AF828825

Nonlinear Partial Differential Equations of Second Order

Translations of
MATHEMATICAL
MONOGRAPHS

Volume 95

Nonlinear Partial Differential Equations of Second Order

Guangchang Dong

American Mathematical Society
Providence, Rhode Island

董 光 昌

非线性二阶偏微分方程

清 华 大 学 出 版 社

Translated from the Chinese by Kai Seng Chou [Kaising Tso]

2010 *Mathematics Subject Classification.* Primary 35Kxx; Secondary 35Jxx, 35Lxx.

ABSTRACT. This is a treatise on nonlinear partial differential equations of second order. With the exception of the first chapter, all the remaining chapters are based on the published or unpublished work of the author.

A priori estimation is the main theme of this book. Emphasis is placed on elliptic and parabolic equations. Nevertheless, some hyperbolic equations are also discussed. Each chapter of this book has its own physical background.

This book can be used as a text for graduate students in mathematics, or as a reference for researchers, teachers, or university seniors.

Library of Congress Cataloging-in-Publication Data

Tung, Kuang-ch'ang.

[Fei hsien hsing erh chieh p'ien wei fen fang ch'eng. English]

Nonlinear partial differential equations of second order/Guangchang Dong; [translated from the Chinese by Kai Seng Chou (Kaising Tso)].

p. cm. — (Translations of mathematical monographs, ISSN 0065-9282; v. 95)

Translation of: Fei hsien hsing erh chieh p'ien wei fen fang ch'eng.

Original Chinese title on verso of t.p.: Fei hsien hsing erh chieh p'ien wei fen fang ch'eng.

Includes bibliographical references.

ISBN 0-8218-4554-3 (alk. paper)

1. Differential equations. Partial. 2. Differential equations. Nonlinear. I. Chou, Kai Seng. II. Title. III. Title: Fei hsien hsing erh chieh p'ien wei fen fang ch'eng. IV. Series.

QA377.T8613 1991 91-27853

515′.353—dc20 CIP

AMS softcover ISBN 978-0-8218-4685-8

∞ The paper used in this book is acid-free and falls within the guidelines established to ensure permanence and durability.
Visit the AMS home page at http://www.ams.org/

10 9 8 7 6 5 4 3 2 20 19 18 17 16 15

Contents

Introduction

There is no routine way of solving nonlinear partial differential equations. Nevertheless, effective approaches which apply to a wide variety of problems are available. The approach we are going to describe belongs to this category. It consists of the following steps:

1. Choose an appropriate function space.
2. Define a family of mappings. In general one uses the associated linear problem to approximate the original nonlinear one.
3. Prove that this family of mappings has a fixed point. Then the nonlinear problem possesses at least one solution.
4. Study various properties of the solution. Usually the solution obtained in Step 3 is rather rough. To know more one needs to study its regularity.

We refer to [49] for a discussion of Banach spaces such as C^α and Sobolev spaces frequently used in Step 1 and on linear problems that appear in Step 2. On the other hand, commonly used fixed point theorems, including the contraction mapping principle and the theorem on topological degree for Leray-Schauder mappings used in Step 3, have been discussed in many texts. For this reason we shall exclude them from this book. So what mainly concerns us is the derivation of various estimates (including a priori ones). They are used throughout Steps 2–4.

In passing, we mention other approaches for solving nonlinear partial differential equations.

A problem in nonlinear partial differential equations may be transformed into another type of problem. A tempting example is to rewrite it into a problem in the calculus of variations [1]. While this approach has many applications, it is subjected to the condition that the original equation must be the Euler-Lagrange equation for some variational problem. This imposes a rather stringent restriction on the coefficients of the equation.

One also employs the method of semigroups to tackle numerous problems. But there are restrictions here too, namely, together with the initial and boundary conditions the equation must admit a semigroup structure. Besides, the domain must be suitably chosen.

There are other approaches as well.

In conclusion the approach of this book, which involves fixed points and a priori estimation, is merely one among many approaches to nonlinear problems. Finally we wish to point out that in dealing with a priori estimation for fully nonlinear equations other than the quasilinear ones a new ingredient is the maximum principle of Aleksandrov's type. It extends the classical maximum principle and is very powerful in deriving a priori estimates.

Notes for the English Translation

The Chinese manuscript of this book was completed in 1986 and published in 1988. Now almost five years have passed and the author would like to take this opportunity to describe his work during this period.

(1) The uniqueness and existence of solutions to nonlinear oblique derivative boundary value problems for fully nonlinear parabolic equations satisfying natural structure conditions. Consider

$$u_t - F(x, t, u, D_x u, D_x^2 u) = 0, \qquad (x, t) \in Q = \Omega \times (0, T); \tag{1}$$

$$u|_{t=0} = \varphi(x), \qquad x \in \Omega; \tag{2}$$

$$G(x, t, u, D_x u) = 0, \qquad (x, t) \in \partial\Omega \times [0, t]; \tag{3}$$

where G satisfies the conditions

$$uG(x, t, u, 0) < 0 \quad \text{for } |u| \geq \mu_3, \tag{4}$$

$$\lambda \leq \frac{\partial G}{\partial p_\gamma}(x, t, u, p) \leq \Lambda \tag{5}$$

(p_γ is a suitable local extension of the inner normal to a neighborhood of the boundary), and

$$\begin{aligned}|G| &+ |G_u| + |G_{uu}| + |G_{px}| + (1 + |p|)^{-1}(|G_x| + |G_{xu}|) \\ &+ (1 + |p|)^2(|G_{xx}| + |G_t|) + (1 + |p|)(|G_p| + |G_{pu}|) + (1 + |p|)^2|G_{pp}| \\ &\leq M_1(u)(1 + |p|).\end{aligned} \tag{6}$$

THEOREM 1. *Suppose that* (4)–(6) *hold and* (1) *satisfies all conditions in* §7 *of Chapter* 9. *Then problem* (1), (2), (3) *has a unique solution.*

This result can be found in [50].

THEOREM 2. *Suppose that* (4)–(6) *hold and* (1) *satisfies all conditions in* §3 *of Chapter* 10. *Then problem* (1), (2), (3), *has a unique solution.*

This result can be found in [51].

(2) The existence, uniqueness, and regularity for viscosity solutions of the first boundary value problem for fully nonlinear parabolic equations satisfying a natural structure condition. When the function $F(x, t, u, p, r)$ in Equation (1) is not convex (or concave) in r, that is

$$\frac{\partial^2 F}{\partial r_{ij} \partial r_{kl}} \eta_{ij} \eta_{kl} \leq 0 \text{ (or } \geq 0), \quad \forall \eta_{ij}, \eta_{kl} \in \mathcal{S}^n$$

no longer holds; classical solutions are not expected. Naturally one turns to viscosity solutions.

Concerning the first boundary value problem

$$u|_{\partial^* Q} = \varphi(x, t) \tag{7}$$

for equation (1), the following results on existence and uniqueness of solutions under a natural structure condition have been obtained.

THEOREM 3 ([52]). *Let λ, μ, μ_i $(i = 1, 2, \ldots)$ be positive constants. Suppose the following conditions are fulfilled:*

$$\lambda I \leq F_r(x, t, z, p, r) \leq \lambda^{-1} I \tag{8}$$

for all $(x, t, z, p, r) \in \Gamma = \overline{Q} \times \mathbb{R} \times \mathbb{R}^n \times \mathcal{S}^n$. (Here I is the $n \times n$ identity matrix);

$$F_z(x, t, z, p, r) \leq 0, \tag{9}$$

$$|F(x, t, 0, 0, 0)| \leq \mu, \tag{10}$$

$$|F(x, t, z, p, 0)| \leq \mu_1 \max_{\overline{Q}} |z|(1 + |p|^2), \tag{11}$$

$$\begin{aligned} &|F| + (1 + |p|)|F_p| + |F_z| + \frac{1}{1 + |p|}|F_x| \\ &\quad \leq \mu_2(\max_{\overline{Q}} |z|)(1 + |p|^2 + |r|), \end{aligned} \tag{12}$$

and

$$|F_{pt}|, |F_{tt}|, |F_{zx}|, |F_{zxx}| \leq \mu_3(1 + |z| + |p| + |r|)^m \tag{13}$$

for some nonnegative m. Then (1) admits a viscosity solution which belongs to $W^{1,0}_\infty(Q) \cap C^{\alpha, \alpha/2}(\overline{Q})$ and satisfies (7).

THEOREM 4 ([52]). *Suppose that for all $(x, t, z, p, r_i) \in \Gamma$, $i = 1, 2$, $r_1 \leq r_2$,*

$$F(x, t, z, p, r_2) - F(x, t, z, p, r_1) \geq \lambda(r_2 - r_1). \tag{14}$$

Moreover suppose that for all $(x, t, z, p, r) \in \Gamma$, $|z| + |p| \leq R$, there exist a positive constant K_R and a positive function $\omega_R(\sigma)$ satisfying $\omega_R(\sigma) \to 0$ as $\sigma \to +0$ and $\omega_R(\sigma)(1 + \sigma)$ is bounded for $\sigma \geq 0$ such that

$$F(x, t, z_2, p, r) - F(x, t, z_1, p, r) \leq K_R(z_2 - z_1), \tag{15}$$

for $z_1 \le z_2$, *and*

$$|F(x, t, z, p, r) - F(y, t, z, p, r)| \le \omega_R(|x-y|^\theta(1+|r|)) \tag{16}$$

for some constant $\theta > 1/2$. *Then the viscosity solutions of* (1) *and* (7) *are unique in* $W_0^{1,0}(Q) \cap C(\overline{Q})$.

Theorem 3 and Theorem 4 are established under some natural structure conditions. For the following regularity result we require a little bit more than the natural structure conditions.

THEOREM 5. *Suppose in addition to* (8)–(12)

$$|F_{pt}|, |F_{tt}|, |F_{zx}|, |F_{zxx}| \le \mu_3(1+|z|+|p|+|r|), \tag{13'}$$

and

$$|F_p(x, t, z, p, r)| \le \mu_4(|z|+|p|) + \mu_5|r|, \tag{17}$$

where μ_5 *is a small, positive parameter. Then for sufficiently smooth initial-boundary value* $\varphi(x, t)$ *a viscosity solution of* (1) *and* (7) *belongs to* $C^{1+\alpha, (1+\alpha)/2}(\overline{Q})$.

The proof of this theorem can be found in [53].

(3) Fully nonlinear degenerate parabolic equations. All equations considered in (1) and (2) are uniformly parabolic. Nevertheless, existence and uniqueness of a viscosity solution can still be studied for some degenerate parabolic equations. In [54] we first proposed the following definition for viscosity sub- and supersolutions. They are the most general ones for a general (including degenerate) parabolic equation.

We call equation (1) parabolic if for all $(x, t, z, p, r) \in \Gamma$, and positive semidefinite $s \in \mathcal{S}^n$

$$F(x, t, z, p, r) \le F(x, t, z, p, r+s). \tag{18}$$

A lower (upper) semicontinuous function u in $\overline{Q}$ is called a viscosity sub- (super-) solution for (1) if for all $\varphi \in C^{2,1}(\overline{Q})$,

$$\varphi_t(x_0, t_0) - F(x_0, t_0, u(x_0, t_0), D_x\varphi(x_0, t_0), D_x^2\varphi(x_0, t_0)) \ge 0 \ (\le 0),$$

whenever $(x_0, t_0) \in Q$ is a local minimum (maximum) for $u - \varphi$. u is called a viscosity solution if it is a viscosity subsolution and supersolution. Let u and v be respectively a subsolution and a supersolution of (1) which are continuous in $\partial^* Q$. The following comparison theorem holds.

THEOREM 6 ([54]). *Suppose for all* $0 < R < \infty$, *there exists* $\gamma_R > -\infty$ *such that*

$$F(x, t, u, p, r) \le F(x, t, v, p, r) - \gamma_R(u-v) \tag{19}$$

for all $(x, t) \in Q$, $R \ge u \ge v \ge -R$, $p \in \mathbb{R}^n$, *and* $r \in \mathcal{S}^n$; *and, for all* $0 < R < \infty$, $(x, t), (y, s) \in Q$, $|u| \le R$, $p \in \mathbb{R}^n$, *and* $r \in \mathcal{S}^n$, *where*

$\omega_R(\sigma) \to 0$ *and* $\overline{\omega}(\sigma) \to 0$ *as* $\sigma \to 0$. *Then*

$$(u - v)_+(x, t) \le e^{(-\gamma_R T)_+} \max_{\partial^* Q}(u - v)_+$$

for all $(x, t) \in \overline{Q}$.

One can deduce the uniqueness of a viscosity solution for (1) and (7) from this comparison principle. Concerning existence we employ the Perron method.

THEOREM 7 ([54]). *Assume that* (18), (19), *and* (20) *hold. Suppose* f *and* g *are viscosity sub- and supersolutions of* (1) *which are continuous in* $\overline{Q}$ *and satisfies* $f|_{\partial^* Q} = g|_{\partial^* Q} = \varphi$. *Then* (1) *and* (7) *admit a unique viscosity solution.*

Uniqueness of a viscosity solution and existence by the Perron method are still possible for certain conditions replacing (19) and (20). See [54].

CHAPTER I

The First Boundary Value Problem for Second-Order Quasilinear Parabolic Equations with Principal Part in Divergence Form [4]

Let Ω be a bounded domain in $\mathbb{R}^n$ and $T > 0$. In $Q = \Omega \times (0, T]$ we consider the following quasilinear equation with principal part in divergence form:

$$\mathscr{L}u = u_t - \frac{d}{dx_i}(a_i(x, t, u, u_{x_k})) + a(x, t, u, u_{x_k}) = 0, \qquad 1 \le k \le n.$$

It is required to fulfill the initial-boundary condition

$$u(x, 0) = \varphi(x), \quad x \in \Omega, \qquad \text{and} \qquad u|_S = \psi,$$

where $S = \partial\Omega \times [0, T]$ and φ, ψ are respectively given functions in Ω and on S. We shall assume that Ω satisfies *Condition* (A):

$$\operatorname{meas}(K(\rho) \cap \Omega) \le (1 - \theta_0)\operatorname{meas} K(\rho), \qquad 0 < \theta_0 < 1, \ \rho < a_0,$$

where $K(\rho) = \{|x - x_0| < \rho : x_0 \in \partial\Omega\}$ and θ_0, a_0 are constants. Requirements on the coefficients a_i and a will be given as we proceed. Our goal is to establish the existence and uniqueness of solutions for this initial-boundary value problem (also known as the *first boundary value problem*).

§1. Uniform and Hölder estimates for the solution

To start with the derivation of the a priori estimates let us assume that a, $\partial a_i/\partial p_j$, $\partial a_i/\partial u$, $\partial a_i/\partial x_i$ are bounded whenever their arguments are bounded. Let $u \in C^{2,1}(\overline{Q})$ be a solution for the first boundary value problem.

THEOREM 1. *Suppose that* a_i $(1 \le i \le n)$ *and* a *satisfy*

$$\frac{\partial a_i}{\partial p_j}\xi_i\xi_j|_{p=0} \ge 0, \qquad p = (p_1, \dots, p_n),$$

and

$$u\left(-\frac{\partial a_i(x, t, u, 0)}{\partial x_i} + a(x, t, u, 0)\right) \ge -b_1u^2 - b_2,$$

for all (x, t) in $\overline{Q}$ where b_1 and b_2 are nonnegative constants. The maximum modulus of the solution can be estimated as

$$\max_{\overline{Q}} |u(x, t)| \leq \min_{b \geq b_1} e^{bT} \left[\max_{\partial^* Q} |u| + \left(\frac{b_2}{b - b_1} \right)^{1/2} \right] \equiv M.$$

Here $\partial^* Q = \Omega \times \{t = 0\} \cup S$.

PROOF. Clearly the theorem is valid when $u(x, t) \equiv 0$. So we may assume $u(x, t) \not\equiv 0$. Rewrite the equation as an equation for v^2, $v = e^{bt} u$ (b is a positive number), as follows:

$$\frac{1}{2}(v^2)_t + bv^2 - \frac{1}{2} \frac{\partial a_i}{\partial p_j} (v^2)_{x_i x_j} + \frac{\partial a_i}{\partial p_j} v_{x_i} v_{x_j}$$
$$- \frac{1}{2} \frac{\partial a_i}{\partial u} (v^2)_{x_i} + \left(a - \frac{\partial a_i}{\partial x_i} \right) v e^{-bt} = 0.$$

If the maximum of v^2 attains somewhere in $\overline{Q} \backslash \partial^* Q$, at this maximum $v_{x_i} = 0$ $(1 \leq i \leq n)$, $(v^2)_t \geq 0$, and $(\partial a_i / \partial p_j)(v^2)_{x_i x_j} \leq 0$. Hence

$$bv^2 \leq \left(\frac{\partial a_i}{\partial x_i} - a \right) \bigg|_{p=0} v e^{-bt} \leq \frac{v}{u} e^{-bt} (b_1 u^2 + b_2) \leq b_1 v^2 + b_2.$$

Taking into account the initial-boundary condition we have

$$|v| \leq \max \left\{ |v|_{\partial^* Q}, \left(\frac{b_2}{b - b_1} \right)^{1/2} \right\},$$

or

$$\max_{\overline{Q}} |u| \leq e^{bT} \left[\max_{\partial^* Q} |u| + \left(\frac{b_2}{b - b_1} \right)^{1/2} \right].$$

The theorem follows after we take the minimum over all $b > b_1$ in the right-hand side of this estimate. □

To get a Hölder estimate we further assume that there are positive constants μ_1 and ν_1 such that for all $(x, t) \in \overline{Q}$, $|u| \leq M$, and $p \in \mathbb{R}^n$,

$$a_i(x, t, u, p) p_i \geq \nu_1 |p|^2 - \mu_1$$

and

$$\sum_i |a_i(x, t, u, p)|(1 + |p|) + |a(x, t, u, p)| \leq \mu_1 (1 + |p|^2).$$

Let $\eta(x, t)$ be a function which belongs to $\overset{\circ}{W}{}_2^1(\Omega)$ for each t. Multiplying $\mathscr{L} u = 0$ by η and then integrating over Ω,

$$\int_\Omega u_t \eta + \int_\Omega (a_i \eta_{x_i} + a\eta) = 0.$$

Choose $\eta = (u(x,t)-k)_+\zeta^2(x)$ where k is a constant and ζ is a cut-off function in $K(\rho) = \{x: |x-x^0| < \rho\}$ $(x^0 \in \Omega)$. To ensure $\eta \in \overset{\circ}{W}{}_2^1$ it suffices to let $k \geq \max_{\overline{K(\rho)\cap S}} u(\cdot, t)$. We have, for arbitrary $\varepsilon > 0$,

$$\begin{aligned}
\frac{1}{2}\frac{\partial}{\partial t}\int_{A_{k,\rho}(t)} (u-k)^2\zeta^2 + \nu_1 \int_{A_{k,\rho}(t)} |\nabla u|^2\zeta^2 \\
\leq \int_{A_{k,\rho}(t)} [\mu_1\zeta^2 + |a_i(u-k)2\zeta\zeta_{x_i}| + |a(u-k)\zeta^2|] \\
\leq \int_{A_{k,\rho}(t)} \{\mu_1\zeta^2 + \mu_1(u-k)[2\zeta|\nabla\zeta|(1+|\nabla u|) + (1+|\nabla u|^2)\zeta^2]\} \\
\leq \mu_1 \int_{A_{k,\rho}(t)} \Big[\zeta^2 + \varepsilon(1+|\nabla u|^2)\zeta^2 + \frac{4}{\varepsilon}(u-k)^2|\nabla\zeta|^2 \\
+ (u-k)(1+|\nabla u|^2)\zeta^2\Big],
\end{aligned}$$

where $A_{k,\rho}(t) = \{x \in K(\rho)\cap\Omega: u(x,t) > k\}$. We can first choose k so that $\max_{\overline{K(\rho)\cap\Omega}}(u-k) \leq \nu_1/4\mu_1$ and then fix $\varepsilon = \nu_1/4\mu_1$. Noticing that $\zeta \leq 1$, we have

$$\begin{aligned}
\frac{\partial}{\partial t}\int_{A_{k,\rho}(t)} (u-k)^2\zeta^2 + \nu_1 \int_{A_{k,\rho}(t)} |\nabla u|^2\zeta^2 \\
\leq \gamma\left(\int_{A_{k,\rho}(t)} (u-k)^2|\nabla\zeta|^2 + \operatorname{meas} A_{k,\rho}(t)\right),
\end{aligned}$$

where

$$k \geq \max\{\sup_{K(\rho)\cap\Omega} u(\cdot,t) - \varepsilon, \ \sup_{K(\rho)\cap S} u(\cdot,t)\}$$

and $\gamma = \gamma(\mu_1, \nu_1, M)$. By the same reasoning we deduce

$$\begin{aligned}
\frac{\partial}{\partial t}\int_{B_{k,\rho}(t)} (u-k)^2\zeta^2 + \nu_1 \int_{B_{k,\rho}(t)} |\nabla u|^2\zeta^2 \\
\leq \gamma\left(\int_{B_{k,\rho}(t)} (u-k)^2|\nabla\zeta|^2 + \operatorname{meas} B_{k,\rho}(t)\right),
\end{aligned}$$

where

$$B_{k,\rho}(t) = \{x \in K(\rho)\cap\Omega: u(x,t) < k\}$$

and $k \leq \min\{\inf_{K(\rho)\cap\Omega} u(\cdot,t) + \varepsilon, \ \inf_{K(\rho)\cap S} u(x,t)\}$. Therefore $u \in \mathscr{B}_2(Q, M, \nu_1, \gamma, \nu_1/4\mu_1, 0)$. (For the definition of the function class $\mathscr{B}_2$ see [49].) From the property of $\mathscr{B}_2$ we deduce a Hölder estimate on the solution u. In other words, there exists α, $0 < \alpha < 1$, such that

$$\|u\|_{\alpha,Q} = \sup_Q |u| + \sup_{(x,t),(x',t')\in Q} \frac{|u(x,t)-u(x',t')|}{(|x-x'|^2 + |t-t'|)^{\alpha/2}}$$

is bounded by a constant depending on M, ν_1, μ_1, θ_0, a_0, and $\|u\|_{\beta,\partial^*Q}$. An interior estimate holds too; namely, for $Q' = \Omega' \times [\varepsilon, T]$, $\overline{\Omega}' \subseteq \Omega$, $\|u\|_{\alpha,Q'}$ can be estimated by M, ν_1, μ_1, ε, and $d(\Omega', \partial\Omega)$.

§2. A uniform bound for $D_x u$

Next we estimate $\nabla u = D_x u = (u_{x_1}, \dots, u_{x_n})$. Suppose that for $(x, t) \in \overline{Q}$, $|u| \le M$, $p \in \mathbb{R}^n$,

$$\nu(M)|\xi|^2 \le \frac{\partial a_i}{\partial p_j}(x, t, u, p)\xi_i\xi_j \le \mu(M)|\xi|^2, \quad \forall \xi \in \mathbb{R}^n,$$

and

$$\sum_i |a_i|(1+|p|) + |a| + \sum_{i,j}\left|\frac{\partial a_i}{\partial p_j}\right|(1+|p|^2) + \sum_i \left|\frac{\partial a_i}{\partial u}\right|(1+|p|) + \sum_{i,k}\left|\frac{\partial a_i}{\partial x_k}\right| \le \mu(M)(1+|p|^2).$$

Also we assume $u|_S = 0$. This can be achieved by replacing u by $u - \psi$ provided ψ can be extended to $\overline{Q}$ as a function in $C^{2,1}(\overline{Q})$.

Under these assumptions we shall derive an estimate on $\max|\nabla u|$. As a primary step we bound $\|\nabla u\|_{L^2(Q)}$. This idea is to look at

$$\int_\Omega \mathscr{L}u(e^{\lambda u} - 1) = 0,$$

where λ is a constant to be specified. By integration by parts

$$\frac{1}{\lambda}\frac{\partial}{\partial t}\int_\Omega (e^{\lambda u} - \lambda u) + \int_\Omega [a_i \lambda u_{x_i} e^{\lambda u} + a(e^{\lambda u} - 1)] = 0.$$

Since

$$\begin{aligned} a_i p_i &= \frac{\partial a_i}{\partial p_j}(x, t, u, \tilde{p})p_i p_j + a_i(x, t, u, 0)p_i \\ &\ge \frac{\nu}{2}|p|^2 - C, \qquad 0 \le \tilde{p}_k \le p_k, \end{aligned}$$

and $|a| \le \mu(1+|p|^2)$,

$$\frac{\lambda\nu}{2}\int_0^t\int_\Omega |\nabla u|^2 e^{\lambda u} \le C\int_0^t\int_\Omega e^{\lambda u}(\lambda + 1 + |\nabla u|^2) + \frac{1}{\lambda}\int_\Omega (\lambda u - e^{\lambda u})|_0^t.$$

Taking $\lambda = 4C/\nu$ we obtain

$$\int_0^t\int_\Omega |\nabla u|^2 \le M_1, \quad \forall t \in [0, T].$$

Next we estimate $\int_\Omega |\nabla u|^{2s+2}$ for $t \in [0, T]$ and $0 \le s \le 3n/2$. For our purpose it suffices to get the estimate up to $s = 3n/2$. As a first step we consider the interior estimate. Let $K(2\rho)$ be in Ω and choose a cut-off function ζ compactly supported in $K(2\rho)$. Consider

$$-\int_0^t \int_{K(\rho)} \mathscr{L}(u) \frac{\partial}{\partial x_k}(|\nabla u|^{2s} u_{x_k} \zeta^2) = 0, \qquad s \geq 0.$$

For now let us assume u_{tx_k} and $D_x^3 u$ are continuous. As a result of integration by parts

$$\frac{1}{2s+2} \int_{K(2\rho)} |\nabla u|^{2s+2} \zeta^2 |_0^t + \int_0^t \int_{K(2\rho)} \left[\frac{da_i}{dx_k} \frac{\partial}{\partial x_i}(|\nabla u|^{2s} u_{x_k} \zeta^2) - a \frac{\partial}{\partial x_k}(|\nabla u|^{2s} u_{x_k} \zeta^2) \right] = 0.$$

Since this equality only involves $D^2 u$, by approximation it continues to hold for u in $C^{2,1}(\overline{Q})$. Substituting

$$\frac{da_i}{dx_k} = \frac{\partial a_i}{\partial p_j} u_{x_j x_k} + \frac{\partial a_i}{\partial u} u_{x_k} + \frac{\partial a_i}{\partial x_k}$$

and

$$\frac{\partial}{\partial x_i}(|\nabla u|^{2s} u_{x_k} \zeta^2) = |\nabla u|^{2s} u_{x_k x_i} \zeta^2 + 2s |\nabla u|^{2s-2} u_{x_l} u_{x_l x_i} u_{x_k} \zeta^2 + |\nabla u|^{2s} u_{x_k} 2\zeta \zeta_{x_i}$$

into this equality and using $2ab \leq \varepsilon a^2 + b^2/\varepsilon$ $(0 < \varepsilon < 1)$, we have

$$\begin{aligned} &\frac{1}{2s+2} \int_{K(2\rho)} |\nabla u|^{2s+2} \zeta^2 |_0^t \\ &\quad + \int_0^t \int_{K(2\rho)} \left(\frac{\partial a_i}{\partial p_j} |\nabla u|^{2s} u_{x_k x_i} u_{x_k x_j} \zeta^2 + 2s \frac{\partial a_i}{\partial p_j} |\nabla u|^{2s-2} u_{x_k} u_{x_k x_j} u_{x_l} u_{x_l x_i} \zeta^2 \right) \\ &\quad \leq C_s \int_0^t \int_{K(2\rho)} \left[\varepsilon |\nabla u|^{2s} \sum_{k,i} u_{x_k x_i}^2 \zeta^2 + \frac{1}{\varepsilon} |\nabla u|^{2s} (1 + |\nabla u|^4) \zeta^2 \right. \\ &\qquad\qquad \left. + \frac{1}{\varepsilon} |\nabla u|^{2s} (1 + |\nabla u|^2) |\nabla \zeta|^2 \right], \end{aligned}$$

where $C_s = C(\mu(M), \nu(M), s)$. Letting $v = \frac{1}{2}|\nabla u|^2$, for ε satisfying $C_s \varepsilon = \nu/2$ we have

$$\begin{aligned} &\frac{1}{s+1} \int_{K(2\rho)} |\nabla u|^{2s+2} \zeta^2 |_0^t \\ &\quad + \nu \int_0^t \int_{K(2\rho)} \left(|\nabla u|^{2s} \sum_{k,i} u_{x_k x_i}^2 \zeta^2 + 4s |\nabla u|^{2s-2} \sum_i v_{x_i}^2 \zeta^2 \right) \\ &\quad \leq C_s \int_0^t \int_{K(2\rho)} (|\nabla u|^{2s+4} \zeta^2 + |\nabla u|^{2s+2} |\nabla \zeta|^2) + C_s' \operatorname{meas} K(2\rho) \frac{1}{\rho^2}. \end{aligned} \tag{1}$$

Since

$$\int_{K(2\rho)} |\nabla u|^{2s+4}\zeta^2 = \int_{K(2\rho)} |\nabla u|^{2s+2} u_{x_i}(u(x,t)-u(x^0,t))_{x_i}\zeta^2$$

$$= -\int_{K(2\rho)} (u(x,t)-u(x^0,t))[|\nabla u|^{2s+2}\Delta u\zeta^2 + |\nabla u|^{2s+2}u_{x_i}2\zeta\zeta_{x_i} + (2s+2)|\nabla u|^{2s}u_{x_l}u_{x_lx_i}u_{x_i}\zeta^2]$$

(x^0 is the center of $K(2\rho)$),

$$\int_{K(2\rho)} |\nabla u|^{2s+4}\zeta^2 \le C_s''\rho^\alpha \int_{K(2\rho)} \left(\varepsilon|\nabla u|^{2s+4}\zeta^2 + \frac{1}{\varepsilon}|\nabla u|^{2s}\sum_{k,i} u_{x_kx_i}^2\zeta^2 + \frac{1}{\varepsilon}|\nabla u|^{2s+2}|\nabla\zeta|^2\right).$$

Notice that we have used the fact u satisfies a C^α-estimate. Fix ε so that $\varepsilon C_s'' = \frac{1}{2}$. It follows that for $\rho \le 1$

$$\int_{K(2\rho)} |\nabla u|^{2s+4}\zeta^2 \le C_s'''\rho^\alpha \int_{K(2\rho)} \left(|\nabla u|^{2s}\sum_{k,i} u_{x_kx_i}^2\zeta^2 + |\nabla u|^{2s+2}|\nabla\zeta|^2\right). \quad (2)$$

Choosing ρ so small that $\rho < 1$ and $C_s'''C_s\rho^\alpha \le \nu/2$ we have, after combining (1) and (2),

$$\frac{1}{s+1}\int_{K(2\rho)} |\nabla u|^{2s+2}\zeta^2\Big|_0^t + \frac{\nu}{2}\int_0^t\int_{K(2\rho)} |\nabla u|^{2s}\sum_{k,i} u_{x_kx_i}^2\zeta^2 \le C_s\int_0^t\int_{K(2\rho)} |\nabla u|^{2s+2}|\nabla\zeta|^2 + C_s', \qquad s \ge 0. \quad (3)$$

Take $\zeta = \zeta(x, \rho+\rho/2^s, \rho+\rho/2^{s+1})$. The right-hand side of (3) is bounded for $s = 0$. One deduces from (2) that $\int_0^t\int_{K(\rho)} |\nabla u|^4\zeta^2$, and hence $\int_0^t\int_{K(3\rho/2)} |\nabla u|^4 dx\,dt$, are bounded. Consequently the right-hand side of (3) ($s = 1$) is bounded. Repeating the same argument finitely many times we conclude that $\int_0^t\int_{K(\rho+\rho/2^{s+1})} |\nabla u|^{2s+4}$ is bounded for $0 \le s \le s_0$. Substituting this into (1) we obtain

$$\int_{K(\rho)} |\nabla u|^{2s+2} \le C_0,$$

for all $\rho \le \rho_0$, and $0 \le s \le s_0$. Here ρ_0 and C_0 only depend on s_0, M, $\mu(M)$, $\nu(M)$, α, and $\max_{\overline{\Omega}}|\nabla\varphi(x)|$.

We can now estimate $|\nabla u|$ in $Q(\rho/2) \equiv K(\rho/2)\times[0,t]$ based on this intermediate result. Setting $w = \zeta^2(x,\rho,\rho/2)|\nabla u|^2$, we consider the equality

$$-\int_0^t\int_{A_{\lambda,\rho}(t)} \mathscr{L}u\frac{\partial}{\partial x_k}[(w-\lambda)\zeta^2u_{x_k}] = 0,$$

where $A_{\lambda,\rho}(t) = \{(x, t) \in K(\rho): w > \lambda\}$ for some $\lambda \geq 0$. By integration by parts, assuming u_{tx_j} and $D_x^3 u$ are continuous,

$$\frac{1}{2}\int_{A_{\lambda,\rho}(t)} (w-\lambda)^2\Big|_0^t + \int_0^t \int_{A_{\lambda,\rho}(t)} \left\{ \frac{da_i}{dx_k}\frac{\partial}{\partial x_i}[(w-\lambda)\zeta^2 u_{x_k}] + a\frac{\partial}{\partial x_k}[(w-\lambda)\zeta^2 u_{x_k}]\right\} = 0.$$

By an approximation argument this equality holds for $u \in C^{2,1}(\overline{Q})$. Since the second integral in this equality is differentiable in t we can differentiate it to obtain

$$\begin{aligned}\frac{1}{2}\frac{\partial}{\partial t}\int_{A_{\lambda,\rho}(t)} (w-\lambda)^2 + \int_{A_{\lambda,\rho}(t)} \Bigg\{ \Bigg[\left(\frac{\partial a_i}{\partial p_j} u_{x_k x_j} + \frac{\partial a_i}{\partial u} u_{x_k} + \frac{\partial a_i}{\partial x_k}\right) \\ \times (w_{x_i}\zeta^2 u_{x_k} + (w-\lambda)u_{x_k x_i}\zeta^2 \\ + (w-\lambda)u_{x_k} 2\zeta\zeta_{x_i})\Bigg] \\ + a\frac{\partial}{\partial x_k}[(w-\lambda)u_{x_k}\zeta^2]\Bigg\} = 0.\end{aligned}$$

From this we further deduce

$$\begin{aligned}\frac{1}{2}\frac{\partial}{\partial t}\int_{A_{\lambda,\rho}(t)} (w-\lambda)^2 + \frac{\nu}{2}\int_{A_{\lambda,\rho}(t)} |\nabla w|^2 + \nu\int_{A_{\lambda,\rho}(t)} (w-\lambda)\sum_{k,i} u_{x_k x_i}^2 \zeta^2 \\ \leq \int_{A_{\lambda,\rho}(t)} \Bigg\{ \frac{\partial a_i}{\partial p_j}[w_{x_i}|\nabla u|^2\zeta\zeta_{x_j} - u_{x_k x_j}(w-\lambda)u_{x_k} 2\zeta\zeta_{x_i}] \\ - \left(\frac{\partial a_i}{\partial u} u_{x_k} + \frac{\partial a_i}{\partial x_k}\right)[w_{x_i}\zeta^2 u_{x_k} + (w-\lambda)u_{x_k x_i}\zeta^2 \\ + (w-\lambda)u_{x_k} 2\zeta\zeta_{x_i}] \\ -a\frac{\partial}{\partial x_k}[(w-\lambda)u_{x_k}\zeta^2]\Bigg\},\end{aligned}$$

or

$$\begin{aligned}\frac{1}{2}\frac{\partial}{\partial t}\int_{A_{\lambda,\rho}(t)} (w-\lambda)^2 + \frac{\nu}{4}\int_{A_{\lambda,\rho}(t)} |\nabla w|^2 \\ \leq C\int_{A_{\lambda,\rho}(t)} (|\nabla u|^6\zeta^2 + |\nabla u|^4|\nabla\zeta|^2) + C' \operatorname{meas} A_{\lambda,\rho}(t).\end{aligned}$$

For $\rho \in [\rho_0/2, \rho_0]$, the constants C and C' only depend on ρ_0. For

$3n/2 - 1 < s \le 3n/2$,

$$\begin{aligned}\int_{A_{\lambda,\rho}(t)} |\nabla u|^6 &\le \left(\int_{A_{\lambda,\rho}(t)} |\nabla u|^{2s+2}\right)^{3/(s+1)} (\operatorname{meas} A_{\lambda,\rho}(t))^{(s-2)/(s+1)} \\ &\le \left(\int_{K(\rho)} |\nabla u|^{2s+2}\right)^{3/(s+1)} (\operatorname{meas} A_{\lambda,\rho}(t))^{(s-2)/(s+1)} \\ &\le C(\operatorname{meas} A_{\lambda,\rho}(t))^{1-2/n+\varepsilon},\end{aligned}$$

holds for some $\varepsilon > 0$. Therefore

$$\frac{1}{2}\frac{\partial}{\partial t}\int_{A_{\lambda,\rho}(t)} (w-\lambda)^2 + \frac{\nu}{4}\int_{A_{\lambda,\rho}(t)} |\nabla w|^2 \le C(\operatorname{meas} A_{\lambda,\rho}(t))^{1-2/n+\varepsilon}.$$

Using the method establishing the boundedness of a weak solution (see §2.5 of [49]) we infer from this inequality a bound on $\operatorname{ess\,sup} w$. In other words, $\operatorname{ess\,sup}|\nabla u|$ is bounded in $Q(\rho/2)$. It remains to estimate $|\nabla u|$ near the lateral boundary.

First we show that $|\nabla u|$ is bounded on the lateral boundary. Since u vanishes on S, it suffices to prove $\partial u/\partial N$ is bounded on S (N is the unit inner normal of S). To this end we assume $\partial\Omega \in C^2$. By a change of variables $x = x(y)$ a small piece of S, S_1, is flattened. The equation $\mathscr{L}u = 0$ is transformed into

$$u_t - \frac{d}{dy_i}\left(a_j \frac{\partial y_i}{\partial x_j}\right) + \tilde{a} = 0,$$

where $\tilde{a} = a + a_j d(\partial y_j/\partial x_i)/dy_i$. For simplicity we still write y as x, $\tilde{a}$ as a, and S_1 as $\partial\Omega_1 \times [0, T]$ where $\partial\Omega_1$ now is contained in $x_n = 0$ and $\partial\Omega\backslash\partial\Omega_1$ belongs to $x_n > 0$. Recall that the equation is given by

$$\mathscr{L}u = u_t - \frac{da_i}{dx_i} + a = u_t - \frac{\partial a_i}{\partial p_j}u_{x_ix_j} - \frac{\partial a_i}{\partial u}u_{x_i} - \frac{\partial a_i}{\partial x_i} + a = 0.$$

Letting $u = \varphi(v)$ the equation becomes

$$\begin{aligned}L(u, v) &= v_t - \frac{\partial a_i}{\partial p_j}v_{x_ix_j} - \frac{\varphi''}{\varphi'}\frac{\partial a_i}{\partial p_j}v_{x_i}v_{x_j} - \frac{1}{\varphi'}\left(\frac{\partial a_i}{\partial u}u_{x_i} + \frac{\partial a_i}{\partial x_i} - a\right) \\ &\equiv v_t - \frac{\partial a_i}{\partial p_j}v_{x_ix_j} + F = 0.\end{aligned}$$

Choose φ with $\varphi' > 0$ and $\varphi'' < 0$. Then

$$\begin{aligned}F &\ge -\frac{\varphi''}{\varphi'}\nu|\nabla v|^2 - \frac{3}{\varphi'}\mu(1 + |\nabla u|^2) \\ &= \left(-\frac{\varphi''}{\varphi'}\nu - 3\mu\varphi'\right)|\nabla v|^2 - \frac{3\mu}{\varphi'}.\end{aligned}$$

We further assume $\varphi(0) = 0$ and $-\varphi''\nu/\varphi' - 3\mu\varphi' \geq 0$. An admissible choice is the function $\varphi(v) = (\nu/3\mu)\log(1+v)$. Then $v = e^{(3\mu/\nu)u} - 1$ and $v|_S = 0$. We have

$$F \geq -\frac{3\mu}{\varphi'} = -\frac{9\mu^2}{\nu}e^{(3\mu/\nu)u} \geq -\frac{9\mu^2}{\nu}e^{3\mu M/\nu} \equiv -C.$$

Consider the function $v + me^{-x_n}$ $(m > 0)$. Since $v|_S = 0$, $v + me^{-x_n}$ attains its maximum on S_1. If we choose $m = (2C/\nu)e^d$ where d is the diameter of Ω, $v + me^{-x_n}$ cannot attain its maximum in Q, for otherwise

$$\begin{aligned}\frac{\partial a_i}{\partial p_j}&(v + me^{-x_n})_{x_i x_j} - (v + me^{-x_n})_t \\ &= F + m\frac{\partial a_n}{\partial p_n}e^{-x_n} \geq -C + m\nu e^{-d} = C > 0,\end{aligned}$$

at this maximum. Therefore $v + me^{-x_n}$ attains its maximum only on S_1. As a result

$$\frac{\partial}{\partial x_n}(v + me^{-x_n})|_{S_1} = \left.\frac{\partial v}{\partial x_n}\right|_{S_1} - m \leq 0,$$

that is, $\sup_{S_1}(\partial v/\partial x_n) \leq m = (18\mu^2/\nu^2)e^{3\mu M/\nu + d}$. From

$$\left.\frac{\partial u}{\partial x_n}\right|_{S_1} = \varphi'(v)\left.\frac{\partial v}{\partial x_n}\right|_{S_1} = \left.\frac{\nu}{3\mu}\frac{\partial v}{\partial x_n}\right|_{S_1},$$

we obtain an upper bound for $\partial u/\partial x_n$ on S_1. To obtain a lower bound we can take $u = \varphi(v) = (\nu/3\mu)\log(1/(1-v))$ and argue as before. Hence $|\nabla u|$ is bounded on S_1. By a finite covering argument $|\nabla u|$ is bounded on S.

Next we estimate $\int_{\Omega\cap K(\rho)}|\nabla u|^{2s+2}$, $0 \leq t \leq T$, $0 \leq s \leq 3n/2$. We distinguish two cases: $s = 0$ and $s > 0$. When $s = 0$, let $M_2 = \sup_S|\nabla u| = \sup_S|\frac{\partial u}{\partial N}|$ and $\Omega_{2\rho} = K(2\rho)\cap\Omega$ where the ball $K(2\rho)$ is centered at $\partial\Omega$. Also let

$$b(x,t) = \begin{cases} 0, & |\nabla u(x,t)|^2 \leq M_2^2, \\ |\nabla u|^2 - M_2^2, & M_2^2 < |\nabla u(x,t)|^2 < M_2^2 + 1, \\ 1, & |\nabla u(x,t)|^2 \geq M_2^2 + 1, \end{cases}$$

and ζ be a cut-off function in $K(2\rho)$. Then $b|_S = 0$ and $b\zeta^2|_{\partial\Omega_{2\rho}} = 0$. From

$$-\int_0^t\int_{\Omega_{2\rho}}\mathscr{L}u\frac{\partial}{\partial x_k}(u_{x_k}b\zeta^2) = 0$$

and

$$\begin{aligned}\int b|\nabla u|_t^2 &= b|\nabla u|^2 - \int|\nabla u|^2 b_t \\ &= b|\nabla u|^2 - \int(b + M_2^2)b_t = b|\nabla u|^2 - \frac{b^2}{2} - M_2^2 b,\end{aligned}$$

we have

$$
\begin{aligned}
&\frac{1}{2}\int_{\Omega_{2\rho}}\left(|\nabla u|^2 b - \frac{b^2}{2} - M_2^2 b\right)\zeta^2\Big|_0^t \\
&\quad + \int_0^t\int_{\Omega_{2\rho}}\left[\left(\frac{\partial a_i}{\partial p_j}u_{x_k x_j} + \frac{\partial a_i}{\partial u}u_{x_k} + \frac{\partial a_i}{\partial x_k}\right)\right. \\
&\qquad\qquad \times (u_{x_k x_i} b\zeta^2 + u_{x_k} b_{x_i}\zeta^2 + 2u_{x_k} b\zeta\zeta_{x_i}) \\
&\qquad\qquad\quad \left. + (a\Delta u b\zeta^2 + u_{x_k} b_{x_k}\zeta^2 + 2u_{x_k} b\zeta\zeta_{x_k})\right] = 0.
\end{aligned}
$$

Therefore

$$
\begin{aligned}
&\frac{1}{2}\int_{\Omega_{2\rho}}|\nabla u|^2 b\zeta^2|_t + \frac{1}{2}\int_0^t\int_{\Omega_{2\rho}}\frac{\partial a_i}{\partial p_j}u_{x_k x_j}u_{x_k x_i}b\zeta^2 + \frac{1}{4}\int_0^t\int_{\Omega_{2\rho}}\frac{\partial a_i}{\partial p_j}b_{x_i}b_{x_j}\zeta^2 \\
&\quad \le \frac{1}{2}\int_{\Omega_{2\rho}}|\nabla u|^2 b\zeta^2|_{t=0} + \frac{1}{2}\int_{\Omega_{2\rho}}\left(\frac{b^2}{2} + M_2^2 b\right)\zeta^2|_0^t \\
&\quad + C\int_0^t\int_{\Omega_{2\rho}}|\nabla u|^6\zeta^2 + C\int_0^t\int_{\Omega_{2\rho}}|\nabla u|^2 b|\nabla\zeta|^2 \\
&\quad + C\int_0^t\int_{\Omega_{2\rho}}|\nabla u|^4 b\zeta^2,
\end{aligned}
$$

where $\Omega'_{2\rho} = \{x \in \Omega_{2\rho} : |\nabla u(x, t)|^2 \le M_2^2 + 1\}$. As $0 \le b \le 1$, the first three terms in the right-hand side of this inequality are bounded. The fourth term is also bounded because $\int_0^T\int_\Omega |\nabla u|^2 \le C$. To estimate the last term we use the Hölder estimate $\max_{\overline{\Omega}_{2\rho}} |u(x, t)| \le C\rho^\alpha$ in the following:

$$
\begin{aligned}
\int_{\Omega_{2\rho}}|\nabla u|^4 b\zeta^2 &= \int_{\Omega_{2\rho}}\sum_k u_{x_k}^2|\nabla u|^2 b\zeta^2 \\
&= -\int_{\Omega_{2\rho}} u(\Delta u|\nabla u|^2 b\zeta^2 + 2u_{x_k}u_{x_i}u_{x_k x_i}b\zeta^2 \\
&\qquad + u_{x_k}|\nabla u|^2 b_{x_k}\zeta^2 + 2u_{x_k}|\nabla u|^2 b\zeta\zeta_{x_k}) \\
&\le C_1\rho^\alpha\int_{\Omega_{2\rho}}\left(\sum_{k,i}u_{x_k x_i}^2 b\zeta^2 + |\nabla b|^2\zeta^2\right) \\
&\quad + C_1\rho^\alpha\int_{\Omega_{2\rho}}|\nabla u|^4 b\zeta^2 + C_1.
\end{aligned}
$$

Let $\rho^\alpha = \min(1/2C_1, \nu/8CC_1)$. Then

$$
\int_{\Omega_{2\rho}}|\nabla u|^2 b\zeta^2|_t + \frac{\nu}{4}\int_0^t\int_{\Omega_{2\rho}}\left(\sum_{k,i}u_{x_k x_i}^2 b\zeta^2 + |\nabla b|^2\zeta^2\right) \le C_2.
$$

Henceforth $\int_{\Omega_{2\rho}}|\nabla u|^2\zeta^2 \le C$ and $\int_{\Omega_\rho}|\nabla u|^2 \le C$.

The derivation of the estimate for $\int_{\Omega\cap K(\rho)} |\nabla u|^{2s+2}$ $(s > 0)$ is simpler. It is sufficient to begin with

$$-\int_0^t \int_{A_{M_2, 2\rho}(t)} \mathscr{L}u \frac{\partial}{\partial x_k}[(|\nabla u|^2 - M_2^2)^s u_{x_k} \zeta^2] = 0,$$

and argue as before. Here $A_{M_2, 2\rho}(t) = \{x \in K(2\rho) \cap \Omega : |\nabla u| > M_2\}$. After accomplishing this we can estimate $\max|\nabla u|$ on $K(\rho/2)\cap\Omega\times[0, T]$ following the same path as we have done in the interior case. Summarizing what has been done we state:

THEOREM 2. *Let u be a $C^{2,1}$-solution of the initial-boundary value problem in $\overline{Q}$ with $\max_{\overline{Q}}|u| \le M$. Suppose that $\partial\Omega \in C^2$ and a_i, a satisfy*

$$\nu(M)|\xi|^2 \le \frac{\partial a_i}{\partial p_j}\xi_i\xi_j \le \mu(M)|\xi|^2, \quad \forall\xi \in \mathbb{R}^n,$$

and

$$\sum_i |a_i|(1 + |p|) + |a| + \sum_{i,j}\left|\frac{\partial a_i}{\partial p_j}\right|(1 + |p|^2) + \sum_i \left|\frac{\partial a_i}{\partial u}\right|(1 + |p|) + \sum_{i,k}\left|\frac{\partial a_i}{\partial x_k}\right|$$
$$\le \mu(M)(1 + |p|^2).$$

Then $\max_{\overline{Q}}|\nabla u| \le M_1$ where M_1 can be estimated in terms of M, $\max_{\overline{\Omega}}|\nabla u(x, 0)|$, μ, ν, and the C^2-modulus of $\partial\Omega$. For Ω' with $\overline{\Omega'} \subseteq \Omega$, $\max_{\overline{\Omega'}\times[0,T]}|\nabla u|$ can be estimated by M, μ, ν, $\max_{\overline{\Omega'}}|u(x, 0)|$, and $d(\Omega', \partial\Omega)$.

§3. A Hölder estimate for $D_x u$

In this section we shall derive a Hölder estimate for ∇u. By differentiating the differential equation $\mathscr{L}u = 0$ in the x_l-direction, we see that $v = u_{x_l}$ satisfies a linear equation with bounded coefficients

$$v_t - \frac{d}{dx_i}\left(\frac{\partial a_i}{\partial p_j}v_{x_j} + \frac{\partial a_i}{\partial u}v + \frac{\partial a_i}{\partial x_l} - \delta_{il}a\right) = 0.$$

Hence $u_{x_l} \in \mathscr{B}_2(Q, M, \gamma_2, \nu_2, \infty, 0)$ where γ_2, ν_2 only depend on $M = \max_{\overline{Q}}|u|$, $M_1 = \max_{\overline{Q}}|\nabla u|$, the maximums of $|a|$, $|a_i|$, $|\partial a_i/\partial x_k|$, $|\partial a_i/\partial u|$, $|\partial a_i/\partial p_k|$, $1 \le i, k \le n$, and $\max_{|\xi|=1}((\partial a_i/\partial p_j)\xi_i\xi_j)^{-1}$ over $(x, t) \in \overline{Q}$, $|u| \le M$ and $|p| \le M_1$. Under the assumptions on a_i and a in Theorem 2 we see that γ_2 and ν_2 only depend on μ, ν, M, and M_1. As a result there exists $\alpha > 0$ depending on these quantities as well as $|u_{x_i}(x, 0)|$

and $d(\Omega', \Omega)$ such that $|u_{x_i}|_{\alpha, Q'}$ is bounded for $Q' = \Omega' \times [0, T]$ and $\overline{\Omega'} \subseteq \Omega$.

We still need to estimate the Hölder norm of ∇u near the boundary. As it is not known whether $\partial u/\partial N$ admits a Hölder estimate on S, it seems difficult to obtain such an estimate directly. To attain our goal we shall first derive Hölder estimates for u_t and u_{x_τ} (τ is a tangential direction).

To show that u_t is bounded and satisfies a Hölder estimate some additional regularity assumptions are required. For $u \in C^{2,1}(\overline{Q})$, let $\max_{\overline{Q}} |u| \leq M$ and $\max_{\overline{Q}} |\nabla u| \leq M_1$. We assume that a_i, a are Lipschitz continuous in t, a_i is differentiable in x, u, and p, and a is differentiable in u and p. For $(x, t), (x, t+h) \in \overline{Q}$, $|u| \leq M$, $|p| \leq M_1$, $1 \leq i, j \leq n$, we then have

$$\max \left| \frac{\partial a_i}{\partial p_j}, \frac{\partial a_i}{\partial u}, \frac{\partial a_i}{\partial x_j}, a, \frac{\partial a}{\partial p_i}, \frac{\partial a}{\partial u}, \frac{a_i(x, t+h, u, p) - a_i(x, t, u, p)}{h}, \frac{a(x, t+h, u, p) - a(x, t, u, p)}{h} \right| \leq C(M, M_1)$$

and

$$\frac{\partial a_i}{\partial p_j} \xi_i \xi_j \geq \nu(M, M_1)|\xi|^2, \quad \forall \xi \in \mathbb{R}^2$$

for some positive $\nu(M, M_1)$.

Lemma 1. *$\max_{\overline{Q}} |u_t|$ can be estimated in terms of $\max_{\partial^* Q} |u_t|$, M, M_1, $C(M, M_1)$, and $\nu(M, M_1)$. Furthermore, $|u_t|_{\alpha, Q'}$ admits a bound for some $\alpha > 0$. This bound and α may further depend on $d(\Omega', \partial\Omega)$. Finally when $\partial\Omega$ satisfies Condition (A) and u_t is Hölder continuous on $\partial^* Q$, $|u_t|_{\alpha, Q}$ admits a bound for some $\alpha > 0$ where α and this bound also depend on $|u_t|_{\beta, \partial^* Q}$ ($\beta > 0$) and a_0, θ_0 in Condition* (A).

Proof. Consider the difference quotient

$$v_h(x, t) = \frac{1}{h}(u(x, t+h) - u(x, t)).$$

It satisfies the equation

$$\frac{\partial v_h}{\partial t} - \frac{d}{dx_i} \frac{\Delta a_i}{h} + \frac{\Delta a}{h} = 0,$$

where

$$\begin{aligned}\frac{\Delta a_i}{h} &= \frac{1}{h}[a_i(x, t+h, u(x, t+h), \nabla u(x, t+h)) \\ &\qquad - a_i(x, t, u(x, t), \nabla u(x, t))] \\ &= \frac{1}{h}\int_0^1 \frac{\partial}{\partial \tau} a_i(x, t+h, \tau u(x, t+h) + (1-\tau)u(x, t), \\ &\qquad \tau \nabla u(x, t+h) + (1-\tau)\nabla u(x, t))d\tau \\ &\quad + \frac{1}{h}[a_i(x, t+h, u(x, t), \nabla u(x, t)) - a_i(x, t, u(x, t), \nabla u(x, t))] \\ &\equiv a_{ij} v_{hx_j} + b_i v_h + d_i\end{aligned}$$

and

$$\frac{\Delta a}{h} = c_i v_{hx_i} + c v_h + d$$

is defined in a similar way. Therefore v_h satisfies a linear parabolic equation whose coefficients are bounded uniformly in h for $0 \le h \le h_0$. (Strictly speaking v_h satisfies the equation in weak sense only.) From linear parabolic theory we conclude v_h is bounded and admits a Hölder estimate uniform in h. Taking $h \to 0$ we prove the lemma. □

Next we consider a Hölder estimate for ∇u in Q. Since we have obtained such an estimate in any subdomain away from the lateral boundary, it remains to derive it near the lateral boundary. As before we use a C^2-change of variables to send a small piece of $\partial\Omega$ to a small piece $\partial\Omega_1$ on $y_n = 0$. Suppose $\partial\Omega_1 \times (0, y_n^0)$ is contained in the new domain. Let ξ be a function compactly supported in $\partial\Omega_1 \times (0, y_n^0)$. Multiply the equation by $\partial\xi/\partial y_l$ $(1 \le l \le n-1)$ and then integrate over Ω. We have

$$\begin{aligned}0 &= \int_\Omega \left[-a_i \frac{\partial}{\partial x_i}\left(\frac{\partial \xi}{\partial y_l}\right) - (a + u_t)\frac{\partial \xi}{\partial y_l}\right] dx \\ &= \int \left[-a_i \frac{\partial^2 \xi}{\partial y_l \partial y_m}\frac{\partial y_m}{\partial x_i} - (a+u_t)\frac{\partial \xi}{\partial y_l}\right] J\, dy \\ &= \int \left[\frac{da_i}{dy_l}\frac{\partial \xi}{\partial x_i} J + a_i \frac{\partial \xi}{\partial y_m}\frac{\partial^2 y_m}{\partial x_i \partial y_l} J + a_i \frac{\partial \xi}{\partial x_i}\frac{\partial J}{\partial y_l} - (a+u_t)\frac{\partial \xi}{\partial y_l} J\right] dy \\ &= \int_\Omega \left\{\left[\frac{\partial a_i}{\partial p_j}\left(u_{y_l y_i} - u_{x_m}\frac{\partial^2 x_m}{\partial y_l \partial x_j}\right) + \frac{\partial a_i}{\partial u} u_{y_l} + \frac{\partial a_i}{\partial y_l}\right] + a_j \frac{\partial^2 y_m}{\partial x_j \partial y_l}\frac{\partial x_i}{\partial y_m}\right. \\ &\qquad \left. + a_i J^{-1}\frac{\partial J}{\partial y_l} - (a+u_t)\frac{\partial x_i}{\partial y_l}\right\}\frac{\partial \xi}{\partial x_i}\, dx, \qquad (4)\end{aligned}$$

where $J = \partial(x_1, \dots, x_n)/\partial(y_1, \dots, y_n)$. Notice that we have used

$$u_{x_j y_l} = u_{x_j x_m}\frac{\partial x_m}{\partial y_l}$$

and

$$u_{y_l x_j} = \left(u_{x_m} \frac{\partial x_m}{\partial y_l}\right)_{x_j} = u_{x_m x_j} \frac{\partial x_m}{\partial y_l} + u_{x_m} \frac{\partial^2 x_m}{\partial y_l \partial x_j},$$

and therefore

$$u_{x_j y_l} = u_{y_l x_j} - u_{x_m} \frac{\partial^2 x_m}{\partial y_l \partial x_j}.$$

Write $v = u_{y_l}$, $a_{ij} = \partial a_i / \partial p_j$, and

$$f_i^l = -\frac{\partial a_i}{\partial p_j} u_{x_m} \frac{\partial^2 x_m}{\partial y_l \partial x_j} + \frac{\partial a_i}{\partial u} u_{y_l} + \frac{\partial a_i}{\partial y_l}$$
$$+ a_j \frac{\partial^2 y_m}{\partial x_j \partial y_l} \frac{\partial x_i}{\partial y_m} + a_i J^{-1} \frac{\partial J}{\partial y_l} - (a + u_t) \frac{\partial x_i}{\partial y_l}.$$

Then (4) can be rewritten as $(d/dx_i)(a_{ij} v_{x_j} + f_i^l) = 0$. For each t, v is a weak solution of a linear elliptic equation with bounded coefficients. Furthermore $u|_{\partial\Omega_1} = 0$ implies $v|_{\partial\Omega_1} = 0$. Using the results in §1.11 of [49] v admits a Hölder estimate near $\partial\Omega_1$. Both the estimate and the Hölder exponent α can be estimated by the quantities specified in Lemma 1.

At last we come to the estimation of u_{y_n}. Its derivation will be based on the characterization of the Hölder condition by some integral inequalities. Let $K_{2\rho}$ be a ball centered at $(y_1^0, \dots, y_{n-1}^0, 0) \in \partial\Omega_1$ and let $\Omega_{2\rho} = K_{2\rho} \cap \{y_n > 0\}$. Let ζ be a nonnegative function compactly supported in $K_{2\rho}$ which equals 1 in K_ρ. Using $\xi = v\zeta^2$ $(v = u_{x_l}, l \neq n)$ as a test function we have

$$\int_\Omega [a_{ij} v_{x_i} v_{x_j} \zeta^2 + a_{ij} v v_{x_j} 2\zeta\zeta_{x_j} + f_i^l (v_{x_i}\zeta^2 + v 2\zeta\zeta_{x_i})] = 0.$$

Consequently

$$\nu \int_{\Omega_{2\rho}} |\nabla v|^2 \zeta^2 \le \varepsilon \int_{\Omega_{2\rho}} |\nabla v|^2 \zeta^2 + \frac{C}{\varepsilon} \int_{\Omega_{2\rho}} v^2 |\nabla\zeta|^2 + C \operatorname{meas} \Omega_{2\rho}.$$

Taking $\varepsilon = \nu/2$, in view of

$$\max_{\overline{\Omega}_{2\rho}} |v| = \max_{\overline{\Omega}_{2\rho}} |u_{y_l}| \le C\rho^\alpha \quad \text{and} \quad |\nabla\zeta| \le \frac{C}{\rho},$$

we have

$$\int_{\Omega_\rho} |\nabla u_{y_l}|^2 \le \int_{\Omega_{2\rho}} |\nabla u_{y_l}|^2 \zeta^2 \le C\rho^{n-2+2\alpha}, \qquad l = 1, 2, \dots, n-1.$$

From the equation $(da_i/dy_k)(\partial y_k/\partial x_i) - (a + u_t) = 0$ we can solve for $u_{y_n y_n}$:

$$u_{y_n y_n} = b_{li} u_{y_l y_i} + b, \qquad b_{li},\ b \text{ bounded},$$

where the summation in l is over $\{1, \ldots, n-1\}$. Consequently

$$\int_{\Omega_\rho} |\nabla w|^2 \le C\rho^{n-2+2\alpha}, \qquad w = u_{y_n}.$$

We extend w as an even function to $y_n < 0$. Then $\int_{K(\rho)} |\nabla w|^2 \le C\rho^{n-2+2\alpha}$.

Recall that $K(\rho)$ is centered on $y_n = 0$. When $K(\rho)$ does not intersect $y_n = 0$, we can obtain the same estimate by choosing $\xi = (v - v^*)\zeta^2$, where v^* is the value of v at the center. In case $\overline{K(\rho)}$ intersects $y_n = 0$, observe that then $K(\rho)$ is contained in a ball centered somewhere at $K(\rho) \cap \{y_n = 0\}$ with radius 2ρ. Hence the estimate still holds. Writing $\sigma(\rho) = \int_{K(\rho)} |\nabla w(y)|$, we have

$$\sigma(\rho) \le C\left(\rho^n \int_{K(\rho)} |\nabla w|^2\right)^{1/2} \le C\rho^{n-1+\alpha}.$$

From Sobolev's decomposition formula [49] we have, for all y in $K(\rho)$,

$$\begin{aligned}
&\left| w(y) - \int_{K(\rho)} \zeta_0(z) w(z)\, dz \right| \\
&\quad \le \left| \int_{K(\rho)} \frac{\omega_i(y, z)}{|y-z|^{n-1}} D_i w(z)\, dz \right| \le C \int_0^{2\rho} \xi^{-n+1} \sigma'(\xi)\, d\xi \\
&\quad = C\left[\xi^{-n+1}\sigma(\xi)\Big|_0^{2\rho} + (n-1)\int_0^{2\rho} \xi^{-n}\sigma(\xi)\, d\xi \right] \le C_1 \rho^\alpha.
\end{aligned}$$

Consequently for any y_1, y_2 in $K(\rho)$,

$$\begin{aligned}
|w(y_2) - w(y_1)| &\le \left| w(y_2) - \int_{K(\rho)} \zeta_0(z) w(z)\, dz \right| \\
&\quad + \left| \int_{K(\rho)} \zeta_0(z) w(z)\, dz - w(y_1) \right| \\
&\le C_2 \rho^\alpha.
\end{aligned}$$

In this way we have derived a Hölder estimate for ∇u in x. That it admits a Hölder estimate in t is given by the following lemma.

Lemma 2. *If* $\operatorname{osc}\{u_{x_k}, K(\rho) \cap \Omega\} \le \mu\rho^\alpha$ *for* $x \in \Omega$, $0 \le t \le T$, *and* $|u(x, t^2) - u(x, t^1)| \le \mu_1 |t^2 - t^1|^\beta$ *for* $x \in \Omega$, $0 \le t^1, t^2 \le T$ *and some* $\beta > 0$, *then* u_{x_k} *admits a Hölder estimate in* t *with exponent* $\delta = \alpha\beta/(1+\alpha)$. *The estimate only depends on* $\alpha, \beta, \mu, \mu_1$, *and* $S = \partial\Omega \times [0, T]$.

Proof. Let x^1 and x^2 be endpoints of a line segment in $\overline{\Omega}$. Writing $x^2 - x^1 = |x^2 - x^1| l$, by the mean value theorem we have

$$u(x^2, t^1) - u(x^1, t^1) = (x^2 - x^1)\frac{\partial u}{\partial l}(x^1 + \theta(x^2 - x^1), t^1), \qquad 0 < \theta < 1,$$

and

$$\left|\frac{u(x^2, t^1) - u(x^1, t^1)}{x^2 - x^1} - \frac{\partial u}{\partial l}(x^1, t^1)\right| \le \mu |x^2 - x^1|^\alpha.$$

Similarly

$$\left|\frac{u(x^2, t^2) - u(x^1, t^2)}{x^2 - x^1} - \frac{\partial u}{\partial l}(x^1, t^2)\right| \le \mu |x^2 - x^1|^\alpha.$$

Therefore

$$\begin{aligned}
&\left|\frac{\partial u}{\partial l}(x^1, t^2) - \frac{\partial u}{\partial l}(x^1, t^1)\right| \\
&\quad \le 2\mu|x^1 - x^2|^\alpha + \left|\frac{u(x^2, t^2) - u(x^1, t^2) - u(x^2, t^1) + u(x^1, t^1)}{x^2 - x^1}\right| \\
&\quad \le 2\mu|x^1 - x^2|^\alpha + 2\mu_1 \frac{|t^2 - t^1|^\beta}{|x^2 - x^1|} = 2(\mu + \mu_1)|t^2 - t^1|^{\alpha\beta/(1+\alpha)},
\end{aligned}$$

provided we choose $|x^2 - x^1| = |t^2 - t^1|^{\beta/(1+\alpha)}$.

In general a segment connecting two points in $\overline{\Omega}$ may lie outside Ω. So we should first employ a C^2 change of variables to send a portion of $\partial\Omega$ into $y_n = 0$ and obtain a Hölder estimate for u_{y_k} $(1 \le k \le n)$ near $y_n = 0$. Going back to the original domain and using a finite covering argument we obtain a Hölder estimate (in t) for ∇u. The proof of the lemma is complete. □

We summarize what has been proved in the following theorem (notice that only the boundedness of u_t is used in the proof of the Hölder estimate for ∇u).

Theorem 3. *Let $u \in C^{2,1}(\overline{Q})$ be a solution of*

$$\begin{cases}
\mathcal{L}u = u_t - \dfrac{da_i}{dx_i}(x, t, u, \nabla u) + a(x, t, u, \nabla u) = 0, \\
u|_S = 0, \\
u|_{t=0} = \varphi(x), \qquad x \in \Omega.
\end{cases}$$

Suppose for $(x, t) \in \overline{Q}$, $|u| \le M$, $|p| \le M_1$, $1 \le i, j \le n$,

$$\max\left|\frac{\partial a_i}{\partial p_j}, \frac{\partial a_i}{\partial u}, \frac{\partial a_i}{\partial x_j}, \frac{\partial a}{\partial p_i}, \frac{\partial a}{\partial u}, \frac{a_i(x, t+h, u, p) - a_i(x, t, u, p)}{h}, \frac{a(x, t+h, u, p) - a(x, t, u, p)}{h}\right| \le C(M, M_1)$$

and

$$\frac{\partial a_i}{\partial p_j}\xi_i\xi_j \ge \nu(M, M_1)|\xi|^2, \quad \forall \xi \in \mathbb{R}^n,$$

where $M = \max_{\overline{Q}}|u|$ *and* $M_1 = \max_{\overline{Q}}|\nabla u|$. *Then there exists* $\alpha \in (0, 1)$ *such that* $|\nabla u|_{\alpha, \overline{Q}}$ *is estimated by* $M, M_1, C(M, M_1), \nu(M, M_1)$, $\max_{\overline{\Omega}}|u_t(x, 0)|$ $(u_t(x, 0) = (da_i/dx_i)(x, 0, \varphi(x), \nabla\varphi(x)) - a(x, 0, \varphi(x), \nabla\varphi(x)))$, *and the* C^2*-norm of* $\partial\Omega$.

§4. Existence and uniqueness of the solution for the first boundary value problem

Consider the problem

$$\begin{cases} \mathscr{L}u = u_t - \dfrac{da_i}{dx_i}(x, t, u, \nabla u) + a(x, t, u, \nabla u) = 0, \\ u|_S = 0, \\ u|_{t=0} = \varphi(x), \qquad x \in \Omega, \end{cases}$$

under the condition $(\partial a_i/\partial p_j)\xi_i\xi_j > 0$, $\forall \xi \neq 0$ in $\mathbb{R}^n$. The equation in this problem is called a quasilinear parabolic equation with principal part in divergence form. It can be rewritten as

$$u_t - \frac{\partial a_i}{\partial p_j} u_{x_i x_j} + A(x, t, u, \nabla u) = 0,$$

where $A = a - (\partial a_i/\partial u)u_{x_i} - (\partial a_i/\partial x_i)$.

A standard method of finding a solution for a nonlinear equation is to approximate it by solutions of a relevant linear equation. Specifically, for a suitably chosen function $w(x, t)$ we may solve the linear problem

$$\begin{cases} v_t - \dfrac{\partial a_i}{\partial p_i}(x, t, w, \nabla w)v_{x_i x_j} + A(x, t, w, \nabla w) = 0, \\ v|_S = 0, \\ v|_{t=0} = \varphi(x), \qquad x \in \Omega, \end{cases}$$

to obtain a solution $v = \Phi(w)$. A solution of the functional equation $\Phi(w) = w$, usually called a fixed point of Φ, is a solution for the nonlinear problem. There are well known tools, such as the contraction mapping principle, Schauder fixed point theorem, and the theorem on Leray-Schauder degree, which can be used to find fixed points of Φ (or to solve $(I - \Phi)w = 0$ where I is the identity operator). Which one of these tools are we going to use? In this and the next several chapters we shall employ these tools so that their main feature can be exposed. For the problem of this chapter we shall use the Leray-Schauder fixed point theorem (also called the topological degree theorem for continuous mappings). This theorem can be stated as follows:

Let $\mathscr{M}$ be a bounded subset of a Banach space B. Suppose a mapping $v = \Phi(w, \tau)$ with parameter τ sends $\overline{\mathscr{M}}$ to $\overline{\mathscr{M}}$ and satisfies the following conditions:

(i) $\Phi(w, \tau)$ is compact, i.e., Φ is continuous in w and $\Phi(\overline{\mathscr{M}}, \tau)$ is compact for each τ;

(ii) Φ is uniformly continuous in τ for $0 \le \tau \le 1$;

(iii) when $\tau = 0$, $(I - \Phi)w = 0$ has finitely many solutions and the degree of $I - \Phi$ (the sum of indices for all solutions) is nonzero; and

(iv) $\partial\mathscr{M}$ does not contain any solution of $(I - \Phi)w = 0$ for all τ.

Under these conditions, the degree of $(I - \Phi)w = 0$ is a homotopy invariant. As a result, $(I - \Phi)w = 0$ is solvable for $\tau = 1$. In other words, the mapping $v = \Phi(w, 1)$ has a fixed point.

To apply this theorem to our problem we define $v = \Phi(w, \tau)$ by solving

$$\begin{cases} v_t - \left[\tau \dfrac{\partial a_i}{\partial p_j}(x, t, w, \nabla w) + (1 - \tau)\delta_{ij}\right] v_{x_i x_j} + \tau A(x, t, w, \nabla w) = 0, \\ v|_S = 0, \\ v|_t = \tau\varphi(x) \end{cases}$$

in the Banach space $H_\delta = \{w : w, w_{x_i} \in C^{\delta, \delta/2}(Q)\}$ whose norm is given by

$$|w|_{H_\delta} = |w|_{\delta, Q} + \sum_1^n |w_{x_i}|_{\delta, Q},$$

where

$$|w|_{\delta, Q} = \sup_Q |w| + \sup_{(x, t), (x', t') \in Q} \frac{|w(x, t) - w(x', t')|}{(|x - x'|^2 + |t - t'|)^{\delta/2}}.$$

δ will be chosen later.

To fix $\mathscr{M}$ it is necessary to look at the set of all fixed points $u^\tau = \Phi(u^\tau, \tau)$. u^τ satisfies:

$$\begin{cases} \mathscr{L}_\tau u \equiv \tau\mathscr{L} u + (1 - \tau)(u_t - \Delta u) = 0, \\ u|_S = 0, \\ u|_{t=0} = \tau\varphi(x), \qquad x \in \Omega. \end{cases}$$

Under the assumptions

$$u\left(-\frac{\partial a_i}{\partial x_i}(x, t, u, 0) + a(x, t, u, 0)\right) \ge -b_1 u^2 - b_2, \qquad b_1, b_2 \ge 0, \quad (5)$$

and

$$\frac{\partial a_i}{\partial p_j}(x, t, u, p)\xi_i\xi_j|_{p=0} \ge 0, \quad \forall \xi \in \mathbb{R}^n,$$

we have

$$\tau u\left(-\frac{\partial a_i}{\partial x_i}(x, t, u, 0) + a(x, t, u, 0)\right) \ge -b_1 u^2 - b_2$$

and

$$\left[\tau\frac{\partial a_i}{\partial p_j} + (1 - \tau)\delta_{ij}\right] \xi_i\xi_j|_{p=0} \ge 0, \quad \forall \xi \in \mathbb{R}^n,$$

for $\tau \in [0, 1]$. Hence for $u^\tau \in C^{2,1}(\overline{Q})$, its maximum norm can be estimated by

$$\max_{\overline{Q}} |u^\tau| \le \min_{b \ge b_1} e^{bT} \left[\max_{\overline{\Omega}} |\varphi(x)| + \left(\frac{b_2}{b - b_1} \right)^{1/2} \right] = M.$$

Next, let us assume

$$\frac{\partial a_i}{\partial p_j}(x, t, u, p)\xi_i \xi_j \ge \nu(|u|)|\xi|^2, \quad \forall \xi \in \mathbb{R}^n,$$

and

$$\sum_i \left(|a_i| + \left| \frac{\partial a_i}{\partial u} \right| \right) (1 + |p|) + \sum_{i,j} \left| \frac{\partial a_i}{\partial p_j} \right| (1 + |p|^2) + |a| + \sum_{i,j} \left| \frac{\partial a_i}{\partial x_j} \right| \le \mu(|u|)(1 + |p|^2), \tag{6}$$

for $(x, t, u, p) \in Q \times [-M, M] \times \mathbb{R}^n$. (6) continues to hold if we replace a_i and a by $a_i^\tau = \tau a_i + (1 - \tau)p_i$ and $a^\tau = \tau a$ respectively. Accordingly $\nu(|u|)$ and $\mu(|u|)$ should be replaced by $\min\{\nu(|u|), 1\}$ and $\max\{\mu(|u|), 1\}$. Henceforth $\max_{\overline{Q}} |\nabla u^\tau| \le M_1$ $(0 \le \tau \le 1)$ where M_1 only depends on $\mu(M)$, $\nu(M)$, C^2-norm of $\partial\Omega$, and $\max_{\overline{\Omega}} |\varphi|$.

To establish the unique solvability more assumptions are needed. A complete statement of the result is:

THEOREM 4. *Assume*

(a) $\partial\Omega \in C^{2+\beta}$ $(\beta > 0)$,
(b) $\varphi \in C^{2+\beta}(\overline{\Omega})$ *and it satisfies the two compatibility conditions* $\varphi|_{\partial\Omega} = 0$ *and* $-da_i/dx_i(x, 0, \varphi(x), \nabla\varphi(x)) + a(x, 0, \varphi(x), \nabla\varphi(x)) = 0$ *for* $x \in \partial\Omega$;
(c) (5) *holds, hence* M *is determined,*
(d) (6) *holds, hence* M_1 *is determined,*
(e) a_i, a, $\partial a_i/\partial p_j$, $\partial a_i/\partial u$, $\partial a_i/\partial x_j$ *are Hölder continuous with exponents* β, $\beta/2$, β, β, *in* $(x, t) \in Q$, $|u| \le M$, *and* $|p| \le M_1$.
(f) a_i *and* a *are Lipschitz continuous in* t; a *is differentiable in* u *and* p *in* $(x, t) \in Q$, $|u| \le M$, *and* $|p| \le M_1$. *The Lipschitz constants,* $|\partial a/\partial u|$, *and* $|\partial a/\partial p_k|$ *are bounded by a constant* C.

Then

$$\begin{cases} \mathscr{L}u = 0, \\ u|_S = 0, \\ u|_{t=0} = \varphi(x) \end{cases}$$

has a unique solution u *in* $C^{2+\beta, 1+\beta/2}(\overline{Q})$. *Moreover,* $u_{x_k t} \in L^2(Q)$ *for* $k = 1, \dots, n$.

PROOF. For a fixed point u^τ of $\Phi(\cdot, \tau)$ in $C^{2,1}(\overline{Q})$ we know that $\max_{\overline{Q}} |u^\tau| \le M$, $\max_{\overline{Q}} |\nabla u^\tau| \le M_1$, and $|u^\tau|_{\alpha, Q} + \sum_1^n |u^\tau_{x_i}|_{\alpha, Q} \le M_2$, where

M, M_1, α, and M_2 are determined by various constants in the theorem. Fix H_δ, $\delta \le \alpha \le 1$, and take $\overline{\mathscr{M}}$ to be $\{w : \max_{\overline{Q}}|w| \le M+\varepsilon,\ \max_{\overline{Q}}|\nabla w| \le M_1+\varepsilon$, and $|w|_{\delta,Q}+\sum_1^n |w_{x_i}|_{\delta,Q} \le M_2+\varepsilon\}$ for some $\varepsilon > 0$. Obviously $\partial\mathscr{M}$ does not contain any fixed points of Φ. When necessary, we can modify the values of a_i and a in $M \le |u| \le M+\varepsilon$ and $M_1 \le |p| \le M_1+\varepsilon$ so that (c)–(f) remain valid for $|u| \le M+\varepsilon$ and $|p| \le M_1+\varepsilon$. This can be simply achieved by a linear extension.

Now we verify all conditions required in the Leray-Schauder theorem. Clearly $\overline{\mathscr{M}}$ is the closure of a bounded, open connected subset in H_δ. In fact, it is convex.

From the interior and boundary estimates for solutions of linear parabolic equations (the two compatibility conditions in (b) are needed for the boundary estimate) we know that for $w \in \overline{M}$, $v = \Phi(w, \tau)$ belongs to $C^{2+\beta\delta,1+\beta\delta/2}(\overline{Q})$ since both $(\partial a_i/\partial p_j)(x, t, w, \nabla w)$ and $A(x, t, w, \nabla w)$ are in $C^{\beta\delta,\beta\delta/2}(\overline{Q})$. As $|w|_{H_\delta} \le M_2+\varepsilon$, there is a constant C such that $|v_t, v_{x_i}, v_{x_ix_j}|_{\beta\delta,\overline{Q}} \le C(M_2+\varepsilon)$. In particular, v_{x_i} is Hölder continuous with exponent 1 in x and v is Lipschitz continuous in t. By Lemma 2, v_{x_i} is Hölder continuous with exponent $1/2$. Therefore, $v, v_{x_i} \in C^{1,1/2}(\overline{Q})$ and v is compactly embedded in H_δ $(\delta < 1)$.

For w^1 and w^2 in H_δ, let $v^1 = \Phi(w^1, \tau)$ and $v^2 = \Phi(w^2, \tau)$. $v = v^2 - v^1$ satisfies:

$$\begin{aligned}
v_t - \left[\tau\frac{\partial a_i}{\partial p_j}(x, t, w^1, \nabla w^1) + (1-\tau)\delta_{ij}\right] v_{x_ix_j}& \\
= \tau\left(\frac{\partial a_i}{\partial p_j}(x, t, w^2, \nabla w^2) - \frac{\partial a_i}{\partial p_j}(x, t, w^1, \nabla w^1)\right) v^2_{x_ix_j}& \\
- \tau(A(x, t, w^2, \nabla w^2) - A(x, t, w^1, \nabla w^1)),& \\
v|_S = 0,& \\
v|_{t=0} = 0.&
\end{aligned}$$

When $|w^2 - w^1|_{H_\delta}$ is small for all τ, the right-hand side of the equation is small in $C^{\beta\delta/2,\beta\delta/4}(\overline{Q})$. (It is merely bounded in $C^{\beta\delta,\beta\delta/2}(\overline{Q})$ and is not sufficient for our purpose.) From linear estimates $|v_t, v_{x_i}, v_{x_ix_j}|_{\beta\delta/2,\overline{Q}}$ and consequently $|v|_{H_\delta}$ are small. Similarly one can show that $\Phi(w, \tau)$ is uniformly continuous in τ for each $w \in \overline{\mathscr{M}}$.

When $\tau = 0$, the problem $v_t - \Delta v = 0$, $v|_S = 0$, and $v|_{t=0} = 0$, has a unique solution which is identically zero. Thus $w = \Phi(w, 0)$ has only one fixed point and its degree is either 1 or -1.

From the linear estimates we also know that the fixed points w^τ satisfy $|w^\tau| \le M$, $|\nabla w^\tau| \le M_1$, and $|w^\tau|_{H_\delta} \le M_2$. Thus they do not lie on $\partial\mathscr{M}$.

Now we can apply the Leray-Schauder theorem to conclude that $\Phi(w, 1)$ has at least one fixed point in $\mathscr{M}$. It is also in $C^{2+\beta\delta, 1+\beta\delta/2}(\overline{Q})$.

In fact this fixed point u belongs to $C^{2+\beta, 1+\beta/2}(\overline{Q})$. For, $u \in C^{2+\beta\delta, 1+\beta\delta/2}(\overline{Q})$ implies $u_{x_i} \in C^{1, 1/2}(\overline{Q})$. This in turn implies

$$(\partial a_i/\partial p_j)(x, t, u, \nabla u), A(x, t, u, \nabla u) \in C^{\beta, \beta/2}(\overline{Q}).$$

From the regularity property of solutions for linear parabolic equations we conclude that u is in $C^{2+\beta, 1+\beta/2}(\overline{Q})$.

We claim the solution is unique in $C^{2,1}(\overline{Q})$. For, recall

$$\int_0^t \int_\Omega (u_t \eta + a_i(x, t, u, \nabla u)\eta_{x_i} + a(x, t, u, \nabla u)\eta) = 0,$$

$$\forall \eta \in C^{1,0}(\overline{Q}), \ \eta|_S = 0.$$

Denote the difference of two solutions u^1 and u^2 by $v = u^2 - u^1$. Since

$$\begin{aligned}\delta a_i &= a_i(x, t, u^2, \nabla u^2) - a_i(x, t, u^1, \nabla u^1) \\ &= v_{x_j} \int_0^1 \frac{\partial a_i}{\partial p_j}(x, t, \tau u^1 + (1-\tau)u^2, \tau \nabla u^1 + (1-\tau)\nabla u^2) d\tau \\ &\quad + v \int_0^1 \frac{\partial a_i}{\partial u}(\cdot) d\tau \\ &= \hat{a}_{ij} v_{x_j} + \hat{b}_i v\end{aligned}$$

and, similarly,

$$\delta a = \hat{c}_i v_{x_i} + \hat{c} v,$$

taking $\eta = v$ we have

$$\frac{1}{2}\int_\Omega v^2 + \int_0^t \int_\Omega [\hat{a}_{ij} v_{x_i} v_{x_j} + (\hat{b}_i + \hat{c}_i) v v_{x_i} + \hat{c} v^2] = 0,$$

or

$$\int_\Omega v^2 + \nu \int_0^t \int_\Omega |\nabla v|^2 \le \gamma \int_0^t \int_\Omega v^2, \quad \gamma \text{ a constant.}$$

Therefore, we deduce $v \equiv 0$ from $v|_{t=0} = 0$.

To conclude the proof we show $u_{x_k t} \in L^2(Q)$. Let

$$v_h = \frac{1}{h}(u(x, t+h) - u(x, t)).$$

Multiplying $0 = (\mathscr{L}u(x, t+h) - \mathscr{L}u(x, t))/h$ with v_h and then using integration by parts we get

$$\frac{1}{2}\frac{\partial}{\partial t}\int_\Omega v_h^2 + \int_\Omega [(\tilde{a}_{ij} v_{hx_j} + \tilde{b}_i v_h + \tilde{d}_i) v_{hx_i} + (\tilde{c}_i v_{hx_i} + \tilde{c} v_h + \tilde{d}) v_h] = 0,$$

where the coefficients $\tilde{a}_{ij}, \ldots$ are bounded in $0 < h \le h_0$. Hence

$$\int_\Omega v_h^2 + C \int_0^t \int_\Omega |\nabla v_h|^2 \le C_1 \int_0^t \int_\Omega v_h^2.$$

In other words, $\int_0^t \int_\Omega |\nabla v_h|^2$ is uniformly bounded in h. Letting $h \to 0$ we conclude $u_{x_k t} \in L^2(\overline{Q})$. The proof of the theorem is complete. □

SOME COMMENTS. 1. A weaker conclusion of Theorem 4 can be drawn when its assumptions are slightly relaxed. Specifically (f) can be waived and the smoothness of the boundary and initial data can be relaxed to $\partial\Omega \in C^2$ and $\max_{\overline{\Omega}} |\nabla\varphi| \le \text{Constant}$. Then the solution

$$u \in C^{2,1}(Q) \cap C^{2+\beta, 1+\beta/2}(\Omega' \times [\delta, T]), \ \delta > 0.$$

To see this we approximate u by the solutions obtained after smoothing out $\partial\Omega$, the coefficients and the boundary-initial data. On the other hand, for the uniqueness of the solution the differentiability of a with respect to u and p is essential.

2. One may also study the unique solvability of the first boundary value problem when a_i and a have different order of growth, or when the equation does not have a divergence structure. See Chapters 9 and 10 for further discussions.

3. In a parallel manner one can also establish the following result for the Dirichlet problem for a quasilinear elliptic equation. Consider

$$\begin{cases} \mathscr{L}u = \dfrac{da_i}{dx_i}(x, u, \nabla u) + a(x, u, \nabla u) = 0, \\ u|_{\partial\Omega} = \varphi, \end{cases}$$

where

$$a(x, u, 0) + \frac{\partial a_i}{\partial x_i}(x, u, 0) \le -b_1 u^2 + b_2 \qquad (b_1, b_2 \ge 0),$$

$$\frac{\partial a_i}{\partial p_j}(x, u, 0)\xi_i\xi_j \ge 0, \quad \forall \xi \in \mathbb{R}^n,$$

$$\nu(|u|)(1 + |p|^{m-2})|\xi|^2 \le \frac{\partial a_i}{\partial p_j}(x, u, p)\xi_i\xi_j \le \mu(|u|)(1 + |p|^{m-2})|\xi|^2, \qquad m \ge 2, \ \forall \xi \in \mathbb{R}^n,$$

and

$$|a(x, u, p)| + \sum_{i,j} \left|\frac{\partial a_i}{\partial x_j}\right| + \sum_i \left(|a_i| + \left|\frac{\partial a_i}{\partial u}\right|\right)(1 + |p|) \le \mu(|u|)(1 + |p|^m).$$

Suppose for $x \in \overline{\Omega}$ and bounded $|u|$, $|\nabla u|$, there exists $\alpha \in (0, 1)$ such that $a_i, a, \partial a_i/\partial p_j, \partial a_i/\partial u, \partial a_i/\partial x_j \in C^\alpha$, $\partial\Omega \in C^{2+\alpha}$, and $\varphi \in C^{2+\alpha}$. Then the Dirichlet problem has a unique solution in $C^2(\Omega) \cap C(\overline{\Omega})$. In fact, it belongs to $C^{2+\alpha}(\overline{\Omega})$.

To prove this result [5] one applies the Leray-Schauder theorem to $\tau\mathscr{L}u + (1-\tau)(1 + |\nabla u|^2)^{(m-2)/2}\Delta u = 0$, $u|_{\partial\Omega} = \tau\varphi$.

4. Concerning boundary value problems for elliptic equations [5] deals with uniform ones (i.e, the ratio of the minimal and maximal eigenvalues of $(\partial a_i/\partial p_j)$ is bounded from both sides by positive constants). Both uniform and nonuniform elliptic equations are discussed in [6]. A well-known example of a nonuniformly elliptic equation is the minimal surface equation

$$\frac{\partial}{\partial x_i}\left(\frac{u_{x_i}}{\sqrt{1+|\nabla u|^2}}\right)=0,$$

which is the Euler-Lagrange equation for the functional $\int(1+|\nabla u|^2)^{1/2}dx$. The eigenvalues of $(\partial a_i/\partial p_j)$ are $1, 1, \dots, 1, 1/(1+|\nabla u|^2)$. A study on the unique solvability of this equation can be found in [6].

6. An unsolved problem. Replace (5) and (6) by

$$\frac{\partial a_i}{\partial p_j}(x, t, u, p)\xi_i\xi_j \geq \nu(|u|)(1+|p|^{m-2})|\xi|^2, \quad \forall\xi\in\mathbb{R}^n,$$

and

$$\sum_i\left(|a_i|+\left|\frac{\partial a_i}{\partial u}\right|\right)(1+|p|)+\sum_{i,j}\left|\frac{\partial a_i}{\partial p_j}\right|(1+|p|^2)+|a|+\sum_{i,j}\left|\frac{\partial a_i}{\partial x_j}\right|$$
$$\leq \mu(|u|)(1+|p|^m), \qquad m>2.$$

How can one generalize the ideas of this chapter to establish the unique solvability of the first boundary value problem?

CHAPTER II

A Periodic Boundary Value Problem for a Nonlinear Telegraph Equation

Let Ω be a bounded domain in $\mathbb{R}^n$ and a_{jk} $(1 \le j, k \le n)$, $a_{jk} = a_{kj}$, be bounded functions defined in Ω. Suppose for some $K > 0$, $a_{jk}\xi_j\xi_k \ge K|\xi|^2$, $\forall \xi \in \mathbb{R}^n$, in Ω. Consider the problem

$$\begin{cases} \mathscr{L}u = u_{tt} - (a_{jk}(x)u_{x_k})_{x_j} - Cu + g(x, t, u)u_t \\ \qquad\qquad + f(x, t, u) = 0, \qquad (x, t) \in \Omega \times [0, 2\pi], \\ u|_{\partial\Omega \times [0, 2\pi]} = 0, \\ u(x, t + 2\pi) = u(x, t), \quad x \in \Omega, \end{cases}$$

where f and g are periodic in t with period 2π and C is a constant.

It is well known that solving a hyperbolic equation with its value prescribed on the entire boundary leads to an ill-posed problem. However, as a natural extension of periodic vibrations for a string or a membrane, a periodic boundary value problem in which solutions periodic in t and attaining a prescribed value on the boundary of the spatial domain are sought becomes well-posed under certain conditions. This observation lays the foundation for the investigation of this chapter.

A large portion of the study of the periodic boundary value problem for nonlinear hyperbolic equations has been devoted to the case of two independent variables [1]:

$$\begin{cases} u_{tt} - u_{xx} + f(x, t, u) = 0, \\ u(-\pi, t) = u(\pi, t) = 0, \\ u(x, t + 2\pi) = u(x, t), \end{cases}$$

where, for instance, f is asymptotically linear or superlinear. An essential reason that accounts for solvability is: The inverse operator $\mathscr{L}^{-1}$ of the self-adjoint operator $\mathscr{L} = \partial^2/\partial t^2 - \partial^2/\partial x^2$ is compact in the function space $L^2[(-\pi, \pi) \times [0, 2\pi]]/\mathscr{M}$. Here $\mathscr{M}$ is the null space of $\mathscr{L}$. The eigenvalues of $\mathscr{L}$ on $L^2[(-\pi, \pi) \times [0, 2\pi]]/\mathscr{M}$ are $m^2 - l^2$ with corresponding eigenfunctions $e^{i(mx+lt)}$ $(m^2 \ne l^2;\ m, l = 0, \pm 1, \pm 2, \dots)$. Since the eigenvalues have no limit points except ∞, $\mathscr{L}^{-1}$ has 0 as its unique limit point.

Expand $u(x, t)$ by the eigenfunctions: $u(x, t) = \sum a_j w_j(x, t)$, where the w_j's are the eigenfunctions. We have

$$\begin{aligned}\mathscr{L}^{-1}u &= \sum_{j\le J} a_j\lambda_j w_j(x, t) + \sum_{j>J} a_j\lambda_j w_j(x, t)\\ &= \mathscr{L}_1^{-1}u + \mathscr{L}_2^{-1}u.\end{aligned}$$

As $\mathscr{L}_1^{-1}$ is of finite rank (and hence compact) and

$$\begin{aligned}\|\mathscr{L}_2^{-1}\| &= \sup\frac{\|\mathscr{L}_2^{-1}u\|}{\|u\|} = \sup\left(\frac{\sum_{j>J}|a_j\lambda_j|^2}{\sum|a_j|^2}\right)^{1/2}\\ &\le \max_{j>J}|\lambda_j| \to 0\end{aligned}$$

as $J\to\infty$, $\mathscr{L}^{-1}$ is compact.

§1. Solvability for higher-dimensional telegraph equations in the nonresonance case

There are very few results for the periodic boundary value problem for nonlinear hyperbolic equations when the number of independent variables is greater than 2. A reason is that the inverse operator of

$$\frac{\partial^2}{\partial t^2} - \frac{\partial}{\partial x_i}\left(a_{ij}(x)\frac{\partial}{\partial x_j}\right)$$

is no longer compact. To illustrate this point, let us consider the special case $\mathscr{L} = \partial^2/\partial t^2 - \Delta$ and $\Omega = \{(x_1, \dots, x_n): -\pi < x_j < \pi,\ j = 1, \dots, n\}$. Nonzero eigenvalues of $\mathscr{L}$ are $\sum_{j=1}^n m_j^2 - l^2$ $(l^2 \ne \sum_{j=1}^n m_j^2)$. Therefore, they may have finite limit points. For example, when $l^2 = m_1^2$ $(l = m_1 = 1, 2, \dots)$, $m_2 = 1$, and $m_3 = \cdots = m_n = 0$, the resulting sequence of eigenvalues converges to 1. Consequently $\mathscr{L}^{-1}$ is not compact. One may not be able to select a convergent subsequence when using the solvability of the associated linear problem to approximate the nonlinear problem. The lack of tool hinders the study in this direction.

Things become better for the periodic boundary value problem for nonlinear hyperbolic equations in higher dimensions when $g \ne 0$ (it may be regarded as a friction term for the telegraph equation). When g is a nonzero constant,

$$\frac{\partial^2}{\partial t^2} - \frac{\partial}{\partial x_j}\left(a_{jk}\frac{\partial}{\partial x_k}\right) + g\frac{\partial}{\partial t}$$

has a compact inverse. As an example, observe that the nonzero eigenvalues of $\mathscr{L} = \partial^2/\partial t^2 - \Delta + g\partial/\partial t$ are $\sum_{j=1}^n m_j^2 - l^2 + igl$ $(i = \sqrt{-1})$. Hence ∞ is the only limit point.

When $g = \text{constant} \ne 0$, a study of solvability for the periodic boundary value problem for the telegraph equation can be found in [8]. Now we generalize the result therein to embrace nonconstant g's. Specifically we assume:

For small $\gamma > 0$, there exists $b_\gamma \in L^2(\Omega \times [0, 2\pi])$ such that $|f(x, t, z)| \leq \gamma|z| + b_\gamma(x, t)$. For example, $f = |z|^{1/2} \leq \gamma|z| + (4\gamma)^{-1}$. $g(x, t, z)$ is continuous in $\Omega \times [0, 2\pi] \times \mathbf{R}$ and periodic in t with period 2π. Also, $0 < \sigma_0 \leq |g(x, t, z)| \leq \sigma_1 < \infty$. Denote the eigenvalues of L by $-(\partial/\partial x_j)\cdot(a_{jk}(x)\partial/\partial x_k)$ by $0 < \lambda_1 \leq \lambda_2 \leq \cdots$ with normalized corresponding eigenfunctions $w_1, w_2, \ldots$.

THEOREM 1. *Under these conditions, the problem*

$$\begin{cases} \mathscr{L}u = 0, \\ u|_{\partial\Omega\times[0,2\pi]} = 0, \\ u(x, t+2\pi) = u(x, t) \end{cases}$$

has a generalized solution u *in* $\overset{\circ}{H}_1(\Omega\times[0, 2\pi])$ *when* $C \neq \lambda_j$ $(j = 1, 2, \ldots)$.

PROOF. For each $U \in L^2(\Omega \times [0, 2\pi])$, we shall solve

$$\begin{cases} \widetilde{\mathscr{L}}u = u_{tt} - \dfrac{\partial}{\partial x_j}(a_{jk}(x)u_{x_k})_{x_j} - Cu + g(x, t, U)u_t + f(x, t, U) = 0, \\ u|_{\partial\Omega\times[0,2\pi]} = 0, \\ u(x, t+2\pi) = u(x, t). \end{cases}$$

First we need some a priori estimates for the solution.

Multiplying the equation by u_t and then integrating it over $\Omega \times [0, 2\pi]$, we have

$$\begin{aligned} \sigma_0\|u_t\|^2 &\leq \int_{\Omega\times[0,2\pi]} g(x, t, u)u_t^2\,dxdt = \int_{\Omega\times[0,2\pi]} -f(x, t, U)u_t\,dxdt \\ &\leq \|f(x, t, U)\|\|u_t\|. \end{aligned}$$

Therefore, $\|u_t\| \leq (1/\sigma_0)\|f(x, t, U)\|$. Expand u by eigenfunctions:

$$u(x, t) = \sum_{j=1}^{\infty}\sum_{k=-\infty}^{\infty} \alpha_{j,k}w_j(x)e^{ikt} \qquad (\alpha_{j,-k} = \overline{\alpha}_{j,k}).$$

Let

$$\begin{aligned} u_0(x, t) &= \sum_{k\neq 0} \alpha_{j,k}w_j(x)e^{ikt}, \\ u_1(x) &= \sum_{\lambda_j > C} \alpha_{j,0}w_j(x), \\ u_2(x) &= \sum_{\lambda_j < C} \alpha_{j,0}w_j(x). \end{aligned}$$

From the assumption we know that $u = u_0 + u_1 + u_2$. We have

$$\|u_0\| \le \left(\sum_{k\neq 0} |\alpha_{j,k}|^2\right)^{1/2} \le \left(\sum_{k\neq 0} k^2 |\alpha_{j,k}|^2\right)^{1/2}$$
$$= \|u_{0t}\| = \|u_t\| \le \frac{1}{\sigma_0}\|f(x, t, U)\|.$$

Let $K_1 = \min_{\lambda_j > C}(\lambda_j - C)$ and $K_2 = \min_{\lambda_j < C}(C - \lambda_j)$. Multiplying $\widetilde{\mathscr{L}} u = 0$ by u_1, by integration we have

$$\begin{aligned} K_1\|u_1\|^2 &\le \int_{\Omega\times[0,2\pi]} \sum (\lambda_j - C)u_1^2 \\ &= \int_{\Omega\times[0,2\pi]} \left(\sum a_{jk}\frac{\partial u_1}{\partial x_j}\frac{\partial u_1}{\partial x_k} - Cu_1^2\right) dxdt \\ &= \int_{\Omega\times[0,2\pi]} -(gu_t + f)u_1 \\ &\le \sigma_1\|u_1\|\|u_t\| + \|f(x, t, U)\|\|u_1\|, \end{aligned}$$

and consequently,

$$\|u_1\| \le \frac{1}{K_1}\left(1 + \frac{\sigma_1}{\sigma_0}\right)\|f(x, t, U)\|.$$

By the same reasoning

$$\|u_2\| \le \frac{1}{K_2}\left(1 + \frac{\sigma_1}{\sigma_0}\right)\|f(x, t, U)\|.$$

Henceforth

$$\|u\|^2 = \|u_0\|^2 + \|u_1\|^2 + \|u_2\|^2 \le K_3^2\|f(x, t, U)\|^2, \tag{1}$$

where

$$K_3^2 = \frac{1}{\sigma_0^2} + \left(\frac{1}{K_1^2} + \frac{1}{K_1^2}\right)\left(1 + \frac{\sigma_1}{\sigma_0}\right)^2.$$

Integrating $u\widetilde{\mathscr{L}} u = 0$ by parts yields

$$\int_{\Omega\times[0,2\pi]} a_{jk}(x)u_{x_j}u_{x_k} - \|u_t\|^2\, dx\, dt = \int_{\Omega\times[0,2\pi]} (Cu - gu_t - f)u\, dx\, dt.$$

From this we deduce

$$\begin{aligned} \|u\|^2_{\overset{\circ}{H}_1(\Omega\times[0,2\pi])} &\le K_4\left[\|u_t\|^2 + \int_{\Omega\times[0,2\pi]} a_{jk}(x)u_{x_j}u_{x_k}\, dx\, dt\right] \\ &\le K_5^2\|f(x, t, U)\|^2. \end{aligned} \tag{2}$$

When g is equal to a constant σ ($\sigma = \pm\sigma_0$, $\sigma_0 > 0$), the problem

$$\begin{cases} \widetilde{\mathscr{L}} u = 0, \\ u|_{\partial\Omega\times[0,2\pi]} = 0, \\ u(x, t + 2\pi) = u(x, t) \end{cases}$$

is solvable. For, since $|f| \le \gamma|U| + b_\gamma(x, t) \in L^2(\Omega \times [0, 2\pi])$, f has the expansion

$$f = \sum f_{j,k} w_j(x) e^{ikt}, \qquad \sum_{j,k} |f_{j,k}|^2 < \infty.$$

Letting $u = \sum \alpha_{j,k} w_j(x) e^{ikt}$, by comparing coefficients in $\widetilde{\mathscr{L}} u = 0$ we have

$$\alpha_{j,k} = \frac{-f_{j,k}}{-k^2 + \lambda_j - C + k\sigma i} \quad \text{and} \quad |\alpha_{j,k}| \le \frac{|f_{j,k}|}{\min(K_1, K_2, k\sigma)}.$$

Therefore

$$\sum |\alpha_{j,k}|^2 < \infty$$

and $u \in L^2(\Omega \times [0, 2\pi])$. From (2) we conclude that $u \in \overset{\circ}{H}_1(\Omega \times [0, 2\pi])$. Hence a generalized solution exists.

Now we can use a continuation argument to extend the solvability to the general case when g is nonconstant. To see this let

$$\begin{cases} \widetilde{\mathscr{L}}_\theta u = u_{tt} - \dfrac{\partial}{\partial x_j}(a_{jk}(x) u_{x_k}) - Cu + [(1-\theta)\sigma + \theta g] u_t = -f, \\ u|_{\partial\Omega \times [0, 2\pi]} = 0, \\ u(x, t + 2\pi) = u(x, t) \end{cases}$$

for $0 \le \theta \le 1$. If the problem is uniquely solvable, it follows from (2) that $\|\widetilde{\mathscr{L}}_\theta^{-1}\| \le K_5$. We can rewrite $\widetilde{\mathscr{L}}_\theta u + \Delta\theta(g - \sigma) u_t = -f$ as

$$\left(I + \Delta\theta \widetilde{\mathscr{L}}_\theta^{-1}(g - \sigma)\frac{\partial}{\partial t}\right) u = -\widetilde{\mathscr{L}}_\theta^{-1} f \qquad (u \in \overset{\circ}{H}_1).$$

In case

$$|\Delta\theta| \left\| \widetilde{\mathscr{L}}_\theta^{-1}(g - \sigma)\frac{\partial}{\partial t} \right\| < 1,$$

the problem is solvable for $[\theta, \theta + \Delta\theta]$. Since

$$\begin{aligned} \left\| \widetilde{\mathscr{L}}_\theta^{-1}(g - \sigma)\frac{\partial}{\partial t} \right\| &= \sup \frac{\|\widetilde{\mathscr{L}}_\theta^{-1}(g - \sigma) u_t\|_{\overset{\circ}{H}_1}}{\|u\|_{\overset{\circ}{H}_1}} \\ &\le K_5 \sup \frac{\|(g - \sigma) u_t\|_{L^2}}{\|u\|_{\overset{\circ}{H}_1}} \\ &\le K_5 \sigma_1 \sup \frac{\|u_t\|_{L^2}}{\|u\|_{\overset{\circ}{H}_1}} \le K_5 \sigma_1, \end{aligned}$$

this inequality holds when $|\Delta\theta| < 1/K_5\sigma_1$. By applying this argument finitely many times this problem is solvable at $\theta = 1$.

From (1) and the assumption on f we know

$$\|u\|_{L^2} \le K_3(\gamma \|U\|_{L^2} + \|b_\gamma\|_{L^2}),$$

when u is a solution of

$$\begin{cases} \widetilde{\mathscr{L}}u = 0, \\ u|_{\partial\Omega\times[0,2\pi]} = 0, \\ u(x, t+2\pi) = u(x, t). \end{cases}$$

Choosing γ smaller such that $K_3\gamma < 1$, we have

$$\|u\|_{L^2} \le \frac{K_3}{1-K_3\gamma}\|b_\gamma\|_{L^2} \quad \text{whenever } \|U\|_{L^2} \le \frac{K_3}{1-K_3\gamma}\|b_\gamma\|_{L^2}.$$

Therefore, the mapping $U \mapsto u$ is bounded on $L^2(\Omega \times [0, 2\pi])$. Furthermore, from (2)

$$\|u\|_{\overset{\circ}{H}_1} \le \frac{K_5}{1-K_3\gamma}\|b_\gamma\|_{L^2}$$

holds. So $U \mapsto u$ is actually compact. By the Schauder fixed point theorem we conclude that the mapping $U \mapsto u$ has a fixed point in the ball $\|u\|_{L^2} \le (K_3/(1-K_3\gamma))\|b_\gamma\|_{L^2}$. Also this fixed point belongs to $\overset{\circ}{H}_1(\Omega\times[0, 2\pi])$. The proof of this theorem is finished. □

§2. A discussion on the resonance case

Next we consider the situation $C = \lambda_j$ for some $j \ge 1$. When this happens, it is said that the equation is at resonance. When the equation is linear, it is well known that solutions exist only if the right-hand side of the equation satisfies a certain orthogonality condition, while in the nonlinear case certain inequalities analogous to the orthogonality condition must be imposed on the nonhomogeneous term of the equation in order to preserve solvability. Usually such conditions are called Landesman-Lazor conditions [9].

Decompose $f(x, t, z) = f^{(1)}(x, t, z) + f^{(2)}(x, t)$, where $f^{(2)} \in L^2(\Omega \times [0, 2\pi])$ and $|f^{(1)}(x, t, z)| \le \gamma|z| + b_\gamma(x, t)$, $\forall\gamma > 0$, $b_\gamma \in L^2(\Omega \times [0, 2\pi])$. For instance, $f^{(1)} = |z|^{1/2} \le \gamma|z| + 1/2\gamma$ satisfies this condition. Further assume

$$f^{(1)}(x, t, z)\operatorname{sgn} z \ge -h(x, t), \qquad h \in L^2(\Omega \times [0, 2\pi]).$$

Let

$$f_+^{(1)}(x, t) = \liminf_{z\to+\infty} f^{(1)}(x, t, z)$$

and

$$f_-^{(1)}(x, t) = \limsup_{z\to-\infty} f^{(1)}(x, t, z)$$

be in $L^2(\Omega \times [0, 2\pi])$. Also we assume a Landesman-Lazor condition

$$\int_{w>0} f_+^{(1)}w + \int_{w<0} f_-^{(1)}w > C_0\|w\|_{L^2(\Omega\times[0,2\pi])}, \tag{3}$$

where w is an eigenfunction with eigenvalue equal to C. The constant C_0 will be chosen below.

Theorem 2. *Under the above assumptions as well as* $g = g(x, u)$, *there exists at least one generalized solution in the resonance case.*

Proof. We approximate the solution by solving

$$\begin{cases} \mathscr{L}u + \varepsilon u = 0, \\ u|_{\partial\Omega\times[0,2\pi]} = 0, \\ u(x, t+2\pi) = u(x, t). \end{cases}$$

When $\varepsilon \neq 0$ and small, solution u_ε exists. Can one find bounds on $\|u_\varepsilon\|_{L^2}$ and $\|u_\varepsilon\|_{\overset{\circ}{H}_1}$ which are independent of ε? From the Fredholm alternative we know that this is impossible when ε tends to zero from both sides. So we consider the case $\varepsilon \to +0$.

For the solution $u_\varepsilon = \sum \alpha^\varepsilon_{j,k} w_j(x) e^{ikt}$, we simplify notation by writing u_ε as u and $\alpha^\varepsilon_{j,k}$ as $\alpha_{j,k}$. Let

$$w = \sum_{\lambda_j = C} \alpha_{j,0} w_j(x).$$

From (1) we have

$$\|u - w\|^2 = \|u_0\|^2 + \|u_1\|^2 + \|u_2\|^2 \leq K_3^2 \|f(x, t, u)\|^2.$$

Therefore

$$\begin{aligned} \|u - w\| &\leq K_3(\gamma\|u\| + \|b_\gamma\| + \|f^{(2)}\|) \\ &\leq K_3\gamma(\|u - w\| + \|w\|) + K_4 \end{aligned}$$

and so

$$\|u - w\| \leq K_5\gamma\|w\| + K_6, \qquad \|u\| \leq (1 + K_5\gamma)\|w\| + K_6,$$

and

$$\|f^{(1)}(x, t, u)\| \leq \gamma\|u\| + \|b_\gamma\| \leq K_7\|w\| + K_8.$$

Now the orthogonality condition implies that

$$0 = \int_{\Omega\times[0,2\pi]} (\mathscr{L}u + \varepsilon u)w = \varepsilon\|w\|^2 + \int_{\Omega\times[0,2\pi]} f(x, t, u)w$$

holds. Notice that its derivation is based on

$$\int_{\Omega\times[0,2\pi]} (u - w)w = 0, \qquad \int_{\Omega\times[0,2\pi]} u_{tt} w = 0,$$

$$\int_{\Omega\times[0,2\pi]} g(x, u)u_t w = . \int_\Omega G(x, u)w\Big|_0^{2\pi} - \int_{\Omega\times[0,2\pi]} G(x, u)w_t = 0,$$

where $G(x, u) = \int g(x, u)\,du$, and

$$\begin{aligned} \int_{\Omega\times[0,2\pi]} \left[(a_{jk}(x)u_{x_k})_{x_j} + Cu\right] w &= \int_{\Omega\times[0,2\pi]} \left[(a_{jk}w_{x_k})_{x_j} + Cw\right] u \\ &= \int_{\Omega\times[0,2\pi]} (-\lambda_j w + Cw)u = 0. \end{aligned}$$

We need to remove the function u appearing in this condition so that it only involves w. To achieve this we use

$$\begin{aligned}
0 &= \varepsilon\|w\|^2 + \int_{\Omega\times[0,2\pi]} f(x,t,u)w \\
&\geq \int_{\Omega\times[0,2\pi]} f^{(1)}(x,t,u)u - \int_{\Omega\times[0,2\pi]} f^{(1)}(x,t,u)(u-w) \\
&\quad + \int_{\Omega\times[0,2\pi]} f^{(2)}w \\
&\geq \int_{\Omega\times[0,2\pi]} \operatorname{sgn} uf^{(1)}|u| - K_3\|f^{(1)}\|^2 - K_3\|f^{(1)}\|\,\|f^{(2)}\| - \|f^{(2)}\|\,\|w\| \\
&= \int_{\Omega\times(0,2\pi]} (\operatorname{sgn} uf^{(1)} + h)(|u| - K_3 \operatorname{sgn} uf^{(1)}) \\
&\qquad - (\|h\| + \|f^{(2)}\|)(\|u\| + K_3\|f^{(1)}\|) - \|f^{(2)}\|\,\|w\| \\
&\geq \int_{\Omega\times[0,2\pi]} (\operatorname{sgn} uf^{(1)} + h)[(1-K_3\gamma)|u| - K_3 b_\gamma] \\
&\qquad - (\|h\| + \|f^{(2)}\|)(\|u\| + K_3\|f^{(1)}\|) - \|f^{(2)}\|\,\|w\| \\
&\geq (1-K_3\gamma)\int_{\Omega\times[0,2\pi]} f^{(1)}(x,t,u)u - K_9\|w\| - K_{10}.
\end{aligned}$$

Taking γ so that $1 - K_3\gamma > 0$, we have

$$\begin{aligned}
0 &\geq \int_{\Omega\times[0,2\pi]} f^{(1)}(x,t,u)u - K_{11}\|w\| - K_{12} \\
&\geq \int_{\{(x,t)\in\Omega\times[0,2\pi]\,:\,u(x,t)>0\}} f_+^{(1)}(x,t)u \\
&\quad + \int_{\{(x,t)\in\Omega\times[0,2\pi]\,:\,u(x,t)<0\}} f_-^{(1)}(x,t)u - K_{11}\|w\| - K_{12}.
\end{aligned}$$

Since

$$\left|\int_{u>0} f_+^{(1)}(u-w)\right| \leq \|f_+^{(1)}\|\,\|u-w\| \leq \|f_+^{(1)}\|(K_5\gamma\|w\| + K_6)$$

and

$$\int_{u>0} f_+^{(1)}w - \int_{w>0} f_+^{(1)}w = \int_{\substack{u>0\\ w\leq 0}} - \int_{\substack{w>0\\ u\leq 0}},$$

when $u > 0$ and $w \leq 0$, $|w| \leq u - w$, we have

$$\left|\int_{\substack{u>0\\ w\leq 0}} f_+^{(1)}w\right| \leq \|f_+^{(1)}\|(K_5\gamma\|w\| + K_6).$$

Estimating the other terms by the same method, we eventually get

$$\begin{aligned}
0 \geq &\int_{\{(x,t)\in\Omega\times[0,2\pi]\,:\,w>0\}} f_+^{(1)}w \\
&+ \int_{\{(x,t)\in\Omega\times[0,2\pi]\,:\,w<0\}} f_-^{(1)}w - [K_{11} + 3K_5\gamma(\|f_+^{(1)}\| + \|f_-^{(1)}\|)]\|w\| \\
&- K_{13}.
\end{aligned}$$

Now, if we assume the Landesman-Lazor condition

$$\int_{w>0} f_{+}^{(1)} z + \int_{w<0} f_{-}^{(i)} z \geq (1+K_{11})\|z\|$$

for any $z(x) = \sum_{\lambda_j = C} \beta_j w_j(x)$, in particular choosing $\beta_j = \alpha_{j,0}^{\varepsilon}$ and $z(x) = w_\varepsilon(x, t)$, we have

$$0 \geq [1 - 3K_5\gamma(\|f_{+}^{(1)}\| + \|f_{-}^{(1)}\|)]\|w_\varepsilon\| - K_{13}.$$

Letting $\gamma = [4K_5(\|f_{+}^{(1)}\| + \|f_{-}^{(1)}\|)]^{-1}$, we obtain $\|w_\varepsilon\| \leq 4K_{13}$, and consequently

$$\|u_\varepsilon\| \leq (1 + K_5\gamma)\|w_\varepsilon\| + K_6 \leq K_{14},$$

$$\|f(x, t, u_\varepsilon)\| \leq \gamma\|u_\varepsilon\| + \|b_\gamma\| + \|f^{(2)}\| \leq K_{15}.$$

From (2)

$$\|u_\varepsilon - w_\varepsilon\|_{\mathring{H}_1} \leq K_5\|f(x, t, u_\varepsilon)\| \leq K_5 K_{15}$$

and

$$\|u_\varepsilon\|_{\mathring{H}_1}^2 \leq \|u_\varepsilon - w_\varepsilon\|_{\mathring{H}_1}^2 + \|w\|_{\mathring{H}_1}^2 \leq K_{16}.$$

Notice that in the last estimate we have used the fact

$$\|w_\varepsilon\|_{L^2} \leq 4K_{13} \Rightarrow \|w_\varepsilon\|_{\mathring{H}_1} \leq K_{17}.$$

Henceforth for suitably small $\varepsilon_0 > 0$, $\{u_\varepsilon\}$ is compact for $0 < \varepsilon < \varepsilon_0$. It follows that a limit function of $\{u_\varepsilon\}$ is a generalized solution for the resonance case $(C = \lambda_j)$. □

One may also consider the one-sided limit $\varepsilon \to -0$. Accordingly the conditions should be changed to

$$f^{(1)}(x, t, z)\operatorname{sgn} z \leq h(x, t), \qquad \int_{w>0} f_{+}^{(1)} w + \int_{w<0} f_{-}^{(1)} w \leq C_0\|w\|_{L^2}.$$

§3. Regularity of a generalized solution

We briefly discuss the regularity of a generalized solution.

THEOREM 3. *When $n = 1$ and $a_{11} = 1$, the generalized solution obtained in Theorems 1 and 2 belongs to $C^\infty([0, \pi] \times [0, 2\pi])$ if both f and g belong to $C^\infty([0, \pi] \times [0, 2\pi] \times \mathbf{R})$.*

PROOF. Let $\sigma_0 \operatorname{sgn} g = \sigma$. The equation can be written as

$$\mathscr{L}_0 u = \left(\frac{\partial^2}{\partial t^2} - \frac{\partial^2}{\partial x^2} + 1 + \sigma\frac{\partial}{\partial t}\right) u = F,$$

where $F = (1 + C)u - (g - \sigma)u_t - f$. F belongs to L^2 as $u \in \mathring{H}_1$. Let

$$F = \sum_{j=1}^{\infty} \sum_{k=-\infty}^{\infty} b_{jk} \sin jx e^{ikt}, \qquad \sum |b_{jk}|^2 < \infty,$$

and

$$u = \sum_{j=1}^{\infty} \sum_{k=-\infty}^{\infty} a_{jk} \sin jx e^{ikt}.$$

As a result of comparing coefficients, we have

$$a_{jk} = \frac{b_{jk}}{j^2 - k^2 + 1 + i\sigma k}.$$

Thus

$$\begin{aligned}
\int_0^\pi (u_t^2 + u_x^2 + u^2)\,dx &\le K_{18} \sum_{j,k_1,k_2} \frac{(1 + j^2 + |k_1 k_2|)|b_{jk_1}||b_{jk_2}|}{|j^2 - k_1^2 + 1 + i\sigma k_1||j^2 - k_2^2 + 1 + i\sigma k_2|} \\
&\le K_{19} \sum_{j,k_2} \sum_{k_1} \frac{(1 + j^2 + k_1^2)|b_{jk_2}|^2}{(j^2 + k_1^2 + 1)^2 + \sigma^2 k_1^2} \\
&\le K_{20} \sum_{j,k_2} |b_{jk_2}|^2 = K_{21}, \quad \forall t \in [0, 2\pi].
\end{aligned}$$

By the same reasoning

$$\int_0^{2\pi} (u_t^2 + u_x^2 + u^2)\,dt \le K_{22}, \quad \forall x \in [0, \pi].$$

Therefore

$$\begin{aligned}
|u(x_2, t) - u(x_1, t)| &= \left| \int_{x_1}^{x_2} u_x(\sigma, t)\,d\sigma \right| \\
&\le |x_2 - x_1|^{1/2} \left| \int_{x_1}^{x_2} u_x^2(\sigma, t)\,d\sigma \right| \le K|x_2 - x_1|^{1/2}
\end{aligned}$$

and, similarly,

$$|u(x, t_2) - u(x, t_1)| \le K|t_2 - t_1|^{1/2},$$

$$\begin{aligned}
|u(x_2, t_2) - u(x_1, t_1)| &\le K(|x_2 - x_1|^{1/2} + |t_2 - t_1|^{1/2}) \\
&\le 2K[(x_2 - x_1)^2 + (t_2 - t_1)^2]^{1/4}.
\end{aligned}$$

In other words, $u \in C^{1/2}$ and $\|u\|_{1/2} \equiv \|u\|_{C^{1/2}} \le 2K$.

Next we estimate the second derivatives. Differentiating the equation

$$u_{tt} - u_{xx} - Cu + gu_t + f = 0$$

in t, we obtain

$$u_{ttt} - u_{txx} - Cu_t + gu_{tt} + g_t u_t + g_u u_t^2 + f_t + f_u u_t = 0.$$

Multiplying it by u_{tt} and then integrating over $[0, \pi] \times [0, 2\pi]$ gives

$$\begin{aligned}
\sigma_0 \|u_{tt}\|^2 &\le \int_0^\pi \int_0^{2\pi} |g| u_{tt}^2 \le \int_0^\pi \int_0^{2\pi} |f_t + (f_u + g_t)u_t + g_u u_t^2||u_{tt}| \\
&\le K\|u_{tt}\| + \frac{\sigma_0}{2}\|u_{tt}\|^2 + K \int_0^\pi \int_0^{2\pi} u_t^4,
\end{aligned}$$

or

$$\|u_{tt}\|^2 \le K + K_3 \int_0^{\pi}\int_0^{2\pi} u_t^4.$$

For each $a \in [0, 2\pi]$, define a continuous cut-off function by

$$\eta(t) = \begin{cases} 1, & |t-a| \le \gamma, \\ 0, & |t-a| \ge 2\gamma, \end{cases}$$

and η is linear in $[a-2\gamma, a-\gamma]$ and $[a+\gamma, a+2\gamma]$. We have

$$\begin{aligned}\int_{a-\gamma}^{a+\gamma} u_t^4 &\le \int_{a-2\gamma}^{a+2\gamma} u_t^4\eta = -\int_{a-2\gamma}^{a+2\gamma}(3uu_t^2u_{tt}\eta + uu_t^3\eta') \\ &= -3\int_{a-2\gamma}^{a+2\gamma}(u(x,t)-u(x,a))u_t^2u_{tt}\eta \\ &\quad - \int_{a-2\gamma}^{a+2\gamma}(u(x,a)+u(x,t))u_t^3\eta' \\ &\le 6\|u\|_{1/2}\gamma^{1/2}\int_{a-2\gamma}^{a+2\gamma} u_t^2|u_{tt}| + \frac{K}{\gamma}\int_{a-2\gamma}^{a+2\gamma}\left[\frac{\gamma u_t^4}{6K} + \left(\frac{6K}{\gamma}\right)^3\right].\end{aligned}$$

Letting $\gamma = \frac{\pi}{m}$ and $a = \gamma, 3\gamma, \dots, (2m-1)\gamma$, and summing up all estimates we have

$$\begin{aligned}\int_0^{2\pi} u_t^4 &\le 2\|u\|_{1/2}\gamma^{1/2}\int_0^{2\pi} u_t^2|u_{tt}| + \frac{1}{3}\int_0^{2\pi} u_t^4 + \frac{K_1}{\gamma^4} \\ &\le \frac{2}{3}\int_0^{2\pi} u_t^4 + 108\gamma\|u\|_{1/2}^2\int_0^{2\pi} u_{tt}^2 + \frac{K_1}{\gamma^4}.\end{aligned}$$

Therefore

$$\int_0^{\pi}\int_0^{2\pi} u_t^4 \le 324\gamma\|u\|_{1/2}^2\int_0^{\pi}\int_0^{2\pi} u_{tt}^2 + \frac{K_2}{\gamma^4}.$$

For small γ we obtain from combining the above two inequalities $\|u_{tt}\|$, $\|u_t^2\| \le K$. As a result,

$$F_t = [(1+C)u - (g-\sigma)u_t - f]_t \in L^2.$$

Following what we have done before, one can show $u_t \in C^{1/2}$ and $|u_t| \le K$.

From the equation

$$u_{ttx} - u_{xxx} - Cu_x + gu_{tx} + g_xu_t + g_uu_tu_x + f_x + f_uu_x = 0$$

we deduce

$$\begin{aligned}\sigma_0\|u_{tx}\|^2 &\le \int_0^{\pi}\int_0^{2\pi}|g|^2u_{tx}^2 \\ &\le \int_0^{\pi}\int_0^{2\pi}|f_x + (g_x+f_u)u_x + g_uu_tu_x||u_{tx}| \\ &\le K\|u_{tx}\| + \frac{\sigma_0}{2}\|u_{tx}\|^2 + K\|u_x\|^2.\end{aligned}$$

Therefore $\|u_{tx}\| \leq K$ and $F_x \in L^2$. This implies $u_x \in C^{1/2}$ and, consequently, $u \in C^{3/2}$. Higher regularity can be obtained by a similar, and actually simpler, argument. The proof of the theorem is complete. □

REMARK. In this chapter we solved a nonlinear problem by the Schauder fixed point theorem. The assumption on $f(x, t, u)$ is sublinear or almost linear in u. Using other fixed point theorems, solvability results have been obtained in [10] where f may contain some superlinear terms in the nonresonance case.

An existing problem is: Can one relax the order of growth for the superlinear term in f in [10]? Also one is interested in extending the result to the resonance case.

CHAPTER III

The Initial Value Problem for a Nonlinear Schrödinger Equation [11]

§1. Background materials

THREE-LINES THEOREM IN ONE COMPLEX VARIABLE. *Let $f(z)$ $(z = x+iy)$ be bounded, analytic in Ω, and continuous in $\overline{\Omega}$, where Ω is a strip $\{z | x_1 < x < x_2\}$. Then*

$$|f(z)| \leq M_1^{(x_2-x)/(x_2-x_1)} M_2^{(x-x_1)/(x_2-x_1)},$$

where M_j $(j = 1, 2)$ satisfies $\sup_{y\in\mathbf{R}} |f(x_j + iy)| \leq M_j$.

PROOF. Let

$$f(z) = F(z) \exp\left(\frac{x_2 - z}{x_2 - x_1} \log M_1 + \frac{z - x_1}{x_2 - x_1} \log M_2\right).$$

Then $\sup_{\partial\Omega} |F(z)| \leq 1$ and $\sup_{\Omega} |F(z)| \leq K$. The function $F_\varepsilon(z) = F(z)e^{\varepsilon z^2}$ $(\varepsilon > 0)$ is analytic in Ω and

$$|F_\varepsilon(z)| \leq Ke^{\varepsilon[\max(x_1^2, x_2^2)-y^2]} \to 0 \quad \text{as } |y| \to \infty.$$

By the maximum principle $|F_\varepsilon(z)|$ can attain its maximum only at $x = x_1$ or $x = x_2$. In other words,

$$\max_{\overline{\Omega}} |F_\varepsilon(z)| \leq \max_{\partial\Omega} |F_\varepsilon(z)| \leq e^{\varepsilon \max(x_1^2, x_2^2)}.$$

Letting $\varepsilon \to 0$ we conclude $\sup_\Omega |F(z)| \leq 1$. □

REMARK 1. Apparently this theorem remains valid when the boundedness of f in Ω is relaxed to $f(z) = O(e^{|y|^2-\delta})$ $(\delta > 0)$.

REMARK 2. This theorem can be generalized to the situation where f is vector-valued, i.e., its range is a Banach space. When f is a finite-dimensional vector $f(z) = \sum_{j=1}^M f_j(z)e_j$ $(e_j \in B)$ we may simply replace the absolute value by a norm. When f is an infinite-dimensional vector, one needs to define a reasonable concept of analyticity.

RIESZ-THORIN INTERPOLATION THEOREM. *Let E_1 (resp. E_2) be a measure space with nonnegative, completely additive measure μ (resp. ν). (For the* *application below* $(E_1, \mu) = (E_2, \nu) = (\mathbb{R}^n$, *Lebesgue measure*).) *Let* T *be a linear mapping that sends simple functions on* E_1 *to* E_2. (*Recall that a simple function is a measurable function that takes finitely many finite values.*) *Suppose that* T *is of type* (α_j, β_j) *with norm* M_j $(j = 1, 2, 1 \le \alpha_j, \beta_j \le \infty)$. *Then for* (α, β) *satisfying*

$$\frac{1}{\alpha} = \frac{1-\lambda}{\alpha_1} + \frac{\lambda}{\alpha_2} \quad \text{and} \quad \frac{1}{\beta} = \frac{1-\lambda}{\beta_1} + \frac{\lambda}{\beta_2} \qquad (0 \le \lambda \le 1),$$

T *is of type* (α, β) *and its norm is less than or equal to* $M_1^{1-\lambda} M_2^{\lambda}$.

Recall that T is of type (α, β) if $|Tf|_{L^\beta} \le M|f|_{L^\alpha}$ for all simple functions f, and its norm is given by $\sup_f(|Tf|_{L^\beta}/|f|_{L^\alpha})$.

PROOF. Clearly the theorem holds when $\alpha = \infty$ or $\beta = 1$. So we may always assume $\alpha < \infty$ and $\beta > 1$ in the following proof.

Let

$$\alpha(z) = \left(\frac{1-z}{\alpha_1} + \frac{z}{\alpha_2}\right)^{-1} \quad \text{and} \quad \beta(z) = \left(\frac{1-z}{\beta_1} + \frac{z}{\beta_2}\right)^{-1}.$$

We have $\alpha = \alpha(\lambda)$ and $\beta = \beta(\lambda)$. Let f and g be simple functions in E_1 and E_2 respectively, normalized so that $|f|_{L^\alpha} = |g|_{L^{\beta/(1-\beta)}} = 1$. We would like to estimate $\int_{E_2} Tf \cdot g \, d\nu$.

For $f = |f|e^{iu}$ and $g = |g|e^{iv}$, define

$$F(z) = \begin{cases} |f|^{\alpha/\alpha(z)} e^{iu}, & f \ne 0, \\ 0, & f = 0, \end{cases}$$

and

$$G(z) = \begin{cases} |g|\dfrac{\beta(\beta(z)-1)}{(\beta-1)\beta(z)} e^{iv}, & g \ne 0, \\ 0, & g = 0. \end{cases}$$

For fixed u and v, F and G are analytic in z, while for each fixed z, they are simple functions. Therefore

$$I(z) = \int_{E_2} T(z)G(z)\, d\nu$$

is analytic in $\Omega = \{z| : 0 < \operatorname{Re} z < 1\}$. To see that I is bounded in $\overline{\Omega}$, denote the nonzero values of f (resp. g) by $C_1, \ldots, C_m$ (resp. $C_1', \ldots, C_M'$) and $\sigma_j = f^{-1}(C_j)$ (resp. $\sigma_j' = g^{-1}(C_j')$). We have

$$F(z) = \sum_j \sigma_j |C_j|^{\alpha/\alpha(z)} e^{iu_j} \qquad (u_j = \arg C_j),$$

$$G(z) = \sum_k \sigma_k' |C_k'|^{\beta(\beta(z)-1)/(\beta-1)\beta(z)} e^{iv_k} \qquad (v_k = \arg C_k'),$$

$$I(z) = \sum_{j,k} |C_j|^{\alpha/\alpha(z)} |C_k'|^{\beta(\beta(z)-1)/(\beta-1)\beta(z)} e^{i(u_j+v_k)} \int_{E_2} T\sigma_j \cdot \sigma_k' \, d\nu$$

and so

$$|I(z)| \leq \sum_{j,k} |C_j|^{\alpha/\alpha(x)} |C_k'|^{\beta(\beta(x)-1)/(\beta-1)\beta(x)} \left| \int_{E_2} T\sigma_j \cdot \sigma_k' \, d\nu \right|$$
$$\leq K \qquad (\operatorname{Re} z = x,\ 0 \leq x \leq 1,\ y \in \mathbb{R}).$$

When $\operatorname{Re} z = x = 0$, from the Hölder inequality and the definition of T

$$|I(z)| \leq |TF(z)|_{L^{\beta_1}} |G(z)|_{L^{\beta_1/(\beta_1-1)}}$$
$$\leq M_1 |F(z)|_{L^{\alpha_1}} |G(z)|_{L^{\beta_1/(\beta_1-1)}}.$$

In case $f = C_j$, $F = |C_j|^{\alpha/\alpha(z)} e^{iu_j}$. Therefore $|F| = |f|^{\alpha/\alpha_1}$ and

$$|F(z)|_{L^{\alpha_1}} = \left(\int_{E_1} |f|^\alpha \, d\mu \right)^{1/\alpha} = 1.$$

Similarly $|G(z)|_{L^{\beta_1/(\beta_1-1)}} = 1$. Hence $|I(z)| \leq M_1$ at $\operatorname{Re} z = 0$. By the same argument $|I(z)| \leq M_2$ at $\operatorname{Re} z = 1$. From three-lines theorem we conclude $|I(\lambda)| \leq M_1^{1-\lambda} M_2^\lambda$ for all normalized simple functions g. Since simple functions are dense in $L^{\beta/(1-\beta)}$, we obtain

$$|Tf|_{L^\beta} = \sup \left| \int_{E_2} Tf \cdot g \, d\nu \right| \leq M_1^{1-\lambda} M_2^\lambda.$$

Hence for an arbitrary f,

$$|Tf|_{L^\beta} \leq M_1^{1-\lambda} M_2^\lambda |f|_{L^\alpha}. \quad \square$$

REMARK. Since simple functions are dense in L^p $(1 \leq p \leq \infty)$ the conclusion of the theorem remains valid for integrable functions.

§2. The initial value problem for the linear Schrödinger equation

Let us begin with the simplest case:

$$\begin{cases} iu_t + u_{xx} = 0, & (x, t) \in \mathbb{R} \times \mathbb{R}^+, \\ u|_{t=0} = \varphi(x), & x \in \mathbb{R}, \end{cases}$$

where u and φ are complex-valued. The solution u can be put into a closed form, which is obtained by formally setting $a = i$ in the well-known formula

$$u = \frac{1}{(4\pi a)^{1/2}} \int_{-\infty}^{\infty} \varphi(\xi) e^{-(\xi - x)^2/4at} \, d\xi$$

for

$$\begin{cases} u_t = au_{xx} & (a > 0), \\ u|_{t=0} = \varphi. \end{cases}$$

To justify this we first prove that the formula holds for $\operatorname{Re} a > 0$.

For $\varphi \in L^2(\mathbb{R})$, let

$$u(x, t) = \int_{-\infty}^{\infty} e^{i\xi x} U(\xi, t) \, d\xi.$$

Then

$$u_{xx} = -\int_{-\infty}^{\infty} e^{i\xi x}\xi^2 U(\xi, t)\, d\xi.$$

From the equation we deduce $U_t + a\xi^2 U = 0$, or $U = U(\xi, 0)e^{-a\xi^2 t}$. Writing $U(\xi, 0) = \Phi(\xi)$, we have

$$\varphi(x) = \int_{-\infty}^{\infty} e^{i\xi x}\Phi(\xi)\, d\xi.$$

As a result,

$$\Phi(\xi) = \frac{1}{2\pi}\int_{-\infty}^{\infty} e^{-i\xi y}\varphi(y)\, dy$$

and

$$\begin{aligned} u(x, t) &= \int_{-\infty}^{\infty} e^{i\xi x}\Phi(\xi)e^{-a\xi^2 t}\, d\xi \\ &= \frac{1}{2\pi}\int_{-\infty}^{\infty}\int_{-\infty}^{\infty} e^{-a\xi^2 t + i\xi(x-y)}\varphi(y)\, dy\, d\xi \\ &= \frac{1}{2\pi}\int_{-\infty}^{\infty} \varphi(y)\, dy \int_{-\infty}^{\infty} e^{-a\xi^2 t + i\xi(x-y)}\, d\xi \\ &= \frac{1}{2\pi}\int_{-\infty}^{\infty} e^{-(x-y)^2/4at}\varphi(y)\, dy \int_{-\infty}^{\infty} e^{-[\xi - i(x-y)/2]^2/at}\, d\xi. \end{aligned}$$

Notice that in exchanging the order of the integration we need $\operatorname{Re} a > 0$. By Cauchy's theorem on contour integration

$$\begin{aligned} \int_{-\infty}^{\infty} e^{-[\xi - i(x-y)/2]^2/at}\, d\xi &= \int_{-\infty}^{\infty} e^{-\xi^2/at}\, d\xi \\ &= \frac{1}{\sqrt{at}}\int_{-\infty}^{\infty} e^{-\eta^2}\, d\eta = \left(\frac{\pi}{at}\right)^{1/2}. \end{aligned}$$

Therefore

$$\begin{aligned} u(x, t) &= \frac{1}{(4\pi at)^{1/2}}\int_{-\infty}^{\infty} e^{-(x-y)^2/4\pi t}\varphi(y)\, dy \\ &= \frac{1}{(\pi a)^{1/2}}\int_{-\infty}^{\infty} e^{-\eta^2/a}\varphi(x + 2\sqrt{t}\eta)\, d\eta. \end{aligned}$$

For smooth φ with compact support,

$$u(x, t) \to \frac{1}{(\pi i)^{1/2}}\int_{-\infty}^{\infty} e^{-\eta^2/i}\varphi(x + 2\sqrt{t}\eta)\, d\eta$$

as $\alpha \to i + 0$. Consequently

$$u(x, t) = \frac{1}{(4\pi it)^{1/2}}\int_{-\infty}^{\infty} e^{-(x-y)^2/4it}\varphi(y)\, dy,$$

where $i^{1/2} = e^{i\pi/4}$. Separating the real and imaginary parts, the initial value problem becomes

$$\begin{cases} u_t^1 + u_{xx}^2 = 0, & u_t^2 - u_{xx}^1 = 0, \\ u^1|_{t=0} = \varphi^1(x), & u^2|_{t=0} = \varphi^2(x) \end{cases}$$

and the solution is given by

$$u^1(x, t) + iu^2(x, t) = \frac{1-i}{\sqrt{8\pi t}} \int_{-\infty}^{\infty} \left[\cos\frac{(x-y)^2}{4t} + i\sin\frac{(x-y)^2}{4t}\right] \times (\varphi^1(y) + i\varphi^2(y))\, dy.$$

When φ is sufficiently smooth, it is not hard to show this formula really represents a classical solution.

Apart from a formula for the solution, conservation laws for this initial value problem are also available.

Multiply the equation by $2\overline{u}$ and then integrate over $[-X, X] \times [0, t]$. The real part of the resulting identity is

$$\int_{-X}^{X} |u|^2 \Big|_0^t = \int_0^t i(\overline{u}u_x - u\overline{u}_x)\Big|_{-X}^{X}.$$

As both $u(X, t)$ and $u_x(X, t)$ tend to zero as $X \to \pm\infty$ when φ has compact support, we have the conservation of mass:

$$\int_{-\infty}^{\infty} |u|^2 = \int_{-\infty}^{\infty} |\varphi|^2.$$

Multiply the equation obtained by differentiating the original equation in x by $2\overline{u}_x$ and then integrate. As above we deduce the conservation of energy:

$$\int_{-\infty}^{\infty} |u_x|^2 = \int_{-\infty}^{\infty} |\varphi_x|^2.$$

More generally, we have

$$\int_{-\infty}^{\infty} |D_x^k u|^2 = \int_{-\infty}^{\infty} |D^k \varphi|^2 \qquad (k = 2, 3, \dots).$$

As in the 1-dimensional case, the initial value problem for the higher-dimensional Schrödinger equation

$$\begin{cases} iu_t + \Delta u = 0, \\ u|_{t=0} = \varphi(x) \end{cases}$$

admits a solution given by the formula

$$u(x, t) = \frac{1}{(4\pi i t)^{n/2}} \int_{\mathbb{R}^n} e^{-\sum_1^n (x_j - y_j)^2/4it} \varphi(y)\, dy \equiv R(t)\varphi$$

when φ has compact support. Also, the conservation law

$$\|R(t)\varphi\|_{L^2} = \|\varphi\|_{L^2}$$

holds. Since smooth functions with compact support are dense in $L^2(\mathbb{R}^n)$, $R(t)$ is an isometry on $L^2(\mathbb{R}^n)$. On the other hand, from the formula for the solution we see that $R(t)$ sends $L(\mathbb{R}^n)$ into $L^\infty(\mathbb{R}^n)$ and, furthermore,

$$\|R(t)\varphi\|_{L^\infty} \le (4\pi t)^{-n/2} \|\varphi\|_L.$$

By the Riesz-Thorin theorem ($\alpha_1 = 1$, $\beta_1 = \infty$, $M_1 = (4\pi t)^{-n/2}$; $\alpha_2 = 2$, $\beta_2 = 2$, $M_2 = 1$)

$$\|R(t)\varphi\|_{L^q} \le (4\pi t)^{n/q-n/2}\|\varphi\|_{L^{\tilde{q}}} \qquad \left(2 < q < \infty, \ \frac{1}{q} + \frac{1}{\tilde{q}} = 1\right)$$

for all $\varphi \in L \cap L^2$. Taking closure in $L^{\tilde{q}}$ this inequality holds for all φ in $L^{\tilde{q}}$.

The connection between the regularity of the solution and its initial value for the Schrödinger equation is unlike other equations. Solutions for elliptic or parabolic equations are much more regular than their initial-boundary data. On the other hand, the solutions for hyperbolic equations are in general much rougher than their initial data. But for the Schrödinger equation the regularity of its solution is more or less the same as its initial value.

By the Duhamel principle one obtains a formula for the solution for the nonhomogeneous problem

$$\begin{cases} iu_t + \Delta u = f(x, t), \\ u|_{t=0} = \varphi(x): \end{cases}$$

$$\begin{aligned} u &= R(t)\varphi + \int_0^t R(t-s)f(s)\,ds \\ &= \int_{\mathbb{R}^n} R(x-\xi, t)\varphi(\xi)\,d\xi + \int_0^t \int_{\mathbb{R}^n} R(x-\xi, t-s)f(\xi, s)\,d\xi\,ds. \end{aligned}$$

Energy inequalities can also be derived. But we shall not go into this.

§3. The initial value problem for a nonlinear Schrödinger equation

Now let us look at the initial value problem for the nonlinear Schrödinger equation,

$$\begin{cases} iu_t + \Delta u = F(u), & (x, t) \in \mathbb{R}^n \times \mathbb{R}^t, \\ u|_{t=0} = \varphi(x), & x \in \mathbb{R}. \end{cases}$$

One may consider two cases:

Case 1. $F(u) = f(|u|^2)u$, where f is real. This case is close to physical reality. Formulas for conservation of mass and energy can be derived.

Case 2. F is a general function.

Here we study the initial value problem from a differential equational point of view. We shall prove

THEOREM 1. *Suppose that* $F(u) \in C^1(\mathbb{R})$, $F'(u) = O(1)$ $(u \in \mathbb{R})$, *and further,* $F(u) = O(|u|^p)$, $F'(u) = O(|u|^{p-1})$ *for small* u, *where* p *satisfies*

$$\frac{1}{2} + \frac{1}{n} + \left(\frac{1}{4} + \frac{3}{n} + \frac{1}{n^2}\right)^{1/2} < p < \frac{n+2}{n-2}$$

(*no restriction on the right-hand side of this inequality when* $n = 1$ *and* 2). *There exists* $\varepsilon > 0$ *such that for* $\varphi \in L(\mathbb{R}^n) \cap W_2^k(\mathbb{R}^n)$,

$$\|\varphi\|_L, \ \|\varphi\|_{L^2}, \ \|D^k\varphi\|_{L^2} < \delta,$$

where δ *is a given small positive number and* $\frac{n}{2} < k \leq \frac{n}{2}+1$, *the initial value problem*

$$\begin{cases} iu_t + \Delta u = F(u), \\ u|_{t=0} = \varphi(x) \end{cases}$$

has a unique generalized solution $u \in C(L^2(\mathbb{R}^n), \mathbb{R}^+)$. *Moreover,*

$$\|u(t)\|_{L^2} \leq K_1,$$
$$\|u(t)\|_{L^{p+1}} \leq K_2(1+t)^{-n(p-1)/2(p+1)},$$
$$\|u(t)\|_{L^\infty} \leq K_3(1+t)^{-\min(1/2, (p-3)/2)}$$

(*a factor* $\log(1+t)$ *should be added to the right-hand side when* $n=1$ *and* $p=4$), *and*

$$\|u(t)\|_{L^{2n/(n-2+2\varepsilon)}} \leq K_4(1+t)^{\varepsilon - \min\{1, n[pn/(n+2-2\varepsilon)-1]\}}$$

for $n \geq 2$, *and* $0 < \varepsilon < 1 - n(p-1)/2(p+1)$ (*a factor* $\log(1+t)$ *should be added to the right-hand side when* $n[pn/(n+2-2\varepsilon)-1] = 1$).

REMARK 1. $u \in C(L^2(\mathbb{R}^n), \mathbb{R}^+)$ means for all $t \in (0, \infty)$,

$$\|u(x, t+\delta t) - u(x, t)\|_{L^2(\mathbb{R}^n)} \to 0 \quad \text{as } \delta t \to 0.$$

REMARK 2. $u \in C(L^2(\mathbb{R}), \mathbb{R}^+)$ is a generalized solution means

(a) it satisfies the equation in the sense

$$i(u(s), v(s))|_0^t + \int_0^t (u(s), iv_s + \Delta v)\, ds = \int_0^t (F(u(s)), v(s))\, ds$$

for all v, with $v, v_t, D_x^2 v \in C(L^2(\mathbb{R}^n), \mathbb{R}^+)$ (here $(\cdot, \cdot)$ is the L^2-product); and

(b) it satisfies $u|_{t=0} = \varphi$ in the $C(L^2(\mathbb{R}^n), \mathbb{R}^+)$-sense.

REMARK 3. The inequalities fulfilled by p can be rewritten as

$$\frac{2(p+1)}{p(p-1)} < n < \frac{2(p+1)}{p-1}.$$

Hence it is possible to have ε with $0 < \varepsilon < 1 - n(p-1)/2(p+1)$.

PROOF. A solution for this nonlinear problem can be regarded as a fixed point $w + W = V$ for the following two problems:

$$\begin{cases} iw_t + \Delta w = 0, \\ w|_{t=0} = \varphi \end{cases}$$

and

$$\begin{cases} iW_t + \Delta W = F(V), \\ W|_{t=0} = 0. \end{cases}$$

We shall first derive estimates for w and W and then decide which fixed point theorem is more appropriate.

Using the Riesz-Thorin theorem we have

$$\begin{aligned}\|w\|_{L^q} = \|R(t)\varphi\|_{L^q} &\le (4\pi t)^{n/q-n/2}\|\varphi\|_{L^{\bar q}} \\ &\le K_5 t^{n/q-n/2}(\|\varphi\|_L + \|\varphi\|_{L^2}), \qquad q \ge 2.\end{aligned}$$

This estimate is not effective for small t. To compensate it we use $\|D_x^k w\|_{L^2} = \|D^k\varphi\|_{L^2}$ and the embedding theorem to derive

$$\begin{aligned}\|w\|_{L^q} &\le \|w\|_{L^\infty}^{(q-2)/q}\|w\|_{L^2}^{2/q} \le (K_6\|w\|_{L^2} + K_7\|D_x^k w\|_{L^2})^{(q-2)/q}\|w\|_{L^2}^{2/q} \\ &= (K_6\|\varphi\|_{L^2} + K_7\|D_x^k\varphi\|_{L^2})^{(q-2)/q}\|\varphi\|_{L^2}^{2/q}.\end{aligned}$$

Combining these two estimates we have

$$\begin{aligned}|||w|||_q &\equiv \sup_{t\ge 0}[(1+t)^{n/2-n/q}\|w\|_{L^q}] \\ &\le K_8(\|\varphi\|_L + \|\varphi\|_{L^2} + \|D^k\varphi\|_{L^2}).\end{aligned}$$

It is not hard to verify that $|||\cdot|||$ is a norm.

Next we treat $W(t) = \int_0^t R(t-s)F(V(s))\,ds$:

$$\begin{aligned}\|W\|_{L^{p+1}} &\le K_{11}\int_0^t (t-s)^{n/(p+1)-n/2}\||V(s)|^p\|_{L^{1+1/p}}\,ds \\ &= K_{11}\int_0^t (t-s)^{n/(p+1)-n/2}\|V(s)\|_{L^{p+1}}^p\,ds \\ &\le K_{12}|||V|||_{p+1}^p\int_0^t (t-s)^{-n(p-1)/2(p+1)}(1+s)^{-np(p-1)/2(p+1)}\,ds \\ &\le K_{13}|||V|||_{p+1}^p(1+t)^{-n(p-1)/2(p+1)}.\end{aligned}$$

Hence

$$|||W|||_{p+1} \le K_{13}|||V|||_{p+1}^p.$$

Since no evidence of the compactness of the mapping $V \mapsto W$ can be seen from this estimate, fixed point theorems such as the Schauder fixed point theorem, or the Leray-Schauder theorem on topological degree, which require compactness, are not applicable. We feel that the contraction mapping principle, which does not use compactness, may work for $|||W|||_{p+1}$. This is suggested by the fact that $|||W|||_{p+1}$ is estimated by the pth power of $|||V|||_{p+1}$. Let

$$V = u^{(m-1)}, \qquad w + W = u^{(m)} \quad (m = 1, 2, \ldots), \qquad u^{(0)} = w.$$

We have

$$|||u^{(m)} - w|||_{p+1} \le K_{13}|||u^{(m-1)}|||_{p+1}^p$$

and

$$|||u^{(m)}|||_{p+1} \le |||w|||_{p+1} + K_{13}|||u^{(m-1)}|||_{p+1}^p.$$

For $|||w|||_{p+1}$ so small that

$$K_{13}|||w|||_{p+1}^{p-1} \leq \frac{1}{2^p},$$

it follows from induction that

$$|||u^{(m)}|||_{p+1} \leq 2|||w|||_{p+1} \qquad (m = 0, 1, 2, \ldots).$$

For this is obviously true when $m = 0$. Assuming that it holds for $m-1$, we have

$$\begin{aligned}|||u^{(m)}|||_{p+1} &\leq |||w|||_{p+1} + K_{13}|||u^{(m-1)}|||_{p+1}^p \\ &\leq |||w|||_{p+1} + K_{13}(2|||w|||_{p+1})^p \leq 2|||w|||_{p+1}.\end{aligned}$$

We further have

$$\begin{aligned}&|||u^{(m+1)} - u^{(m)}|||_{p+1} \\ &= \sup_{t\geq 0}\left\{(1+t)^{n/2-n/(p+1)}\left\|\int_0^t R(t-s)[F(u^{(m)}) - F(u^{(m-1)})]\,ds\right\|_{L^{p+1}}\right\} \\ &\leq K_{14}\sup_{t\geq 0}\left\{(1+t)^{n/2-n/(p+1)}\right. \\ &\quad\left.\times\int_0^t (t-s)^{n/(p+1)-n/2}\|(u^{(m)} - u^{(m-1)})F'(\theta u^{(m)} + (1-\theta)u^{(m-1)})\|_{L^{1+1/p}}\,ds\right\} \\ &\leq K_{15}|||w|||_{p+1}^{p-1}|||u^{(m)} - u^{(m-1)}|||_{p+1} \qquad (0<\theta<1).\end{aligned}$$

Notice that we have used the Riesz-Thorin theorem, the Hölder inequality, $F'(u) = O(|u|^{p-1})$, and $|||u^{(m)}|||_{p+1} \leq 2|||w|||_{p+1}$. From this estimate we see that for small $|||w|||_{p+1}$ so that $K_{15}|||w|||_{p+1}^{p-1} = \delta_1 < 1$,

$$|||u^{(m+1)} - u^{(m)}|||_{p+1} \leq \delta_1|||u^{(m)} - u^{(m-1)}|||_{p+1}.$$

Hence $u^{(m)}$ converges to some u in the norm $|||\cdot|||$. In order to guarantee $K_{15}|||w|||_{p+1}^{p-1} < 1$ and $K_{13}|||w|||_{p+1}^{p-1} \leq 1/2^p$, in view of the estimate $|||w|||_{p+1} \leq K_8(\|\varphi\|_L + \|\varphi\|_{L^2} + \|D^k\varphi\|_{L^2})$ it suffices to restrict the initial value φ to $\|\varphi\|_L$, $\|\varphi\|_{L^2}$, $\|D^k\varphi\|_{L^2} < \delta$ for suitably small δ.

To derive further properties of the limit function u, first we note that $|||u|||_{p+1} \leq K_2$, i.e., $\|u\|_{L^{p+1}} \leq K_2(1+t)^{-n(p-1)/2(p+1)}$. Multiplying $iW_t + \Delta W = F(V)$ by $2\overline{W}$ and then integrating over $\mathbb{R}^n \times (0, t)$, the imaginary part of the resulting identity is

$$\begin{aligned}\|W(t)\|_{L^2}^2 &= 2\,\mathrm{Im}\int_0^t ds\int_{\mathbb{R}^n}\overline{W}F(V)\,dx \\ &\leq 2\int_0^t \|W(s)\|_{L^{p+1}}\|F(V(s))\|_{L^{1+1/p}}\,ds \\ &\leq K_{16}|||V|||_{p+1}^{2p}\int_0^t (1+s)^{-n(p-1)/2}\,ds.\end{aligned}$$

As $n(p-1) > 2(1+\frac{1}{p}) > 2$,

$$\|W(t)\|_{L^2} \le K_{17}|||V|||_{p+1}^p.$$

Substituting $V = u^{(m-1)}$ and $W = u^{(m)} - w$ into this estimate, we obtain

$$\|u - w\|_{L^2} \le K_{17}|||u|||_{p+1}^p$$

after passing to the limit. Hence $u \in L^2(\mathbb{R}^n)$, $\forall t > 0$, and

$$\|u\|_{L^2} \le \|\varphi\|_L + \|\varphi\|_{L^2} + K_1\|u\|_{p+1}^p \le \widetilde{K}_1.$$

From $W(t) = \int_0^t R(t-s)F(V(s))\,ds$ we have, for $n = 1$,

$$\begin{aligned}
|W(t)| &\le 4\pi \int_0^t (t-s)^{-1/2}\|F(V(s))\|_L\,ds \\
&\le K_{18}\int_0^t (t-s)^{-1/2}\||V(s)|^p\|_L\,ds \\
&\le K_{18}\|V\|_{L^2}^{2/(p-1)}|||V|||_{p+1}^{(p-2)(p+1)/(p-1)} \int_0^t (t-s)^{-1/2}(1+s)^{1-p/2}\,ds \\
&\le K_{19}\|V\|_{L^2}^{2/(p-1)}|||V|||_{p+1}^{(p-2)(p+1)/(p-1)} \\
&\quad \times (1+t)^{-\min(1/2,(p-3)/2)} \cdot \begin{cases} 1, & p > 3, p \ne 4, \\ \log(1+t), & p = 4. \end{cases}
\end{aligned}$$

For $n \ge 2$ and $0 < \varepsilon < 1 - n(p-1)/2(p+1)$, from the Riesz-Thorin theorem we have

$$\begin{aligned}
&\|W(t)\|_{L^{2n/(n-2+2\varepsilon)}} \le K_{20}\int_0^t (t-s)^{\varepsilon-1}\|F(V(s))\|_{L^{2n/(n+2-2\varepsilon)}}\,ds \\
&\le K_{21}\int_0^t (t-s)^{\varepsilon-1}\||V(s)|^p\|_{L^{2n/(n+2-2\varepsilon)}}\,ds \\
&\le K_{22}\|V\|_{L^2}^{2(p+1)/(n+2-2\varepsilon)[1-n(p-1)/2(p+1)-\varepsilon]}|||V|||_{p+1}^{[2(p+1)/(p-1)](pn/(n+2-2\varepsilon)-1)} \\
&\quad \times \int_0^t (t-s)^{\varepsilon-1}\cdot(1+s)^{-n[pn/(n+2-2\varepsilon)-1]}\,ds \\
&\le K_{23}\|V\|_{L^2}^{[2(p+1)/(n+2-2\varepsilon)][1-n(p-1)/2(p+1)-\varepsilon]}|||V|||_{p+1}^{(2(p+1)/(p-1))[pn/(n+2-2\varepsilon)-1]} \\
&\quad \times (1+t)^{\varepsilon\min\{1,n[pn/(n+2-2\varepsilon)-1)]\}} \\
&\quad \times \begin{cases} 1, & n(p-1) > 2,\ n\left(\dfrac{pn}{n+2-2\varepsilon} - 1\right) \ne 1, \\ \log(1+t), & n\left(\dfrac{pn}{n+2-2\varepsilon} - 1\right) = 1. \end{cases}
\end{aligned}$$

Letting $W = u^{(m)} - w$ and $V = u^{(m-1)}$ in this estimate, we conclude an estimate on $\|u\|_{L^\infty}$ $(n = 1)$ or $\|u\|_{L^{2n/(n-2+2\varepsilon)}}$ after taking the limit.

To show that u takes φ as its initial value, we need

$$|||u(t+\delta t) - u(t)|||_{p+1} = O(|\delta t|),$$
$$\|u(t+\delta t) - u(t)\|_{L^2} = O(|\delta t|^{1/2}).$$

These two estimates can be derived by the same reasoning as above. Hence

$$u \in C^1(L^{p+1}(\mathbb{R}^n), \mathbb{R}^+) \cap C^{1/2}(L^2(\mathbb{R}^n), \mathbb{R}^+)$$

and it takes the initial value in the sense $|||u(\delta t) - \varphi|||_{p+1} \to 0$ or

$$\|u(\delta t) - \varphi\|_{L^2} \to 0 \quad \text{as } \delta t \to 0.$$

Next we show that u satisfies the equation. For $v, v_t, D_x^2 v \in C(L^2(\mathbb{R}^n), \mathbb{R})$, multiply the equation

$$\begin{cases} iu_t^{(m)} + \Delta u^{(m)} = F(u^{(m-1)}), \\ u^{(m)}(0) = \varphi \end{cases}$$

by $\overline{v}$ and then integrate. We have

$$i(u^{(m)}, v)|_0^T + \int_0^T (u^{(m)}, iv_t + \Delta v)\, dt = \int_0^T (F(u^{(m-1)}), v)\, dt$$

(it is not hard to see that $D_x^i u^{(m)}$, $i = 1, 2$, belongs to $L^2(\mathbb{R}^n)$). Letting $m \to \infty$,

$$i(u, v)|_0^T + \int_0^T (u, iv_t + \Delta v)\, dt = \int_0^T (F(u), v)\, dt, \quad \forall T \in \mathbb{R}^+.$$

To show the uniqueness of this generalized solution, we argue as follows. Suppose u and $u+U$ are two generalized solutions in $C(L^2(\mathbb{R}^n), \mathbb{R}^+)$. Then $U|_{t=0} = 0$ and

$$\begin{aligned} i(U(T), v(T)) &+ \int_0^T (U, iv_t + \Delta v)\, dt \\ &= \int_0^T ([F(u+U) - F(u)], v)\, dt. \end{aligned} \tag{1}$$

If we can take $v = U$, then after taking the imaginary part we have

$$\begin{aligned} \frac{1}{2}\|U(T)\|_{L^2}^2 &= \operatorname{Im} \int_0^T ([F(u+U) - F(u)], U)\, dt \\ &\le \max_{0<\theta<1} |F'(u+\theta U)| \int_0^T \|U(t)\|_{L^2}^2\, dt \\ &\le K_{24} \int_0^T \|U(t)\|_{L^2}^2, \quad \forall T \in \mathbb{R}^+ \end{aligned}$$

since $F'(u) = O(1)$ $(u \in \mathbb{R})$. From this we deduce $U(x, t) = 0$, $\forall (x, t) \in \mathbb{R}^n \times \mathbb{R}^+$.

Unfortunately we cannot take $v = U$ directly since U merely belongs to $C(L^2(\mathbb{R}^n), \mathbb{R}^+)$. To overcome this difficulty, we smooth out U as follows.

The function

$$j(x) = \begin{cases} C \exp[(|x|^2/\lambda^2 - 1)^{-1}], & |x| < \lambda, \\ 0, & |x| \ge \lambda \end{cases}$$

discussed in [49] satisfies conditions $j \in C_0^\infty(\mathbb{R}^n)$, $\int_{\mathbb{R}^n} j(x)\,dx = 1$, $j(x)$ is even in x, $j(x) = O(\lambda^{-n})$, and $Dj(x) = O(\lambda^{-n-1})$. Similarly we define $k(t)$ which satisfies $k \in C_0^\infty(\mathbb{R})$, $\int_{\mathbb{R}} k(t)\,dt = 1$, $k(t) = 0$ for $|t| \geq \mu$, $k(t)$ is even in t, $k(t) = O(\mu^{-1})$, and $k'(t) = O(\mu^{-2})$.

Extend $U(x, t)$ to $t < 0$ by setting it equal to zero. Since $U(x, 0) = 0$, the extended function belongs to $C(L^2(\mathbb{R}^n), \mathbb{R})$.

Set $v(x, t) = \int_{\mathbb{R}^{n+1}} U(y, \tau) j(x-y) k(t-\tau)\,dy\,d\tau$ in (1). Taking the imaginary part and then letting λ and μ tend to zero, we have

$$\|U(T)\|^2 + \lim_{\lambda\to 0} I_1 + \lim_{\mu\to 0} I_2 = \operatorname{Im} \int_0^T ([F(u+U) - F(u)], U)\,dt,$$

where

$$I_1 = \int_0^T dt \int_{\mathbb{R}^n\times\mathbb{R}^n} \Delta_x j(x-y)[U_1(y, t)U_2(x, t) - U_1(x, t)U_2(y, t)]\,dx\,dy,$$

$$I_2 = \int_{\mathbb{R}^n} dx \int_0^T dt \int_{t-\mu}^{t+\mu} [U_1(x, t)U_1(x, \tau) + U_2(x, t)U_2(x, \tau)] k_t(t-\tau)\,d\tau,$$

$$U_1 = \operatorname{Re} U \quad \text{and} \quad U_2 = \operatorname{Im} U.$$

Observe that the integrands of I_1 and I_2 are odd. We have $I_1 = 0$ and

$$\begin{aligned} I_2 &= \int_{\mathbb{R}^n} dx \left(\int_{T-\mu} dt \int_T^{t+\mu} d\tau + \int_0^\mu \int_{-t} d\tau \right) \\ &\quad \times [U_1(x, t)U_1(x, \tau) + U_2(x, t)U_2(x, \tau)] k_t(t-\tau)\,d\tau \\ &= \|U(T)\|^2 \int_{T-\mu}^T dt \int_T^{t+\mu} k_t(t-\tau)\,d\tau + o(1) \\ &= -\frac{1}{2}\|U(T)\|^2 + o(1). \end{aligned}$$

As a result, we still have

$$\frac{1}{2}\|U(T)\|^2 = \operatorname{Im} \int_0^T ([F(u+U) - F(u)], U)\,dt, \quad \forall T \in \mathbb{R}^+.$$

The proof of Theorem 1 is complete. □

THEOREM 2. *The solution $u(x, t)$ found in Theorem 1 is asympotically equal to a solution $u_+(x, t)$ of the linear equation $iu_t + \Delta u = 0$ in the L^2-sense (for large t).*

PROOF. More precisely, u_+ is given by

$$u_+(x, t) = R(t)\varphi + \int_0^\infty R(t-s)F(u(s))\,ds.$$

It solves

$$\begin{cases} iv_t + \Delta v = 0, \\ v|_t = \varphi + \int_0^\infty R(-s)F(u(s))\,ds. \end{cases}$$

We would like to show

$$\|u(x,t) - u_+(x,t)\|_{L^2} = \left\| \int_t^\infty R(t-s)F(u(s))\,ds \right\|_{L^2} \to 0 \qquad (t \to \infty)$$

(as a by-product $u_+(x,t) \in L^2 \ \forall t$). Since

$$\left\| \int_t^\infty R(t-s)F(u(s))\,ds \right\|_{L^2} \le \int_t^\infty \|F(u(s))\|_{L^2}\,ds,$$

the theorem follows if we can prove

$$\|F(u(s))\|_{L^2} = O((1+s)^{-1-\eta}) \tag{2}$$

for some $\eta > 0$.

When $n = 1$,

$$\begin{aligned} \|F(u(s))\|_{L^2}^2 &\le K\||u(s)|^{2p}\|_L \le K\|u(s)\|_{L^{p+1}}^{p+1}\|u(s)\|_{L^\infty}^{p-1} \\ &\le \widetilde{K}(1+s)^{-(p+1)/2-(p-1)\min\{1/2,(p-3)/2\}} \\ &\quad \times \begin{cases} 1, & p \ne 4, \\ \log^{p-1}(1+s), & p = 4. \end{cases} \end{aligned}$$

As for $p \ge 4$,

$$\frac{p-1}{2} + (p-1)\min\left\{\frac{1}{2}, \frac{p-3}{2}\right\} = p - 1 > 2$$

and for $p < 4$, $p(p-1) > 2(p+1)$, i.e., $(p-1)(p-2) > 4$,

$$\frac{p-1}{2} + (p-1)\min\left\{\frac{1}{2}, \frac{p-3}{2}\right\} = \frac{(p-1)(p-2)}{2} > 2.$$

Consequently (2) always holds.

When $n \ge 2$,

$$\begin{aligned} \|F(u(s))\|_{L^2}^2 &\le K\||u(s)|^{2p}\|_{L^2} \\ &\le K\|u(s)\|_{L^{p+1}}^{\alpha}\|u(s)\|_{L^{2n/(n-2+2\varepsilon)}}^{2p-\alpha} \le \widetilde{K}(1+s)^{-\beta}, \end{aligned}$$

where

$$\alpha = \frac{1-(n-2+2\varepsilon)p/n}{1/(p+1)-(n-2+2\varepsilon)/n} = \frac{2p(1-\varepsilon)-n(p-1)}{1-\varepsilon-n(p-1)/2(p+2)}$$

and

$$\beta = \frac{n(p-1)}{2(p+1)}\alpha + (2p-\alpha)\left[-\varepsilon + \min\left\{1, n\left(\frac{pn}{n+2-2\varepsilon} - 1\right)\right\}\right].$$

When $n(pn/(n+2-2\varepsilon)-1)>1$, i.e., $p>(1+1/n)(1+(2-2\varepsilon)/n)$,

$$\begin{aligned}\beta &= 2p(1-\varepsilon)-\alpha\left[1-\varepsilon-\frac{n(p-1)}{2(p+1)}\right]\\ &= 2p(1-\varepsilon)-[2p(1-\varepsilon)-n(p-1)]\\ &= n(p-1)>\frac{2(p+1)}{p}>2;\end{aligned}$$

when $p\le(1+1/n)(1+(2-2\varepsilon)/n)$,

$$\begin{aligned}&\left[1-\varepsilon-\frac{n(p-1)}{2(p+1)}\right](\beta-2)\\ &=\frac{np(p-1)}{2(p+1)}\frac{n^2+4}{n+2}+\frac{1}{2(p+1)}[n^2-2n-4-(n^2-2n+4)p]+O(\varepsilon)\\ &=\left[\frac{np(p-1)}{2(p+1)}-1\right]\frac{n^2+4}{n+2}+\frac{1}{2(p+1)(n+2)}\\ &\quad\times\left\{n^2(n-2)\left[\left(1+\frac{1}{n}\right)\left(1+\frac{2-2\varepsilon}{n}\right)-p\right]+(n-2)^2\right\}+O(\varepsilon).\end{aligned}$$

From our assumption

$$1-\varepsilon-\frac{n(p-1)}{2(p+1)}>0,\qquad \frac{np(p-1)}{2(p+1)}-1>0,$$

and

$$\left(1+\frac{1}{n}\right)\left(1+\frac{2-2\varepsilon}{n}\right)-p\ge 0.$$

Hence, for sufficiently small ε, $\beta-2>0$.

We have seen that (2) holds in either case. The proof of the theorem is finished. □

We refer to [11] for a discussion on the regularity of the solution.

Two points in the approach of this chapter are worth mentioning. The first point is, as suggested by the linear estimate, to adapt the fixed point theorem—the contraction mapping principle—which does not require compactness. The second is to replace the space $L^2\cap L^\infty$ used in [12] and [13] by a more appropriate L^{p+1}-space so that assumptions on $F(u)$ can be less strict.

The study for various problems (initial, mixed, boundary value problems, etc.) on various domains (bounded, or exterior of a bounded, domain, the whole space) of the Schrödinger equation with general right side $F(u, D_x u)$ can be found in [14], [15], and [16].

Open problems. 1. Study the problem for

$$p<\frac{1}{2}+\frac{1}{n}+\left(\frac{1}{4}+\frac{3}{n}+\frac{1}{n^2}\right)^{1/2}.$$

Probably for some T_0 the solution exists for $t<T_0$. But, for $t\ge T_0$, there are examples showing that the solution no longer exists. This is a question of blow-up.

2. Consider the regularity of the solution for

$$\begin{cases} iu_t + \Delta u = |u|^2 u, \\ u|_{t=0} = \varphi(x). \end{cases}$$

Prove that $u \in C^\infty$ for $t > 0$ if $\varphi \in L^2$ but $D\varphi \notin L^2$. Generalize this to general right-hand $F(u)$.

3. Solve the problem with data prescribed at ∞. For the solution of

$$\begin{cases} iu_t + \Delta u = F(u), \quad F \text{ satisfying the conditions in Theorem 1}, \\ u|_{t=0} = \varphi, \end{cases}$$

u satisfies the decay property $u \approx u_+ \approx R(t)\varphi + \int_0^\infty R(t-s)F(u(s))\,ds$ for large t. If now we are given $u|_{t=\infty} = u_+(x)$, that is, the state of u at ∞ is known, can we solve for u? Also do we still have uniqueness?

4. Study the analogous problem for $(\alpha + i\beta)u_t + \Delta u = F(u)$.

Summing up, there are relatively scant results for nonlinear Schrödinger equations. A lot of work remains to be done.

CHAPTER IV

Multi-Dimensional Subsonic Flows Around an Obstacle

§1. Introduction

Mathematically speaking, the problem of a subsonic flow passing an obstacle means to solve a quasilinear elliptic equation for the velocity potential of which the normal derivative on the boundary of the obstacle vanishes and of which the derivative at infinity is prescribed. This problem was studied thoroughly [17] in the two-dimensional case. In this chapter we consider this problem in higher dimensions. When the dimension is greater than three, the problem loses its physical meaning. However, it does no harm in adapting the same terminologies as in the three-dimensional case.

The existence and uniqueness of the solution for $n = 3$ was established in the case: Mach number < 0.7 [18]. In [19] we improve and generalize the result therein to any Mach number. Certain simplifications of [19] can be found in [20]. In [21] the result of [19] was generalized to the presence of external force. In addition, a variational proof of the existence of a solution, which clears the way for practical computation, was given. In the following we shall present the result of [19], incorporating the simplified proof in [20].

Consider the quasilinear boundary value problem on a domain Ω which is the exterior of a bounded, closed hypersurface Γ:

$$\begin{cases} \dfrac{\partial}{\partial x_i}\left(\rho \dfrac{\partial \varphi}{\partial x_i}\right) = 0, \\ \nabla\varphi|_\infty = u^\infty = (U, 0, \dots, 0), \\ \left.\dfrac{\partial \varphi}{\partial N}\right|_\Gamma = 0, \end{cases} \tag{1}$$

where φ is the velocity potential (i.e., $\nabla\varphi = (u_1, \dots, u_n)$ is the velocity of the flow), $q = (\sum_1^n u_i^2)^{1/2}$ is the speed of the flow, $\rho = \rho(q)$ is the density of the gas, and $\nabla\varphi|_\infty$ indicates the uniform outflow at infinity. Without loss of generality we may assume the outflow is along the positive x_1-direction $(U \geq 0)$. $(\partial\varphi/\partial N)|_\Gamma = 0$ (N is the unit outer normal) means the normal velocity on the boundary of the obstacle is equal to zero. Finally ρ is a given function of q. In the boundary value problem only the constant U is prescribed.

The dimensionless density of a perfect gas is given by

$$\rho(q) = \left(1 - \frac{\gamma - 1}{2}q^2\right)^{1/(\gamma-1)},$$

where γ, $1 < \gamma < 2$, stands for the ratio between the isobaric and isochroic specific heats. We shall assume ρ satisfies: $\rho > 0$, $\rho(0) = 1$, $\rho'(q) \leq 0$, $\rho'(0) = 0$, and $\rho \in C^2[0, q_{\lim}]$ for some $q_{\lim} \leq \infty$. $q_{\lim}$ is equal to $(2/(\gamma - 1))^{1/2}$ for a perfect gas.

The equation we are going to study is not always elliptic. We write it in the form

$$\left(\rho\delta_{ij} + \frac{\rho'}{q}\varphi_{x_i}\varphi_{x_j}\right)\varphi_{x_ix_j} = 0.$$

To compute the eigenvalues of the quadratic form $(\rho\delta_{ij} + (\rho'/q)\varphi_{x_ix_j})\xi_i\xi_j = \rho|\xi|^2 + (\rho'/q)(\varphi_{x_i}\xi_i)^2$, we rotate the axes so that $\nabla\varphi$ is equal to $(q, 0, \ldots, 0)$, then the quadratic form becomes $\rho|\eta|^2 + \rho' q\eta_1^2$, $\eta = (\eta_1, \ldots, \eta_n)$. Therefore the eigenvalues are $\lambda_1 = \rho + \rho' q$, $\lambda_2 = \cdots = \lambda_n = \rho$. The equation is elliptic when $\lambda_1 > 0$. The corresponding point $(x, \varphi(x))$ is called an elliptic point. It is called a hyperbolic point when $\lambda_1 < 0$.

From Bernoulli's law $(q^2/2) + \int dp/\rho = c$ (p is the pressure) and the definition of the speed of sound $c^2 = dp/d\rho$ we have $c^2 = -\rho q(dq/d\rho) = -\rho q/\rho'(q)$. On the other hand, recalling that the Mach number of a flow is given by $M = q/c$, we have

$$M^2 = \frac{q^2}{c^2} = -\frac{\rho' q}{\rho} = 1 - \frac{\lambda_1}{\rho}.$$

Hence $M < 1$ (> 1) when $\lambda_1 > 0$ (< 0) and elliptic (hyperbolic) points occur where the flow is *subsonic* ($M < 1$) or *supersonic* ($M > 1$). $\lambda_1 = 0$ or $M = 1$ corresponds to a sonic point and $q = q_c$ is called the critical speed. For a perfect gas $q_c = (2/\gamma + 1)^{1/2}$.

A solution of the boundary value problem is called a subsonic (supersonic) flow according to whether all points in $\overline{\Omega}$ are subsonic or supersonic. A flow is called transonic if there are subsonic and supersonic points in $\overline{\Omega}$. Up to now there are very few results concerning the existence and uniqueness of solutions for the complicated transonic flows, although some methods for practical numerical computations presupposing existence are available. This state of research for transonic flows also presents problems for the study for subsonic flows. This is because even for the outflow $q_n = U < q_c$ and $M_\infty < 1$ it does not necessarily imply $\max_{\overline{\Omega}} M < 1$. We shall show that this is true when U is sufficiently small. But, for how large U does $\max_{\overline{\Omega}} M < 1$ continue to hold? We do not have a satisfactory answer.

The condition $\nabla\varphi|_\infty = (U, 0, \ldots, 0)$ asserts not only the flow comes in in a uniform way but also the stronger fact that it is uniform at ∞ in the whole space. This does not reflect what really happens. In reality there are wakes after passing the obstacle. Therefore we should keep in mind that assuming the flow is uniform at infinity is merely an approximation to reality.

The obstacle Γ may be composed of one or several closed hypersurfaces. We shall always assume $\Gamma \in C^{2+\tau_0}$.

Since it is not known for how large U one still has $\max_{\overline{\Omega}} M < 1$ we shall consider an auxiliary problem first:

$$\begin{cases} \dfrac{\partial}{\partial x_i}\left(\rho \dfrac{\partial \varphi}{\partial x_i}\right) = 0, \\ \dfrac{\nabla\varphi}{|\nabla\varphi|}\Big|_\infty = (1, 0, \dots, 0), \\ \sup\limits_{\overline{\Omega}} |\nabla\varphi| = Q \qquad (< q_c), \\ \dfrac{\partial\varphi}{\partial N}\Big|_\Gamma = 0. \end{cases} \tag{2}$$

Here only the constant Q is given. We shall use the Leray-Schauder theorem on topological degree to prove that this problem is solvable. We consider the linearization of the problem. It turns out that the map $\varphi \mapsto \Phi$ is not compact if we take $(\partial/\partial x_i)(\rho(q)(\partial\Phi/\partial x_i)) = 0$ to be the linearized equation. So instead we use

$$\left(\rho(q)\delta_{ij} + \frac{\rho'}{q}\varphi_{x_i}\varphi_{x_j}\right)\Phi_{x_i x_j} = 0.$$

Under $\partial\Phi/\partial N|_\Gamma = 0$, $[\nabla\Phi/|\nabla\Phi|]_\infty = (1, 0, \dots, 0)$ and $\sup_{\overline{\Omega}}|\nabla\Phi| = Q$ it can be shown that $\varphi \mapsto \Phi$ is compact. Setting

$$\Phi = Q(\Psi + x_1)/\sup_{\overline{\Omega}}|\nabla(\Psi + x_1)|,$$

since $\nabla\Psi|_\infty = 0$ implies $\Psi|_\infty = 0$ we have

$$\begin{cases} \left(\rho(q)\delta_{ij} + \dfrac{\rho'}{q}\varphi_{x_i}\varphi_{x_j}\right)\Psi_{x_i x_j} = 0, \\ \Psi|_\infty = 0, \\ \dfrac{\partial\Psi}{\partial N}\Big|_\Gamma = -\dfrac{\partial x_1}{\partial N}\Big|_\Gamma = -\cos(N, x_1)|_\Gamma. \end{cases}$$

As a first step we need to study and solve this linear problem. Necessary knowledge will be developed in the next section.

§2. Background material for the linear problem

For $\Psi \in C^k(\overline{\Omega})$ or $C^{k+\tau}(\overline{\Omega})$ $(0 < \tau < 1)$, we write:

$$M_{m,k}(\Psi) = \sup_{x\in\overline{\Omega}} \max\{1, d_x^{m+k}\}|D^k\Psi(x)|,$$

$$M_{m,k+\tau}(\Psi) = \sup_{x,y\in\overline{\Omega}} \max\{1, d_{xy}^{m+k+\tau}\}\frac{|D^k\Psi(x) - D^k\Psi(y)|}{|x-y|^\tau},$$

where m is nonnegative, $d_x = d(x, \partial\Omega)$, $d_{xy} = \min\{d_x, d_y\}$, and $D^k\Psi$ is any kth order derivative of Ψ. Notice in the definition of $M_{m,k}(\Psi)$ and $M_{m,k+\tau}(\Psi)$ supremums are to be taken over all kth order derivatives. Define norms

$$||\Psi||_{m,k} = \sum_0^k M_{m,j}(\Psi),$$

$$||\Psi||_{m,k+\tau} = ||\Psi||_{m,k} + M_{m,k+\tau}(\Psi).$$

Let $C_{m,k}(\overline{\Omega})$ ($C_{m,k+\tau}(\overline{\Omega})$ resp.) be the Banach space $\{\Psi \in C^k(\overline{\Omega})$: $||\Psi||_{m,k}(||\Psi||_{m,k+\tau}$ resp.$) < \infty\}$.

Consider the linear problem

$$\begin{cases} \mathscr{L}(\Psi) = b_{ij}(x)\dfrac{\partial^2\Psi}{\partial x_i \partial x_j} + b_i(x)\dfrac{\partial\Psi}{\partial x_i} + b(x)\Psi = f(x), \\ \left.\dfrac{\partial\Psi}{\partial N}\right|_\Gamma = \Psi_0, \\ \Psi|_\infty = 0. \end{cases} \tag{3}$$

LEMMA 1. *Suppose for some positive constants K and σ,*

$$\begin{gathered} \frac{1}{\sigma}|\xi|^2 \le b_{ij}\xi_i\xi_j \le \sigma|\xi|^2, \quad \forall\xi \in \mathbb{R}^n, \\ ||b_{ij}||_{0,\tau}, \ ||b_i||_{1,\tau}, \ ||b||_{2,\tau} \le K, \\ f \in C_{n,\tau}(\overline{\Omega}), \qquad \Psi_0 \in C^{1+\tau}(\Gamma). \end{gathered} \tag{4}$$

Then for $0 < \tau \le \tau_0$ a solution $\Psi \in C_{n-2,2+\tau}$ of (3) *satisfies*

$$||\Psi||_{n-2,2+\tau} \le C(M_{n-2,0}(\Psi) + ||\Psi_0||_{C^{1+\tau}(\Gamma)} + ||f||_{n,\tau}), \tag{5}$$

where C only depends on n, Ω, K, σ, and τ_0.

PROOF. Although the domain is unbounded, the derivation of the interior estimate is similar to the Schauder estimate presented in [49]. For estimates near the boundary we refer to [6]. □

LEMMA 2. *In addition to* (4) *suppose that*

$$||b_{ij}(x) - b_{ij}(\infty)||_{0,\tau}, \ ||b_i||_{0,1+\tau}, \ ||b||_{0,2+\tau} \le K. \tag{6}$$

Then for $f \in C_{n+\tau,0}(\overline{\Omega}) \cap C_{n,\tau}(\overline{\Omega})$ and $0 < \tau \le \tau_0$, a solution $\Psi \in C_{n-2,2+\tau}(\overline{\Omega})$ of (3) *satisfies:*

$$M_{n-2,2+\tau}(\Psi) \le C_1(||\Psi_0||_{C^{1+\tau}(\Gamma)} + ||f||_{n,\tau} + ||f||_{n+\tau,0}), \tag{7}$$

where C_1 only depends on n, σ, Ω, K, and τ_0.

PROOF. In view of Lemma 1 (7) can be deduced from

$$M_{n-2,0}(\Psi) \le C_2(||\Psi_0||_{C^{1+\tau}(\Gamma)} + ||f||_{n,\tau} + ||f||_{n+\tau,0}). \tag{8}$$

On the other hand, (8) can be deduced from

$$M_{0,0}(\Psi) \le C_3(\|\Psi_0\|_{C^{1+\tau}(\Gamma)} + \|f\|_{n,\tau} + \|f\|_{n+\tau,0}). \tag{9}$$

For, assuming $\|\Psi_0\|_{C^{1+\tau}(\Gamma)} + \|f\|_{n,\tau} + \|f\|_{n+\tau,0} \le 1$, it follows from (9) that $M_{0,0}(\Psi) \le C_3$. Let $R = (B_{ij}(\infty)x_ix_j)^{1/2}$, where (B_{ij}) is the inverse matrix of $(b_{ij}(\infty))$. By (4) and (6)

$$\frac{1}{\sigma}|x| \le R \le \sigma|x| \quad \text{and} \quad b_{ij}(\infty)\frac{\partial^2 R^{2-n}}{\partial x_i \partial x_j} = 0.$$

Therefore

$$\mathscr{L}(R^{2-n} - R^{2-n-\tau/2}) = -\frac{\tau}{2}\left(n - 2 + \frac{\tau}{2}\right)R^{-n-\tau/2} + O(R^{-n-\tau}).$$

Taking R_0 sufficiently large so that $\{x\colon R > R_0\} \subseteq \Omega$ and $R_0^{2-n} - R_0^{2-n-\tau/2} < 1$, we have

$$\mathscr{L}\left[R^{2-n} - R^{2-n-\tau/2} \pm \frac{1}{C_3}(R_0^{2-n} - R_0^{2-n-\tau/2})\Psi\right]$$
$$\le -\frac{\tau}{2}\left(n - 2 + \frac{\tau}{2}\right)R^{-n-\tau/2} + O(R^{-n-\tau}) + \frac{|f|}{C_3} < 0.$$

Clearly the choice of R_0 only depends on Ω, K, τ, n, and σ. By the maximum principle we conclude

$$|\Psi| \le \frac{C_3}{R_0^{2-n} - R_0^{2-n-\tau/2}}(R^{2-n} - R^{2-n-\tau/2}), \qquad R \ge R_0.$$

Combining with $M_{0,0}(\Psi) \le C_3$ we obtain $M_{n-2,0}(\Psi) \le C_2$. In other words, (8) holds.

Now we prove by contradiction that (9) holds.

If on the contrary (9) is invalid, then for each $k = 1, 2, \ldots,$ there exist b_{ij}^k, b_i^k $(1 \le i, j \le n)$, b^k, f^k, Ψ_0^k, Ψ^k fulfilling the hypotheses in Lemma 2 but

$$M_{0,0}(\Psi^k) > k(\|\Psi_0^k\|_{C^{1+\tau}(\Gamma)} + \|f^k\|_{n,\tau} + \|f^k\|_{n+\tau,0}).$$

Without loss of generality we may assume $M_{0,0}(\Psi^k) = 1$. Consequently

$$\|\Psi_0^k\|_{C^{1+\tau}(\Gamma)} + \|f^k\|_{n,\tau} + \|f^k\|_{n+\tau,0} \le \frac{1}{k}.$$

Following the path used to obtain (8) from (9), we have $M_{n-2,0}(\Psi^k) \le C_2$. By Lemma 1, $M_{n-2,2+\tau}(\Psi^k) \le C_1$.

By passing to a subsequence (using the same notation) b_{ij}^k, b_i^k, and b^k uniformly converge to b_{ij}^0, b_i^0, and b^0 on every compact subset of $\overline{\Omega}$. Similarly Ψ^k uniformly converges to Ψ^0 up to second order. b_{ij}^0, b_i^0, and b^0

satisfy all hypotheses in Lemma 2 and $\|\Psi^0\|_{n-2,2+\tau} \le C_2$. Hence Ψ_0 satisfies

$$\begin{cases} b_{ij}^0 \Psi^0_{x_i x_j} + b_i^0 \Psi^0_{x_i} + b^0 \Psi^0 = 0, \\ \left.\dfrac{\partial \Psi^0}{\partial N}\right|_\Gamma = 0, \\ \Psi^0|_\infty = 0. \end{cases}$$

But then from the strong maximum principle $\Psi^0 = 0$. This is in contradiction with $M_{0,0}(\Psi^k) = 1$ and $M_{n-2,0}(\Psi^k) \le C_2$. The proof of Lemma 2 is complete. □

THEOREM 3. *Under the hypotheses in Lemma 2, the linear problem* (3) *admits a* $C^{2+\tau}(\overline{\Omega})$*-solution which is unique in* $C^2(\Omega) \cap C^1(\overline{\Omega})$. *Moreover, it satisfies* (7).

PROOF. The uniqueness of the solution follows from the maximum principle. For existence we use a continuation argument. We solve

$$\begin{cases} [(1-\theta)\mathscr{L}_\infty + \theta \mathscr{L}]\Psi = \theta f, \qquad 0 \le \theta \le 1, \\ \left.\dfrac{\partial \Psi}{\partial N}\right|_\Gamma = \Psi_0, \\ \Psi|_\infty = 0, \end{cases}$$

where $\mathscr{L}_\infty = b_{ij}(\infty)(\partial^2/\partial x_i \partial x_j)$. When $\theta = 0$, this problem becomes an oblique derivative problem in an exterior domain for the equation $\Delta\Psi = 0$ (after a linear transformation). Its uniqueness and solvability has been established in [22].

From the estimate in Lemma 2 we see that solvability at $\theta = 0$ implies the same property at $\theta = 1$. Theorem 3 is thus established. □

§3. Solution to the auxiliary problem

Consider the auxiliary problem (2). We need to modify the function $\rho(q)$ in the range $q > Q$ so that the equation becomes uniformly elliptic and the theorem on the topological degree of Leray-Schauder can be used. Since in (2) we have $\sup_{\overline{\Omega}} |\nabla\varphi| = Q$, the solution of the modified problem solves the original problem.

Here is one of the admissible modifications. Writing $\overline{Q} = (Q + q_c)/2$, we let

$$s = s(q) = \frac{1}{2}\left[1 + \tanh \frac{(\overline{Q} - Q)(2q - Q - \overline{Q})}{2(q - Q)(\overline{Q} - q)}\right],$$

where tanh stands for the hyperbolic tangent function. For $Q \le q \le \overline{Q}$, s

is a smooth, monotonic increasing function. Also, $s(Q) = s'(Q) = s''(Q) = \cdots = s'(\overline{Q}) = s''(\overline{Q}) = \cdots = 0$ and $s(\overline{Q}) = 1$. Let

$$\overline{\rho} = \begin{cases} \rho, & 0 \le q \le Q, \\ \rho(1-s) + \rho^* s, & Q < q \le \overline{Q}, \\ \rho^*, & \overline{Q} \le q < \infty, \end{cases}$$

where $\rho^* = \max_{Q \le \rho \le \overline{Q}} \rho(q)$. $\overline{\rho}$ satisfies our requirement. For, from $\rho \in C^2$ $(0 \le q \le \overline{Q})$ we know $\overline{\rho} \in C^2$ $(0 \le q \le \infty)$. $\rho > 0$ $(0 \le q \le \overline{Q})$ implies $\overline{\rho} > 0$ $(0 \le q < \infty)$ and $\rho + \rho' q > 0$ $(0 \le q \le \overline{Q})$ implies $\overline{\rho} + \overline{\rho}' q = (\rho + \rho' q)(1-s) + \rho^* s + q(\rho^* - \rho)s' \ge \min\{\rho + \rho' q, \rho^*\} > 0$ $(Q \le q \le \overline{Q})$. Hence there exists a positive constant K such that $\overline{\rho} + \overline{\rho}' q \ge K > 0$ $(0 \le q < \infty)$. The matrix $(\overline{a}_{ij}(D\varphi)) = (\overline{\rho}(D\varphi)\delta_{ij} + (\overline{\rho}'/|D\varphi|)\varphi_{x_i}\varphi_{x_j})$ has eigenvalues $\lambda_1 = \overline{\rho} + \overline{\rho}' q$, $\lambda_2 = \cdots = \lambda_n = \overline{\rho}$, and

$$\sigma|\xi|^2 \le \sum \overline{a}_{ij}(D\varphi)\xi_i\xi_j \le \frac{1}{\sigma}|\xi|^2, \quad \forall \xi \in \mathbb{R}^n, \quad \forall q \in [0, \infty),$$

for some positive constant $\sigma = \sigma(Q)$. To simplify notations from now on we write $\overline{\rho}$ and $\overline{a}_{ij}$ as ρ and a_{ij}. Now the following equation

$$\frac{\partial}{\partial x_i}\left(\rho \frac{\partial \varphi}{\partial x_i}\right) = a_{ij}(D\varphi)\varphi_{x_i x_j} = 0$$

is uniformly elliptic for $0 \le q < \infty$.

Assume that $\varphi \in C^{1+\tau}(\overline{\Omega})$, $|D\varphi(x) - D\varphi(\infty)| \le C_1 d_x^{-\tau}$, $D\varphi(\infty) = (U, 0, \ldots, 0)$, $(\varphi(x) - Ux_1)_{x=\infty} = 0$, and $|D\varphi(x) - D\varphi(y)| \le C_2 \min\{1, d_{xy}^{-\tau}\}|x-y|^\tau$, where $\tau = \tau(Q, \Gamma) \le \tau_0$ and τ_0 is the exponent for $\Gamma \in C^{2+\tau_0}$. The choice of τ will be made in due course. We define a norm for φ by

$$||\varphi||_E = |U| + \sup_{\overline{\Omega}} |\varphi - Ux_1| + \inf C_1 + \inf C_2.$$

The collection of all functions with finite $||\cdot||_E$ forms a Banach space E.

For $\varphi \in E$, we solve

$$\begin{cases} a_{ij}(D\varphi)\Psi_{x_i x_j} = 0, \\ \left.\dfrac{\partial \Psi}{\partial N}\right|_\Gamma = -\cos(N, x_1)|_\Gamma, \\ \Psi|_\infty = 0. \end{cases} \tag{10}$$

From Theorem 3 there exists a unique solution Ψ of (10) in $C_{n-2, 2+\tau}(\overline{\Omega})$. In particular $\sup_\Omega |\nabla(\Psi(x) + x_1)|$ is finite since $\nabla\Psi = O(d_x^{1-n})$. Let

$$\Phi(x) = \frac{\theta Q(\Psi(x) + x_1)}{\sup_\Omega |\nabla(\Psi(x) + x_1)|}, \qquad 0 \le \theta \le 1.$$

Φ satisfies

$$a_{ij}(D\varphi)\Phi_{x_i x_j} = 0, \qquad \left.\frac{\partial \Phi}{\partial N}\right|_{\Gamma} = 0,$$

$$\sup_{\Omega}|\nabla\Phi| = \theta Q, \qquad \left.\frac{\nabla\Phi}{|\nabla\Phi|}\right|_{\infty} = (1, 0, \dots, 0).$$

$\Psi \in C_{n-2,2+\tau}(\overline{\Omega})$ implies that $\Psi \in E$. Hence $\Phi \in E$ and $\Phi = T(\varphi, \theta)$ ($\theta \in [0, 1]$) defines a linear mapping from E into itself. We are going to show that this linear mapping has a fixed point for each θ. When $\theta = 1$, this fixed point is a solution to our auxiliary problem.

We verify various requirements in the Leray-Schauder theorem. From Theorem 3 $||\Psi||_{n-2,2+\tau} \le K||\varphi||_E$. Hence the image of T is compact. To show that T is continuous in φ, we assume on the contrary that there exists a sequence $\{\varphi_k\}$ in E such that $||\varphi_k - \varphi||_E \to 0$ for some $\varphi \in E$ but $||\Phi_k - \Phi||_E \ge \delta$. As

$$||\Psi_k||_{n+2,2+\tau} \le K||\varphi_k||_E \le K(||\varphi||_E + 1),$$

$\{\Psi_k\}$ is compact in $C_{n-2,2+\tau/2}(\overline{\Omega})$. We can find a subsequence $\{\Psi_{k'}\}$ and Ψ_* such that

$$||\Psi_{k'} - \Psi_*||_{n-2,2+\tau/2} \to 0 \quad \text{as } k' \to \infty.$$

Taking the limit in $a_{ij}(D\varphi_{k'})\partial^2\Psi_{k'}/\partial x_i \partial x_j = 0$ we have

$$a_{ij}(D\varphi)\frac{\partial^2 \Psi_*}{\partial x_i \partial x_j} = 0.$$

By the same reasoning $\partial\Psi_*/\partial N|_\Gamma = -\cos(N, x_1)$ and $\Psi_*|_\infty = 0$. From the uniqueness of (10) $\Psi_* = \Psi$. As $||\Psi_{k'} - \Psi||_{n-2,2+\tau/2} \to 0$ and $D\Phi(\infty) = (\theta Q/\sup|\nabla(\Psi(x) + x_1)|, 0, \dots, 0)$, $||\Phi_{k'} - \Phi||_E \to 0$ as $k' \to \infty$, giving a contradiction.

$T(\varphi, \theta)$ is uniformly continuous in θ simply because it is linear in θ.

When $\theta = 0$, $\Phi(x) = 0$. Hence $\varphi = T(\varphi, 0)$ only allows $\varphi \equiv 0$ as its solution. In other words, the index of the solutions at $\theta = 0$ is ± 1.

Next we show the solutions of $\varphi - T(\varphi, \theta) = 0$ are bounded in E, or equivalently, $||\varphi||_E = |U| + \sup_{\overline{\Omega}}|\varphi - Ux_1| + \inf C_1 + \inf C_2$ is bounded. U is already bounded due to the fact that $U = D\varphi(\infty)$ and $0 \le U \le \sup|\nabla\varphi| = \theta Q \le Q$ ($0 \le \theta \le 1$). In the following we shall use a priori interior and near-boundary estimates to bound $\inf C_2$ and then use the estimates at ∞ to bound $\inf C_1$ and $\varphi - Ux_1$.

We use the result on Hölder estimates in [5]. (It is essentially a special case for a time dependent estimate in [49].) For $u \in W^{1,m}(\Omega)$ ($1 < m \le n$) and $M = \operatorname{ess\,sup}|u|$, denote

$$A_{k,\rho} = \{x \in \Omega \cap |x - x_0| < \rho : u(x) > k\}$$

and

$$B_{k,\rho} = \{x \in \Omega \cap |x - x_0| < \rho : u(x) < k\},$$

where x_0 is fixed. We say u belongs to $\mathscr{B}_m(\Omega, M, \gamma, \delta, 1/q)$ if for all x_0 in $\overline{\Omega}$, $\rho > 0$, and $\sigma \in (0, 1)$,

$$\int_{A_{k,(1-\sigma)\rho}} |\nabla u|^m \leq \gamma \left\{ \frac{1}{\sigma^m \rho^m (1 - n/q)} \sup_{A_{k,\rho}} [(u(x) - k)^m + 1] \right\} \operatorname{meas}^{1-m/q} A_{k,\rho},$$

where k satisfies $k \geq \max\{\sup_{K_\rho \cap \Omega} u(x) - \delta, \sup_{K_\rho \cap \partial\Omega} u(x)\}$ ($K_\rho = \{x: |x - x_0| < \rho\}$) and

$$\int_{B_{k,(1-\sigma)\rho}} |\nabla u|^m \leq \gamma \left\{ \frac{1}{\sigma^m \rho^m (1 - n/q)} \sup_{B_{k,\rho}} [(k - u(x))^m + 1] \right\} \operatorname{meas}^{1-m/q} B_{k,\rho},$$

where k satisfies $k \leq \min\{\inf_{K_\rho \cap \Omega} u(x) + \delta, \inf_{K_\rho \cap \partial\Omega} u(x)\}$. It is proved in [5] that every element u in $\mathscr{B}_m(\Omega, M, \gamma, \delta, 1/q)$ admits Hölder estimates in the interior of the domain as well as near its boundary. For the interior estimate Hölder exponent only depends on m, M, γ, δ, q, and the distance of the subdomain to the boundary. For the near-boundary estimate it is further required that $\partial\Omega$ satisfies an exterior cone condition and $u|_{\partial\Omega}$ satisfies a Hölder condition. The Hölder exponent and modulus depend on both conditions.

A fixed point of $\Phi = T(\varphi, \theta)$ is an element $\varphi \in E$ so that $\Psi = \varphi - Ux_1$ belongs to $C_{n-2,2+\tau}(\Omega)$ and

$$\begin{cases} a_{ij}\varphi_{x_i x_j} = 0, \\ \left.\dfrac{\partial \varphi}{\partial N}\right|_\Gamma = 0, \\ \nabla\varphi|_\infty = (U, 0, \dots, 0), \\ \sup\limits_\Omega |\nabla\varphi| = q \leq Q. \end{cases}$$

From Schauder estimates for linear equations we know $\varphi \in C^{3+\tau}(\Omega) \cap C^{2+\tau}(\overline{\Omega})$.

Differentiating the equation in x_l, yields

$$0 = \frac{\partial}{\partial x_l}\frac{\partial}{\partial x_i}\left(\rho \frac{\partial \varphi}{\partial x_i}\right) = \frac{\partial}{\partial x_i}\left(a_{ij}\frac{\partial u}{\partial x_j}\right), \qquad u = \frac{\partial \varphi}{\partial x_l}.$$

Let $\{x: |x - x_0| \leq \mu\} \subseteq \Omega$ and $A_{\eta,\mu} = \{x \in \Omega: |x - x_0| < \mu,\ u(x) > \eta\}$. Multiplying the above equation by $\zeta^2(x)(u - \eta)^+$, where ζ is a cut-off function in $|x - x_0| < \mu$. After an integration

$$\int_{A_{\eta,\mu}} \zeta^2 |\nabla u|^2 \leq K \int_{A_{\eta,\mu}} (u - \eta)^2 |\nabla\zeta|^2.$$

Choose ζ so that it is equal to 1 in $|x - x_0| \leq (1 - \theta)\mu$, $0 < \theta < 1$, equal to

0 on $|x - x_0| = \mu$, and linear along each radius in between. This inequality becomes

$$\int_{A_{\eta,(1-\theta)\mu}} |\nabla u|^2 \leq K \operatorname{meas} A_{\eta,\mu} \frac{1}{(\theta\mu)^2} \max_{A_{\eta,\mu}} (u(x) - \eta)^2.$$

A similar estimate holds on $B_{\eta,(1-\theta)\mu}$. Henceforth

$$u \in \mathscr{B}_2(|x - x_0| < \mu, Q, K, \infty, \infty)$$

and $|u(x) - u(y)| \leq K d_{xy}^{\tau_1} |x - y|^{\tau_1}$.

Notice that the proof of the last estimate in [5] requires an additional condition $|x-y| \leq K_1 d_{xy}$. However, this estimate is still valid when $|x-y| \geq K_1 d_{xy}$. For,

$$|u(x) - u(y)| \leq |u(x)| + |u(y)| \leq 2Q \leq \frac{2Q}{K_1^{\tau_1}} d_{xy}^{-\tau_1} |x - y|^{\tau_1}.$$

Next we consider the estimate near the boundary. Since $\Gamma \in C^{2+\tau_0}$, a small piece of Γ has a parametrization $x_i = x_i(\xi_1, \ldots, \xi_{n-1}) \in C^{2+\tau_0}$. Take a point P_0 on this piece. For a suitable choice of $(\xi_1, \ldots, \xi_{n-1})$, $\xi_i = \xi_i(x_1, \ldots, x_n) \in C^{2+\tau_0}$ on a small ball ω centered at P_0, $\{\xi_1, \ldots, \xi_n\}$ are orthonormal at P_0, and ξ_n is in the normal direction on $\omega \cap \Gamma$. Such a choice can be made as follows: first we pick $\xi_1, \ldots, \xi_{n-1}$ so that they are orthonormal at P_0. As

$$\cos(\nu, x_i) = \lambda_i(\xi_1, \ldots, \xi_{n-1}) \in C^{1+\tau_0}, \qquad i = 1, 2, \ldots, n,$$

$x_i = x_i(\xi_1, \ldots, \xi_{n-1}) + \xi_n \lambda_i(\xi_1, \ldots, \xi_{n-1})$ in general merely belongs to $C^{1+\tau_0}$ and does not suit our purpose. Instead we define

$$x_i = x_i(\xi_1, \ldots, \xi_{n-1}) + \xi_n^{2-n} \int_{\xi_1}^{\xi_1+\xi_n} \cdots \int_{\xi_{n-1}}^{\xi_{n-1}+\xi_n} \lambda_i(\sigma_1, \ldots, \sigma_{n-1}) d\sigma_1 \cdots d\sigma_{n-1},$$

and take its inverse as our choice. Writing $\alpha_{jk} = (\partial\xi_j/\partial x_i)(\partial\xi_k/\partial x_i)$, then

$$\alpha_{jk}|_{P_0} = \delta_{jk}, \qquad j, k = 1, \ldots, n,$$
$$\alpha_{jn}|_{\omega\cap\Gamma} = 0, \qquad j = 1, \ldots, n-1.$$

Thus

$$q^2 = \sum_i u_{x_i}^2 = \sum_i \left(\frac{\partial\varphi}{\partial x_i}\right)^2 = \frac{\partial\varphi}{\partial\xi_j}\frac{\partial\varphi}{\partial\xi_k}\frac{\partial\xi_j}{\partial x_i}\frac{\partial\xi_k}{\partial x_i} = \alpha_{jk}\overline{u}_j\overline{u}_k, \qquad \overline{u}_j = \frac{\partial\varphi}{\partial\xi_i}.$$

The equation $\partial/\partial x_i(\rho\partial\varphi/\partial x_i) = 0$ is the Euler-Lagrange equation for the variational problem $\delta \iint \rho q \, dq \, dx = 0$. For,

$$0 = \delta \iint \rho q \, dq \, dx = \int \rho q \, \delta q \, dx$$
$$= \int \rho \sum \frac{\partial\varphi}{\partial x_i} \delta \frac{\partial\varphi}{\partial x_i} dx = -\int \frac{\partial}{\partial x_i}\left(\rho\frac{\partial\varphi}{\partial x_i}\right)\delta\varphi \, dx.$$

After a change of coordinate, we have

$$\delta \iint \rho q\, dq\, J\, d\xi = 0, \qquad J = \frac{\partial(x_1, \dots, x_n)}{\partial(\xi_1, \dots, \xi_n)},$$

or

$$\frac{\partial}{\partial \xi_i}\left(\rho \alpha_{ij} \overline{u}_j J\right) = 0.$$

Differentiating this equation in ξ_l gives

$$\frac{\partial}{\partial \xi_i}\left(J \beta_{ij} \frac{\partial \overline{u}_l}{\partial \xi_j} + \beta_i\right) = 0,$$

where

$$\beta_{ij} = \rho \alpha_{ij} + \frac{\rho'}{q} \alpha_{ik} \alpha_{jm} \overline{u}_k \overline{u}_m$$

and

$$\beta_i = \frac{\partial}{\partial \xi_l}(J \alpha_{ij}) \rho \overline{u}_j + \frac{J \rho'}{2q} \alpha_{ij} \frac{\partial \alpha_{km}}{\partial \xi_l} \overline{u}_j \overline{u}_k \overline{u}_m.$$

Therefore

$$\beta_{ij}|_{P_0} = \left(\rho \delta_{ij} + \frac{\rho'}{q} \overline{u}_i \overline{u}_j\right)\Big|_{P_0}.$$

When the ball is sufficiently small,

$$\frac{\sigma}{2}|\lambda|^2 \le \beta_{ij} \lambda_i \lambda_j \le \frac{2}{\sigma}|\lambda|^2, \quad \forall \lambda \in \mathbb{R}^n.$$

The images of ω, $\omega \cap \Omega$, $\omega \cap \Gamma$ are written as $\widetilde{\omega}$, $\widetilde{\Omega}$, $\widetilde{\Gamma}$. Then $(\partial \varphi / \partial N)|_\Gamma = 0$ becomes $\overline{u}_n|_{\widetilde{\Gamma}} = 0$.

We multiply the equation for $\overline{u}_l$ by $\zeta^2(\overline{u}_l - \eta)_+$, where ζ is a cut-off function centered at ξ_0 and supported in $\widetilde{\omega}$. When $l = n$ we require $\eta \ge 0$. After an integration by parts we have

$$\int_{\widetilde{\omega} \cap \{\overline{u}_l > \eta\}} \left(J \beta_{ij} \frac{\partial \overline{u}_l}{\partial \xi_j} + \beta_i\right) \frac{\partial}{\partial \xi_i}[\zeta^2(\overline{u}_l - \eta)]\, d\xi = 0.$$

The boundary integrals vanish due to the following reason: When $i \ne n$, $\zeta = 0$ outside $\widetilde{\omega}$; when $i = n$ and $l = n$, $[(\overline{u}_n - \eta)_+]_{\widetilde{\Gamma}} = (-\eta)_+ = 0$; when $i = n$, $l \ne n$, if further $j = n$, then $\partial \overline{u}_l / \partial \xi_n|_{\widetilde{\Gamma}} = \partial \widetilde{u}_n / \partial \xi_l|_{\widetilde{\Gamma}} = 0$ and so $J \beta_{nn} \partial \overline{u}_l / \partial \xi_n|_{\widetilde{\Gamma}} = 0$; if $j \ne n$, then $\alpha_{jn}|_{\widetilde{\Gamma}} = 0$, and $\beta_{jn}|_{\widetilde{\Gamma}} = (\rho \alpha_{nj} + \frac{\rho'}{q} \alpha_{nk} \alpha_{jm} \overline{u}_k \overline{u}_m)_{\widetilde{\Gamma}} = (\frac{\rho'}{q} \alpha_{nn} \overline{u}_n \alpha_{jm} \overline{u}_m)_{\widetilde{\Gamma}} = 0$ imply $J \beta_{nj} \partial \overline{u}_l / \partial \xi_j|_{\widetilde{\Gamma}} = 0$. Similarly $b_n|_\Gamma = 0$ $(i = n, l \ne n)$.

Extend $\overline{u}_l$ to $\widetilde{\omega} \backslash \widetilde{\Omega}$ by setting

$$\overline{u}_l(\xi_1, \dots, \xi_n) = \begin{cases} \overline{u}_l(\xi_1, \dots, \xi_{n-1}, -\xi_n), & l \ne n, \\ 0, & l = n. \end{cases}$$

$\overline{u}_l$ belongs to $W_2'(\widetilde{\omega})$ since $\overline{u}_n|_{\widetilde{\Gamma}} = 0$. Integrating the identity above we have

$$\int_{A_{\eta,(1-\theta)\mu}} |\nabla \overline{u}_l|^2 \, d\xi \le K \operatorname{meas} A_{\eta,\mu} \left\{ \frac{1}{(\theta\mu)^2} \sup_{\xi \in A_{\eta,\mu}} [(\overline{u}_l(\xi) - \eta)^2 + 1] \right\}, \quad 0 < \theta < 1,$$

where η is an arbitrary number when $l \neq \eta$ and is nonnegative when $l = n$.

By the same method we obtain a similar estimate on $B_{\eta,(1-\theta)\mu}$. Therefore, $\overline{u}_l \in \mathscr{B}_2(\widetilde{\omega}, Q, K, \infty, \infty)$ $(l = 1, 2, \dots, n-1)$ and $\overline{u}_n \in \mathscr{B}_2(\widetilde{\omega}, Q, K, 0, \infty)$. It follows from [5] that the near-boundary estimate

$$|u(x) - u(y)| \le K|x - y|^{\tau_2} \qquad (u = u_l, l = 1, 2, \dots, n)$$

holds for some $\tau_2 = \tau_2(Q, \Gamma) > 0$.

To obtain an estimate of u near ∞, recall that u_k $(1 \le k \le n)$ satisfies

$$\frac{\partial}{\partial x_i}\left(a_{ij}\frac{\partial u}{\partial x_j}\right) = 0.$$

Letting $v_k(x) = u_k(x) - u_k(\infty)$, v_k satisfies

$$\frac{\partial}{\partial x_i}\left(a_{ij}\frac{\partial v}{\partial x_j}\right) = 0. \tag{11}$$

Since $v_k = \partial\Psi/\partial x_k$ and $\Psi \in C_{n-2,2+\tau}(\Omega)$ we have

$$v_k = O(d_x^{1-n}). \tag{12}$$

Here $O(d_x^{1-n})$ depends on v_k. Suppose the origin is located inside Γ. We let

$$\zeta(x) = \begin{cases} |x|/R - 1, & R \le |x| \le 2R, \\ 1, & 2R < |x| < \frac{1}{2}\overline{R}, \\ 2(1 - |x|/R), & \frac{1}{2}\overline{R} \le |x| \le \overline{R}, \end{cases}$$

where R is the diameter of Γ and $\overline{R}$ is a large positive number. Multiplying (11) by $\zeta^2 v$ and then integrating by parts, by (12) we have

$$\begin{aligned} \int \zeta^2 |\nabla v|^2 &\le K_1 \int v^2 |\nabla\zeta|^2 \\ &\le K_1 \int_{R \le |x| \le 2R} v^2 |\nabla \zeta|^2 + O\left(\overline{R}^{-2} \int_{\overline{R}/2}^{\overline{R}} r^{-n+1} dr\right). \end{aligned}$$

Letting $\overline{R} \to \infty$

$$\int_{2R}^{\infty} |\nabla v|^2 \le 2^n K_1 Q^2 R^{n-2} = K_2.$$

Write $x/|x|^2 = z$, $x^0/|x^0|^2 = z^0$, $v(z/|z|^2) = w(z)$. This inequality implies $\int_{|z|<1/2R} |\nabla w|^2 < \infty$. As w is also bounded, $w(t) \in W_2^1$ $(|z| < 1/2R)$. By

Sobolev's decomposition formula

$$\begin{aligned}
|v(x^0)| &= |w(z^0)| \\
&\le \left| \int_{|z|\le 2|z^0|} \zeta_0(z) w(z)\,dz + \int_{|z|\le 2|z^0|} \frac{w_i(z^0, z) D_{z_i} w}{|z^0 - z|^{n-1}}\,dz \right| \\
&\le K_3 |z^0|^n + K_4 \int_{|z|\le 2|z^0|} \frac{|\nabla w|}{|z - z^0|^{n-1}}\,dz \\
&= K_3 |x^0|^{-n} + K_4 \int_{|x|\ge |x^0|/2} \frac{|\nabla v|}{|x - x^0|^{n-1}} \left(\frac{|x^0|}{|x|}\right)^{n-1} dx \\
&\le K_3 |x^0|^{-n} + K_5 \left(\int_{\substack{|x|\ge |x^0|/2 \\ |x-x^0|>\rho}} + \int_{|x-x^0|\le\rho} \right) \frac{|\nabla v|}{|x - x^0|^{n-1}}\,dx.
\end{aligned}$$

Notice that we have used

$$\left| \frac{x}{|x|} \cdot x^0 - \frac{x^0}{|x^0|} \cdot x \right|^2 = |x^0|^2 + |x|^2 - 2x^0 \cdot x = |x - x^0|^2.$$

Taking $|x^0|$ so that $|x^0| \ge 4R$, for $\rho = o(|x^0|)$ we have

$$\begin{aligned}
&\int_{\substack{|x|\ge |x^0|/2 \\ |x-x^0|>\rho}} \frac{|\nabla v|}{|x - x^0|^{n-1}}\,dx \\
&\quad\le \left(\int_{|x|\ge |x^0|/2} |\nabla v|^2 dx \int_{|x-x^0|\ge\rho} \frac{dx}{|x - x^0|^{2n-2}} \right)^{1/2} \\
&\quad\le K_6 \rho^{1-n/2}.
\end{aligned}$$

Now, multiply (11) by $(v(x) - v(x^0))\zeta^2$, where ζ is a cut-off function vanishing outside $|x - x^0| \ge 2\rho$, and then integrate. As a result

$$\int_{|x-x^0|\le 2\rho} \zeta^2 |\nabla v|^2 dx \le K_7 \int_{|x-x^0|\le 2\rho} |v(x) - v(x^0)|^2 |\nabla\zeta|^2 dx. \tag{13}$$

By virtue of the interior estimate

$$|v(x) - v(x^0)| \le K_8 d_{xx^0}^{-\tau_1} |x - x^0|^{\tau_1},$$

and the choice of ζ:

$$\zeta(x) = \begin{cases} \rho_1^{1-(n+\tau_1)/2}, & |x - x^0| < \rho_1, \\ |x - x^0|^{1-(n+\tau_1)/2}, & \rho_1 \le |x - x^0| \le \rho, \\ \rho^{1-(n+\tau_1)/2}\left(2 - \dfrac{|x - x^0|}{\rho}\right), & |x - x^0| > \rho, \end{cases}$$

in (13), we have

$$\int_{|x-x^0|\le\rho} \frac{|\nabla v|^2\,dx}{|x - x^0|^{n-2+\tau_1}} \le K_8 |x^0|^{-2\tau_1} \rho^{\tau_1}$$

after letting $\rho_1 \to 0$. Therefore,

$$\int_{|x-x^0|\le\rho} \frac{|\nabla v|\,dx}{|x-x^0|^{n-1}} \le \left(\int_{|x-x^0|\le\rho} \frac{|\nabla v|^2\,dx}{|x-x^0|^{n-2+\tau_1}} \int_{|x-x^0|\le\rho} \frac{dx}{|x-x^0|^{n-\tau_1}}\right)^{\frac{1}{2}} \le K_{10}|x^0|^{-\tau_1}\rho^{\tau_1}.$$

This implies

$$|v(x^0)| \le K_{11}(|x^0|^{-n} + \rho^{1-n/2} + |x^0|^{-\tau_1}\rho^{\tau_1}).$$

Choose ρ so that $\rho^{n/2-1+\tau_1} = |x^0|^{\tau_1}$. We obtain

$$|v(x^0)| \le K_{12}|x^0|^{-(n/2-1)\tau_1/(n/2-1+\tau_1)}.$$

In other words,

$$|u(x) - u(\infty)| = |v(x)| \le K_{12}|x|^{-(n/2-1)\tau_1/(n/2-1+\tau_1)}. \tag{14}$$

Take $\tau = \min\{\tau_1, \tau_2, (n/2-1)\tau_1/(n/2-1)+\tau_1\}$. By combining this estimate with the interior and near-boundary estimates we obtain

$$|D\varphi(x) - D\varphi(\infty)| \le K_{13}d_x^{-\tau},$$
$$|D\varphi(x) - D\varphi(y)| \le K_{14}\min\{1, d_{xy}^{-\tau}\}|x-y|^{\tau}.$$

Thus $\inf C_1 \le K_{13}$ and $\inf C_2 \le K_{14}$.

Finally, for a fixed point $\varphi(x) = \theta Q(\Psi(x) + x_1)/\sup|\nabla(\Psi(x) + x_1)|$, Ψ solves (10) and belongs to $C_{n-2,2+\tau}(\overline{\Omega})$. Since $|D\varphi(x) - D\varphi(\infty)| \le C_1 d_x^{-\tau}$,

$$\nabla\varphi|_\infty = \left(\frac{\theta Q}{\sup|\nabla(\Psi(x)+x_1)|}, 0, \dots, 0\right) = (U, 0, \dots, 0).$$

From $\nabla(\Psi(x)+x_1)|_\infty = (1, 0, \dots, 0)$ we also have $\sup|\nabla(\Psi(x)+x_1)| \ge 1$. As

$$|\varphi - Ux_1| = \left|\frac{\theta Q\Psi(x)}{\sup|\nabla(\Psi(x)+x_1)|}\right| \le Q|\Psi(x)|$$

and $|\Psi(x)|$ is bounded in $\overline{\Omega}$ (because $\Psi \in C_{n-2,2+\tau}(\overline{\Omega})$), $\sup_{\overline{\Omega}}|\varphi - Ux_1|$ is bounded.

In conclusion the fixed points of $\varphi - T(\varphi, \theta)$ $(0 \le \theta \le 1)$ are uniformly bounded. By enlarging the domain a bit we may assume there are no fixed points on the boundary. All conditions for applying the Leray-Schauder theorem on the topological degree are fulfilled. We deduce

THEOREM 4. *The auxiliary problem* (2) *has a solution in* $C^{2+\tau}(\Omega) \cap C^{1+\tau}(\overline{\Omega})$.

§4. Resolution of the problem of a flow passing an obstacle and elementary properties of the solution

THEOREM 5. *There is at most one subsonic* $(\sup|\nabla\varphi| < q_c)$ *solution, up to the addition of a constant, in* $C^{2+\tau}(\Omega)\cap C^{1+\tau}(\overline{\Omega})$ *for Problem* (1).

PROOF. From the study of the fixed points in the previous section we know that $\varphi - Ux_1$ is continuous at ∞. We may fix φ by requiring $(\varphi - Ux_1)|_\infty = 0$.

The difference of two solutions φ_1 and φ_2, $\overline{\varphi} = \varphi_2 - \varphi_1$, satisfies the equation

$$a_{ij}(u_1^{(2)}, \dots, u_n^{(2)})\frac{\partial^2\overline{\varphi}}{\partial x_i \partial x_j} + b_k\frac{\partial\overline{\varphi}}{\partial x_k} = 0,$$

where

$$u_i^{(m)} = \frac{\partial\varphi^{(m)}}{\partial x_i} \qquad (m = 1, 2;\ i = 1, \dots, n),$$

$$b_k = \frac{1}{u_k^{(2)} - u_k^{(1)}}(a_{ij}(u_1^{(2)}, \dots, u_n^{(2)}, u_{k+1}^{(1)}, \dots, u_n^{(1)})$$
$$- a_{ij}(u_1^{(2)}, \dots, u_{k-1}^{(2)}, u_k^{(1)}, \dots, u_n^{(1)}))\frac{\partial^2\varphi^{(1)}}{\partial x_i \partial x_j},$$

and

$$\left.\frac{\partial\overline{\varphi}}{\partial N}\right|_\Gamma = 0, \qquad \overline{\varphi}(\infty) = 0.$$

By virtue of the maximum principle a nonconstant $\overline{\varphi}$ can neither attain a positive maximum nor a negative minimum in $\overline{\Omega}$. As $\overline{\varphi}(\infty) = 0$, $\overline{\varphi} = 0$ in $\overline{\Omega}$. □

THEOREM 6. *There exists a positive number* q_c^∞ *so that Problem* (1) *has a subsonic solution if* $0 \le U \le q_c^\infty$. *Furthermore, the function* $Q(U)$ *belongs to* $C[0, q_c^\infty)$ *and* $\lim_{U\uparrow q_c} Q(U) = q_c$.

PROOF. The solutions of Problem (1) and the auxiliary problem are determined respectively by the constants U and Q. We may call the corresponding solution a solution for U or for Q. We can also determine $Q(U)$ and $U(Q)$, where $U(Q)$ is in general multi-valued—uniqueness of solutions for the auxiliary problem has not been established—and $Q(U)$ is at most one-valued in the subsonic range.

Fix $Q_0 < q_c$. We claim for each U, $0 < U \le \inf U(Q_0)$, there corresponds a solution with $0 < Q(U) < Q_0$. Suppose on the contrary that there is $\widetilde{U}$, $0 < \widetilde{U} \le \inf U(Q_0)$ which does not correspond to any solution for $0 < Q \le Q_0$. In case there exists a sequence $\{(U_i, Q_i)\}$, $\lim U_i = U$ and $\sup_i Q_i \le Q_0$, then the sequence of solutions $\{\varphi_i\}$ is compact in $\|\cdot\|_{n-2,2+\tau/2}$. One of its subsequences, $\{\varphi_{i_j}\}$, converges to some $\widetilde{\varphi}$. But then $\lim Q_{i_j} = \widetilde{Q} \le Q_0$, i.e., $\widetilde{\varphi}$ is a solution corresponding to $(\widetilde{U}, \widetilde{Q})$. This is impossible. Hence, there must be $\widetilde{\varepsilon} > 0$, $\widetilde{\varepsilon} < \widetilde{U}$, such that each U in the interval $(\widetilde{U} - \widetilde{\varepsilon}, \widetilde{U})$ does not correspond to any solution with $0 < Q \le Q_0$. However, on the other hand, applying the Leray-Schauder theorem to the set $\{\varphi \in E: -\infty < U < \widetilde{U} - \widetilde{\varepsilon}/2\}$ one knows that for each Q, $0 < Q \le Q_0$, (2) has at least one solution with $0 < U \le \widetilde{U} - \widetilde{\varepsilon}/2$. Thus we have arrived at a contradiction since $\inf U(Q_0) \le \widetilde{U} - \widetilde{\varepsilon}/2 < \widetilde{U}$.

Next we claim $Q \in C[0, \inf U(Q_0)]$. For any sequence $\{U_i\}$, $U_i \to \widetilde{U} \le \inf U(Q_0)$, we know from Theorem 5 that $Q(U_i) \le Q_0$. If $Q(U_{i_j}) \to \overline{Q} \ne Q(\widetilde{U})$ for some subsequence $\{U_{i_j}\}$, by compactness there is a solution corresponding to $(\widetilde{U}, \widetilde{Q})$. But this is in conflict with Theorem 5. Hence $\{Q(U_i)\}$ must tend to $Q(\widetilde{U})$ as $i \to \infty$.

Let $\sup_{Q<q_c} \inf U(Q) = q_c^\infty$. Because $\inf U(Q) \le Q$, it follows that q_c^∞ is finite ($q_c^\infty < q$) and $Q(U) \in C[0, q_c^\infty)$. In view of $\sup_{U \le \inf U(Q_0)} Q(U) = Q_0$, $\forall Q_0 < q_c$, letting $Q_0 \uparrow q_c$ we have $\sup_{U<q_c^\infty} Q(U) = q_c$. If $\lim_{U \uparrow q_c^\infty} \inf Q(U) = \widetilde{Q} < q_c$, the continuous function $Q(U)$ oscillates in $[\widetilde{Q}, q_c]$ as U tends to $q_c^\infty - 0$. Therefore, every point in $[\widetilde{Q}, q_c]$ is a limit point of $Q(U)$. In particular, there exists $\{U_m\}$, $U_m \uparrow q_c^\infty$ and $\{U'_m\}$, $U'_m \uparrow q_c^\infty$, such that $Q(U_m) \to \widetilde{Q}$ and $Q(U'_m) \to (\widetilde{Q} + q_c)/2$ as $m \to \infty$. The corresponding sequences of solutions $\{\varphi_m\}$ and $\{\varphi'_m\}$ subconverge to φ and φ' respectively. But this is impossible as φ and φ' would then be solutions for $(q_c^\infty, \widetilde{Q})$ and $(q_c^\infty, (\widetilde{Q}+q_c^\infty)/2)$ respectively. The proof of the theorem is complete. □

§5. Further properties of the solution

So far we have assumed the gas flows in from infinity along the x_1-direction. Now we remove this requirement. Consider

$$\begin{cases} \dfrac{\partial}{\partial x_i}\left(\rho \dfrac{\partial \varphi}{\partial x_i}\right) = 0, \\ \left.\dfrac{\partial \varphi}{\partial N}\right|_\Gamma = 0, \\ \nabla \varphi|_\infty = u^\infty, \end{cases}$$

where $u^\infty = (u_1^\infty, \dots, u_n^\infty)$ is a given vector. Since both the equation and the boundary conditions are invariant under rotations, this problem has a unique subsonic solution (up to the addition of a constant) φ in $C^{2+\tau}(\Omega) \cap C^{1+\tau}(\overline{\Omega})$ provided $|u^\infty|$ is not too large. Denote $\varphi = u_i^\infty x_i + \psi$ where $\psi(\infty) = 0$. Regarding ψ as a solution for the linear problem we have, as before,

$$\|\psi\|_{n-2, 2+\tau} \le K_{16} \|u^\infty\|. \tag{15}$$

The solution depends continuously on u^∞ in the following sense.

THEOREM 7. *For a fixed $Q < q_c$, there exists a positive constant $K_2 = K_2(Q)$ such that for $u^{\infty,1}$, $u^{\infty,2}$ satisfying*

$$|u^\infty| \le \inf_{|l|=1} U(Ql_1, \dots, Ql_n),$$

the functions $\psi^{(1)}$, $\psi^{(2)}$ determined from their corresponding solutions $\varphi^{(1)}$, $\varphi^{(2)}$ satisfy

$$\|\psi^{(2)} - \psi^{(1)}\|_{n-2,2+\tau} \le K_{17}|u^{\infty,2} - u^{\infty,1}|.$$

PROOF. Let $\overline{\varphi} = \psi^{(2)} - \psi^{(1)}$. Then $\overline{\varphi}$ satisfies the equation in Theorem 5 as well as $\partial\overline{\varphi}/\partial N|_\Gamma = -(u_i^{\infty,2} - u_i^{\infty,1})\cos(N, x_i)|_\Gamma$, $\overline{\varphi}(\infty) = 0$. The condition at ∞ follows from (15). By (15) it is not hard to see that the b_i's satisfy the condition in Theorem 3. Thus by Theorem 3 we deduce this theorem. □

THEOREM 8. *For a nonzero subsonic flow the maximum of its speed attains on Γ and nowhere else.*

PROOF. Suppose $q_{\max} > 0$ attains at a point p in Ω. After rotating the axes we may assume $\nabla u|_p = (q_{\max}, 0, \dots, 0)$. So u_1 attains a positive maximum at p. As u_1 satisfies an elliptic equation, by the strong maximum principle u_1 must be a constant, i.e., $u_1 = q_{\max}$. Hence $u_2 = \dots = u_n = 0$. However it follows from $\partial\varphi/\partial N|_\Gamma = 0$ that $q_{\max} = 0$, which gives a contradiction.

If $q_{\max} > 0$ attains at ∞, as above we may assume $\nabla u|_\infty = (q_{\max}, 0, \dots, 0)$. Then $\|\varphi - q_{\max}x_1\| < \infty$. Perform a linear transformation so that $a_{ij}(\infty)$ becomes δ_{ij} and $\partial/\partial x_i(a_{ij}\partial u/\partial x_j) = 0$ becomes $\partial/\partial y_i(c_{ij}\partial u/\partial y_j) = 0$. Let $v = |y|^{n-2}(u_1 - q_{\max})$. From the above estimate we have $v = O(1/|y|)$ for large y. By the inversion $Y_i = y_i/|y|^2$, $1 \le i \le n$, the equation becomes

$$\frac{\partial}{\partial Y_i}\left(\tilde{c}_{ij}(Y)\frac{\partial v}{\partial Y_j}\right) + B_i\frac{\partial v}{\partial Y_i} + Bv = 0,$$

where $\tilde{c}_{ij}(0) = \delta_{ij}$, $B_i = O(|Y|^{n-2})$, and $B = O(|Y|^{n-3})$. Introduce another transform

$$v = \begin{cases} \dfrac{\sin(C|Y|)}{Y}w, & n = 3, \\ (1 - C|Y|^2)w, & n > 3. \end{cases}$$

The equation can further be written as

$$\frac{\partial}{\partial Y_i}\left(\tilde{c}_{ij}(Y)\frac{\partial w}{\partial Y_j}\right) + \tilde{B}_i\frac{\partial w}{\partial Y_i} + \tilde{B}w = 0,$$

where $\tilde{B}(0) < 0$ for sufficiently large C. By the strong maximum principle w must be a constant in case it takes a nonnegative maximum or a nonpositive minimum at $Y = 0$. So there are two cases to be considered. In the first case $w(0)$ is not a nonnegative maximum or a nonpositive minimum. As $w(0) = 0$, w takes both positive and negative values near the origin. So u_1 may take values greater or less than $q_{\max}$ near ∞. In particular, $q > q|_\infty$ somewhere near ∞. But which contradicts the fact $q|_\infty = q_{\max}$. In the other case w is a constant, $w = 0$, and $u_1 = q_{\max}$ which also leads to a contradiction. □

THEOREM 9. *For a subsonic flow near* ∞, $u|_\infty$ *must exist and* $u - u_\infty = O(|x|^{-\alpha})$ *for some* $\alpha > 0$. *Also a subsonic flow in the whole space must be uniform.*

PROOF. The first assertion follows from the estimate near ∞ derived earlier. The second assertion is a consequence of the maximum principle. □

OPEN PROBLEMS. 1. A monotonicity property for the solution. Show that $Q(U)$ is increasing (or strictly increasing) for U in $[0, q_c^\infty)$.

2. Relax the regularity requirement $\Gamma \in C^{2+\tau_0}$. Establish solvability when the obstacle (flying object) has corners.

3. Prove Problem (1) still admits a unique solution when there are sonic points on Γ (still subsonic in Ω).

4. Study the time-dependent problem for Problem (1).

5. Give a theoretical estimate on the ratio q_c^∞ / q_c in terms of the geometry of the obstacle.

6. Problems on practical computations for q_c^∞ / q_c.

CHAPTER V

The Initial-Boundary Value Problem for Degenerate Quasilinear Parabolic Equations

In this chapter we treat the initial and initial-boundary value problems for equations such as $u_t = \Delta(u^m)$ $(m > 1)$ which describe the motion of fluids in porous media [23].

Writing out the equation

$$u_t = mu^{m-1}\Delta u + m(m-1)u^{m-2}|\nabla u|^2,$$

we see that it is parabolic on $u > 0$ and degenerate at $u = 0$. The case $u < 0$ will not be considered since it is not realistic.

Under certain assumptions it is still possible to establish the unique solvability for the initial-boundary value problem or the Cauchy problem of this degenerate parabolic equation. However, its solution possesses a property which is in sharp contrast with the solution of a nondegenerate parabolic equation. Consider the equation

$$\begin{cases} u_t = u_{xx}, \\ u|_{t=0} = \phi(x), \end{cases}$$

whose solution is given by

$$u(x, t) = \frac{1}{(4\pi t)^{1/2}} \int_{-\infty}^{\infty} \phi(\xi)e^{-(\xi-x)^2/4t} d\xi.$$

For $\phi(x) > 0$ in $|x| < a$ and vanishing outside $|x| = a$, one reads from this formula that $u > 0$ for any $t > 0$. In other words, u becomes positive everywhere after an arbitrarily small increment in time provided it is nonzero initially. This means that the speed of propagation of the solution is infinity.

It can be shown that this property remains valid for nonlinear nondegenerate parabolic equations. However, for the problem

$$\begin{cases} u_t = (u^m)_{xx} \qquad (m > 1), \\ u|_{t=0} = \phi(x), \end{cases}$$

where ϕ is as before, we can establish the existence of a function $\sigma(t) > 0$ so that $u(x, t) = 0$ for $x > \sigma(t)$, where $\sigma(0) = a$ and $\sigma(t)$ is monotonic increasing. So the solution of a degenerate parabolic equation has finite speed of propagation. We shall establish this fact in the next chapter. The domains of dependence and influence are like those of a hyperbolic equation. On the other hand, when $\phi > 0$ in the entire space the solution is nondegenerate and behaves like a solution of a nonlinear parabolic equation.

Work on these topics began with a single spatial variable for which a typical example is the equation $u_t = (u^m)_{xx}$. Soon equations such as $u_t = a(u)_{xx} + b(u)_x$, $u_t = \Delta(u^m)$, $u_t = \Delta(a(u)) + b_i(u)u_{x_i}$, etc. were studied. Finally the work was extended to more general equations. See [24]–[28].

§1. Formulation of the problem and a Hölder estimate for the solution

Let Ω be a bounded domain in $\mathbb{R}^n$. For $T > 0$ we write $Q = \Omega \times (0, T]$ and $\Gamma = \partial\Omega \times [0, T] \cup \Omega \times \{t = 0\}$. Consider

$$\begin{cases} u_t = (a_{ij}(x, t, u)u_{x_j})_{x_j} + b_i(x, t, u)u_{x_i} + c(x, t, u)u, & (x, t) \in Q, \\ u|_{t=0} = u_0(x), \quad x \in \Omega, \\ u|_{\partial\Omega\times[0,T]} = \psi(x, t)|_{\partial\Omega\times[0,T]}, \end{cases}$$

where the coefficients of the equations, the domain Ω, and the initial-boundary values $u_0(x)$, $\psi(s, t)$ $(s \in \partial\Omega)$ satisfy

(1) a_{ij}, b_i, c, $(\partial/\partial x_j)a_{ij}$, and $(\partial/\partial x_i)b_i$ are in $C(\overline{Q} \times \mathbb{R})$ $(i, j = 1, \dots, n)$.

(2) $(1/\Lambda)\nu(|r|)|\xi|^2 \le a_{ij}(x, t, r)\xi_i\xi_j \le \Lambda\nu(|r|)|\xi|^2$, for all $\xi \in \mathbb{R}^n$ and $(x, t, r) \in Q \times \mathbb{R}$. Here Λ is a constant and the function $\nu(r)$ $(0 \le r < \infty)$ satisfies $\nu \in C[0, \infty)$, $\nu(0) = 0$, $\nu(r) > 0$ $(r > 0)$ and $\exists\delta > 0$, $m > 1$ such that

$$1 \le \frac{r\nu(r)}{\int_0^r \nu(s)ds} \le m \qquad (0 < r \le \delta).$$

So ν is a little bit more general than a power of r—its exponent is between 0 and $m - 1$.

(3) $|b_i|$, $|\partial b_i/\partial x_i|$, $|c| \le \Lambda$ in $\overline{Q} \times \mathbb{R}$.

(4) there exist constants $a_0 > 0$ and $\theta_0 \in (0, 1)$ such that $\text{meas}\{K(\rho) \cap \Omega\} \le (1 - \theta_0)K(\rho)(\rho \le a_0)$, where the ball $K(\rho)$ is centered on $\partial\Omega$.

(5) $u_0(x) \ge 0$ and $\psi(s, t) \ge 0$ $(s \in \partial\Omega)$.

However, in deriving a priori estimates we shall assume $u_0(x) > 0$ in $\overline{\Omega}$, $\psi > 0$ in $\partial\Omega \times [0, T]$, and they satisfies the compatibility condition $u_0(s) = \psi(s, 0)$. Further we require that Γ satisfies a Hölder condition and that u satisfy a Hölder condition on Γ.

Add

$$\begin{cases} 0, & |u| \ge \varepsilon, \\ ((\varepsilon^2 - u^2)^m u_{x_i})_{x_i}, & |u| < \varepsilon \qquad (\varepsilon \text{ to be specified}), \end{cases}$$

to the right-hand side of the equation and let $u \in C^{2,1}(Q) \cap C(\overline{Q})$ be a solution of this modified problem. By setting $u = e^{\lambda t}v$, λ a constant, the equation is transformed into

$$v_t = (a^{\varepsilon}_{ij} v_{x_j})_{x_i} + b_i v_{x_i} + (c - \lambda) v,$$

where $a^{\varepsilon}_{ij} = a_{ij} + \delta_{ij}(\varepsilon^2 - u^2)^m$. At an interior minimum, $v_t \le 0$, $v_{x_i} = 0$, and $(a^{\varepsilon}_{ij} v_{x_j})_{x_i} \ge 0$. This implies $(c - \lambda)v \le 0$. Hence if we take $\lambda > \sup_{Q\times\mathbf{R}} c(x, t, r)$, $\lambda = 2\Lambda$ for example, v cannot attain a negative minimum in Q. Also a zero minimum is excluded by the strong maximum principle. So v and u are positive in $\overline{Q}$. Now, if we take $\lambda < \inf_{Q\times(0,\infty)} c$, $\lambda = -2\Lambda$ for example, v cannot attain a positive minimum in Q. We have

$$u \ge e^{-2\lambda T} \min\left\{ \inf_{\Omega} u_0, \inf_{\partial\Omega\times[0,T]} \psi \right\} \equiv M_1.$$

By a similar reasoning (choosing $\lambda > \max\{0, \sup_{Q\times(0,\infty)} c\}$) v cannot attain a positive maximum in Q. Hence

$$u \le e^{2\Lambda T} \max\left\{ \sup_{\Omega} u_0, \sup_{\partial\Omega\times[0,T]} \psi \right\} \equiv M.$$

The modified equation becomes the original one when $\varepsilon \le M_1$. Hence a $C^{2,1}(Q) \cap C(\overline{Q})$-solution of the original problem satisfies $0 < M_1 \le u \le M$.

In Chapter I we discussed Hölder estimates on u which depend on M and M_1. Now we need to consider Hölder estimates independent of M_1, for in the proof of existence in the degenerate case we shall construct a sequence of approximate solutions $\{u_k\}$ with $\min_{\Gamma} u_k \to 0$ as $k \to \infty$.

Rewrite the equation as an equation for $e^{-\Lambda t}u$ and maintain the same notation. Notice that now the equation satisfies $c \le 0$.

By a suitable change of the values of ν in $\delta \le r < \infty$, for instance setting $\nu(r) = \nu(\delta)$ $(r > \delta)$ and enlarging Λ, we may assume ν satisfies

$$\frac{1}{\Lambda}\nu(r)|\xi|^2 \le a_{ij}(x, t, r)\xi_i\xi_j \le \Lambda\nu(r)|\xi|^2,$$

$$1 \le \frac{r\nu(r)}{\int_0^r \nu(s)\,ds} \le m,$$

$$|b_i|, \left|\frac{\partial b_i}{\partial x_i}\right| \le \Lambda, \qquad -\Lambda \le c(x, t, r) \le 0 \quad \text{for } (x, t) \in Q, r \in (0, \infty).$$

Denote the inverse of $w = \int_0^u \nu(s)\,ds$ by $u = \Phi(w)$. The above inequality for ν becomes

$$\frac{1}{mw} \le \frac{\Phi'(w)}{\Phi(w)} \le \frac{1}{w} \qquad (w > 0).$$

Integrating from w_1 to w_2 $(w_1 < w_2)$ yields

$$\left(\frac{w_2}{w_1}\right)^{1/m} \le \frac{\Phi(w_2)}{\Phi(w_1)} \le \frac{w_2}{w_1}.$$

Therefore

$$\frac{1}{m} \le \frac{\Phi'(w_1)}{\Phi'(w_2)} \le m\left(\frac{w_2}{w_1}\right)^{1-1/m}.$$

Similarly we have

$$\frac{1}{m} \le \frac{\nu(u_2)}{\nu(u_1)} \le m\left(\frac{u_2}{u_1}\right)^{m-1} \qquad (0 < u_1 < u_2).$$

Let K_ρ be a ball centered at $x_0 \in \overline{\Omega}$ with radius ρ. For $0 \le t \le T$ write

$$A_{k,\rho}(t) = \{x \in K_\rho \cap \Omega\colon\ w(x, t) > k\},$$
$$B_{k,\rho}(t) = \{x \in K_\rho \cap \Omega\colon\ w(x, t) < k\},$$

where $w = \int_0^u \nu(s)\,ds$. Let ζ be a cut-off function in K_ρ. For $k \ge \sup_{K_\rho \cap \partial\Omega} w(x, t)$, we multiply the equation by $\zeta^2(w-k)_+$ and integrate over Ω. Using

$$\begin{aligned}\int_\Omega \zeta^2 u_t (w-k)_+ dx &= \int_\Omega \zeta^2 \Phi'(w)(w-k)_+ w_t\, dx\\ &= \frac{\partial}{\partial t}\int_\Omega \zeta^2 \int_k^w \Phi'(\tau)(\tau - k)_+\, d\tau\, dx\\ &= \frac{\partial}{\partial t}\int_\Omega \zeta^2 \int_0^{w-k} \Phi'(\tau + k)\tau_+\, d\tau\, dx\\ &= \frac{\partial}{\partial t}\int_{A_{k,\rho}(t)} \zeta^2 \chi_k(w-k)\, dx,\end{aligned}$$

where

$$\chi_k(s) = \int_0^s \Phi'(\tau + k)\tau\, d\tau,$$

we have

$$\begin{aligned}&\frac{\partial}{\partial t}\int_{A_{k,\rho}(t)} \zeta^2 \chi_k(w-k) + \int_{A_{k,\rho}(t)} \zeta^2 a_{ij} w_{x_i} w_{x_j}\\ &\quad = -2\int_{A_{k,\rho}(t)} \zeta(w-k) a_{ij}\zeta_{x_i} u_{x_j} + \int_{A_{k,\rho}(t)} \zeta^2 (w-k)(b_i u_{x_i} + cu).\end{aligned}$$

Set $h_i(x, t, r) = \int_0^r b_i(x, t, \Phi(k+\tau))\Phi'(k+\tau)\,\tau\, d\tau$. We have

$$|h_i| \le \Lambda \int_0^r \Phi'(k+\tau)\,\tau\, d\tau = \Lambda\chi_k(r), \qquad \left|\frac{\partial h_i}{\partial x_i}\right| \le \Lambda\chi_k(r),$$

and

$$\begin{aligned}&\int_{A_{k,\rho}(t)} \zeta^2 (w-k) b_i u_{x_i}\\ &\quad = \int_{A_{k,\rho}(t)} \zeta^2 \frac{d}{dx_i} h_i(x, t, w-k)\\ &\quad\quad - \int_{A_{k,\rho}(t)} \zeta^2 \left(\int_0^{w-k} \frac{\partial b_i}{\partial x_i}(x, t, \Phi(k+\tau))\Phi'(k+\tau)\tau\, d\tau\right) dx.\end{aligned}$$

From integration by parts and

$$a_{ij}w_{x_i}u_{x_j} \geq \frac{a_{ij}w_{x_i}w_{x_j}}{\nu(u)} \geq \frac{1}{\Lambda}\frac{\nu(e^{\lambda t}u)}{\nu(u)} \geq \frac{1}{m\Lambda},$$

we have

$$\frac{\partial}{\partial t}\int_{A_{k,\rho}(t)} \zeta^2\chi_k(w-k) + \frac{1}{m\Lambda}\int_{A_{k,\rho}(t)} \zeta^2|\nabla w|^2$$
$$\leq 2\Lambda\int_{A_{k,\rho}(t)} \zeta|\nabla\zeta|\,|\nabla w|(w-k) + 2n\Lambda\int_{A_{k,\rho}(t)} \zeta|\nabla\zeta|\chi_k(w-k)$$
$$+ n\Lambda\int_{A_{k,\rho}(t)} \zeta^2\chi_k(w-k).$$

Recall that $c \leq 0$. By the Schwarz inequality we further deduce

$$\frac{\partial}{\partial t}\int_{A_{k,\rho}(t)} \zeta^2\chi_k(w-k) + \frac{1}{2m\Lambda}\int_{A_{k,\rho}(t)} \zeta^2|\nabla w|^2$$
$$\leq \gamma\left[\int_{A_{k,\rho}(t)} |\nabla\zeta|^2(w-k)^2 + \Phi'(k)\int_{A_{k,\rho}(t)} \zeta^2\chi_k(w-k)\right],$$

where $\gamma = \gamma(n, \Lambda, m, M)$. Notice that we have used the fact

$$\frac{\chi_k(s)}{s^2} = \frac{1}{s^2}\int_0^s \Phi'(k+\tau)\,\tau\,d\tau \leq m\Phi'(k)\frac{\int_0^s \tau\,d\tau}{s^2} = \frac{m}{2}\Phi'(k).$$

The following estimate can then be obtained:

$$\frac{\partial}{\partial t}\left[e^{-\gamma\Phi'(k)t}\int_{A_{k,\rho}(t)} \zeta^2\chi_k(w-k)\right] + \frac{1}{2m\Lambda}e^{-\gamma\Phi'(k)t}\int_{A_{k,\rho}(t)} \zeta^2|\nabla w|^2$$
$$\leq \gamma e^{-\gamma\Phi'(k)t}\int_{A_{k,\rho}(t)} |\nabla\zeta|^2(w-k)^2. \tag{1}$$

For $k \leq \inf_{K(\rho)\cap\partial\Omega} w(x, t)$, multiply the equation by $\zeta^2(w-k)_-$ and then integrate over Ω. Writing $\widetilde{\chi}_k(s) = \int_0^s \Phi'(k-\tau)\,\tau\,d\tau$, analogously we obtain

$$\frac{\partial}{\partial t}\left[e^{-\gamma\Phi'(k)t}\int_{B_{k,\rho}(t)} \zeta^2\widetilde{\chi}_k(k-w)\right] + \frac{1}{2m\Lambda}e^{-\gamma\Phi'(k)t}\int_{B_{k,\rho}(t)} \zeta^2|\nabla w|^2$$
$$\leq \gamma e^{-\gamma\Phi'(k)t}\left[\int_{B_{k,\rho}(t)} |\nabla\zeta|^2(k-w)^2 + \operatorname{meas} B_{k,\rho}(t)\right], \tag{2}$$

where $\gamma = \gamma(n, \Lambda, m, M)$. Notice we have used

$$\frac{\widetilde{\chi}_k(k-w)}{(k-w)^2\Phi'(k)} = \frac{1}{(k-w)^2}\int_0^{k-w} \frac{\Phi'(k-w)}{\Phi'(k)}\,\tau\,d\tau$$
$$\leq \frac{m}{(k-w)^2}\int_0^{k-w}\left(1-\frac{\tau}{k}\right)^{1/m-1}\tau\,d\tau$$
$$\leq \frac{m}{k-w}\int_0^{k-w}\left(1-\frac{\tau}{k-w}\right)^{1/m-1}d\tau = m^2.$$

Also the term meas $B_{k,\rho}(t)$ is derived from $\int_\Omega \zeta^2(w-k)_-c(x,t,u)u\,dx$.

All functions satisfying inequalities (1) and (2) comprise the *extended function class* $\mathscr{B}_2(Q, M, m, \gamma)$. To investigate functions in this class we need to look at χ_k and $\widetilde{\chi}_k$ first. For $0 < w_1 \le w_2 \le 2w_1$ we have

$$\frac{1}{m} \le \frac{\Phi'(w_1)}{\Phi'(w_2)} \le 2^{1-1/m} m < 2m.$$

Therefore, when $\mu > k \ge \mu/2 > 0$, $H \le \mu - k$, and $0 < \beta < 1$,

$$\frac{\chi_k(H)}{\Phi'(\mu)} = \int_0^H \frac{\Phi'(k+\tau)}{\Phi'(\mu)} \tau\,d\tau,$$

$$\frac{1}{m} \le \frac{\Phi'(k+\tau)}{\Phi'(\mu)} \le 2m.$$

Consequently

$$\frac{H^2}{2m} \le \frac{\chi_k(H)}{\Phi'(\mu)} \le mH^2,$$

$$\frac{\chi_k(H)}{\chi_k(\beta H)} - 1 = \frac{\int_{\beta H}^H (\Phi'(k+\tau)/\Phi'(k+\beta H))\tau\,d\tau}{\int_0^{\beta H} (\Phi'(k+\tau)/\Phi'(k+\beta H))\tau\,d\tau} \le m^2 \frac{\int_{\beta H}^H \tau\,d\tau}{\int_0^{\beta H} \tau\,d\tau}$$
$$= \frac{m^2(1-\beta^2)}{\beta^2},$$

and hence

$$\frac{\chi_k(H)}{\chi_k(\beta H)} \le 1 + \frac{m^2(1-\beta^2)}{\beta^2}$$

holds.

When $k > H > 0$ and $0 < \beta < 1$, from

$$\frac{1}{m} \le \frac{\Phi'(w_1)}{\Phi'(w_2)} \le m\left(\frac{w_2}{w_1}\right)^{1-1/m} \qquad (0 < w_1 < w_2),$$

one obtains

$$\frac{1}{m}\int_0^H \tau\,d\tau \le \int_0^H \frac{\Phi'(k-\tau)}{\Phi'(k)} \tau\,d\tau \le m\int_0^H \left(1-\frac{\tau}{k}\right)^{1/m-1} \tau\,d\tau$$
$$\le mH\int_0^H \left(1-\frac{\tau}{H}\right)^{1/m-1} d\tau.$$

This means

$$\frac{H^2}{2m} \le \frac{\widetilde{\chi}_k(H)}{\Phi'(k)} \le m^2H^2,$$

$$\frac{\widetilde{\chi}_k(H)}{\widetilde{\chi}_k(\beta H)} - 1 = \frac{\int_{\beta H}^H (\Phi'(k-\tau)/\Phi'(k-\beta H))\tau\,d\tau}{\int_0^{\beta H} (\Phi'(k-\tau)/\Phi'(k-\beta H))\tau\,d\tau}$$
$$\le m^2 \frac{\int_{\beta H}^H (k-\tau)^{1/m-1}\tau\,d\tau}{\int_0^{\beta H} (k-\tau)^{1/m-1}\tau\,d\tau}. \tag{3}$$

For $k/2 \le H \le k$, the right-hand side of (3) does not exceed

$$\frac{m^2 H \int_{\beta H}^{H} (H-\tau)^{1/m-1}\,d\tau}{(2H)^{1/m-1}\int_0^{\beta H}\tau\,d\tau} \le \frac{4m^3}{\beta^2}(1-\beta)^{1/m};$$

for $H < k/2$, it does not exceed

$$\frac{m^2 \int_{\beta H}^{H} (k/2)^{1/m-1}\tau\,d\tau}{\int_0^{\beta H} k^{1/m-1}\tau\,d\tau} \le \frac{2m^2}{\beta^2}(1-\beta^2).$$

Thus

$$\frac{\widetilde{\chi}_k(H)}{\widetilde{\chi}_k(\beta H)} \le 1 + \frac{4m^3(1-\beta)^{1/m}}{\beta^2}.$$

For a fixed $\rho \in (0, 1]$, let $Q_\rho = \{(x, t): |x - x^0| < \rho,\ t^0 - a\eta\rho^2 < t < t^0\}$, where $\eta = \Phi'(\rho^\varepsilon)$ $(0 < \varepsilon \le \frac{1}{2})$. The constants ε and a will be specified in the lemmas below. Denote

$$\operatorname{meas} K(1) = \kappa_n, \qquad \sup_{Q_\rho \cap Q} w = \mu, \qquad \inf_{Q_\rho\cap Q} w = \widetilde{\mu},$$

$$\omega = \mu - \widetilde{\mu}, \quad \rho_i = \left[1 - \frac{i}{3}(1 - 2^{-1/n})\right]\rho \qquad (i = 1, 2, 3),$$

$$\xi = \frac{1}{m^2}\Phi'(\mu), \qquad \eta_k = \frac{1}{m}\Phi'(k).$$

Also assume $\mu > 2\rho^\varepsilon$, $\rho^\varepsilon \le k \le \mu$. Then we have $\xi \le \eta_k \le \eta$.

LEMMA 1. *There exist constants β, a, $b \in (0, 1)$ such that*

$$\operatorname{meas}\{K(\rho_1)\backslash A_{k+\beta H, \rho_1}(t)\} \ge b\kappa_n\rho_1^n, \qquad t \in [t^0 - a\xi\rho^2, t^0],$$

when

$$k \ge \max\{\mu/2, \sup_{Q_\rho\cap\Gamma} w\}, \qquad H = \mu - k > 0,$$

and $\operatorname{meas} A_{k,\rho}(t^0 - a\xi\rho^2) \le \frac{\kappa_n}{2}\rho_1^n$;

$$\operatorname{meas}\{K(\rho_1)\backslash B_{k-\beta H, \rho_1}(t)\} \ge b\kappa_n\rho_1^n, \qquad t \in [t^0 - a\eta_k\rho^2, t^0],$$

when $k \le \inf_{Q_\rho\cap\Gamma} w$, $H = k - \widetilde{\mu} \ge \rho^\varepsilon$, *and* $\operatorname{meas} B_{k,\rho_1}(t^0 - a\eta_k\rho^2) \le \frac{\kappa_n}{2}\rho_1^n$.

PROOF. For $\sigma \in (0, 1)$, let

$$\zeta(x, s, s(1-\sigma)) = \begin{cases} 1, & |x - x^0| \le s(1-\sigma), \\ \dfrac{s - |x - x^0|}{\sigma s}, & s(1-\sigma) < |x - x^0| < s, \\ 0, & |x - x^0| \ge s. \end{cases}$$

Substituting $\zeta = \zeta(x, \rho_1, \rho_1(1-\sigma))$ into (1) and then integrating from $t^0 - a\xi\rho^2$ to t, where $t^0 - a\xi\rho^2 \le t \le t^0$ yields

$$e^{-\gamma\Phi'(k)t}\int_{A_{k,\rho_1}(t)} \zeta^2\chi_k(w-k)$$
$$\le e^{-\gamma\Phi'(k)(t^0-a\xi\rho^2)}\left[\int_{A_{k,\rho_1}(t^0-a\xi\rho^2)} \zeta^2\chi_k(w-k) + \gamma a\xi\rho^2\frac{H^2}{(\sigma\rho_1)^2}\kappa_n\rho_1^n\right].$$

By virtue of the hypotheses of the lemma we have

$$\int_{A_{k,\rho_1}(t)} \zeta^2\chi_k(w-k) \le e^{\gamma\Phi'(k)a\xi\rho^2}\left(\chi_k(H)\frac{\kappa_n}{2}\rho_1^n + 2\gamma a\xi\frac{H^2}{\sigma^2}\kappa_n\rho_1^n\right)$$

and

$$\int_{A_{k,\rho_1}(t)} \zeta^2\chi_k(w-k) \ge \int_{A_{k+\beta H,\rho_1(1-\sigma)}(t)} \chi_k(w-k)$$
$$\ge \chi_k(\beta H)\operatorname{meas} A_{k+\beta H,\rho_1(1-\sigma)}(t).$$

Therefore

$$\operatorname{meas} A_{k+\beta H,\rho_1(1-\sigma)}(t)$$
$$\le e^{\gamma\Phi'(k)a\xi\rho^2}\left(\frac{\chi_k(H)}{\chi_k(\beta H)}\frac{1}{2}\kappa_n\rho_1^n + 2\gamma a\sigma^{-2}\frac{\xi H^2}{\chi_k(\beta H)}\kappa_n\rho_1^n\right)$$
$$\le e^{\gamma\Phi'(k)a\xi\rho^2}\left\{\frac{1}{2}\left[1+\frac{m^2(1-\beta^2)}{\beta^2}\right] + \frac{4a\gamma}{m\beta^2\sigma^2}\right\}\kappa_n\rho_1^n.$$

Notice that we have used $\chi_k(\beta H)/\Phi'(\mu) \ge (\beta H)^2/(2m)$. Fix $\beta = \beta(m) \in (0, 1)$ satisfying

$$\frac{1}{2}\left[1+\frac{m^2(1-\beta^2)}{\beta^2}\right] \le \frac{3}{4}.$$

Since $k \ge \mu/2 \ge \rho^\varepsilon$,

$$\Phi'(k) \le m\Phi'(M)\left(\frac{M}{k}\right)^{1-1/m} \le m\Phi'(M)M^{1-1/m}\rho^{\varepsilon(1/m-1)},$$

$\mu \ge 2\rho^\varepsilon$, and $\xi = (1/m^2)\Phi'(\mu) \le 1/m\Phi'(M)M^{1-1/m}(2\rho^\varepsilon)^{1/m-1}$, $\gamma\Phi'(k)a\xi\rho^2 \le a\gamma\Phi'(M)^2M^{2-2/m}$. So we may choose small a, b_1, σ_0 depending only on n, M, m, γ, and T such that

$$\frac{1}{2}\left[1+\frac{m^2(1-\beta^2)}{\beta^2}\right]e^{a\gamma\Phi'(k)\xi\rho^2} < (1-b_1)(1-\sigma_0)^n.$$

By further restricting a,

$$e^{a\gamma\Phi'(k)\xi\rho^2}\left\{\frac{1}{2}\left[1+\frac{m^2(1-\beta^2)}{\beta^2}\right] + \frac{4a\gamma}{m\beta^2\sigma_0^2}\right\} \le (1-b_1)(1-\sigma_0)^n.$$

In conclusion, for $t \in [t^0 - a\xi\rho^2, t^0]$, we have

$$\operatorname{meas} A_{k+\beta H, \rho_1(1-\sigma_0)}(t) \le (1-b_1)(1-\sigma_0)^n,$$

and hence

$$\begin{aligned}\operatorname{meas}\{K(\rho_1)\backslash A_{k+\beta H, \rho_1}(t)\} &\ge \operatorname{meas}\{K(\rho_1(1-\sigma_0))\backslash A_{k+\beta H, \rho_1(1-\sigma_0)}(t)\} \\ &\ge (1-\sigma_0)^n \kappa_n \rho_1^n - (1-b_1)(1-\sigma_0)^n \kappa_n \rho_1^n \\ &= b_1(1-\sigma_0)^n \kappa_n \rho_1^n.\end{aligned}$$

The first inequality of the lemma follows by taking $b = b_1(1-\sigma_0)^n$. By a similar argument one can prove the second inequality. □

LEMMA 2 (interior estimate). *Let $Q_\rho \subseteq Q$. For each $\theta_1 > 0$ there exists $s = s(\theta_1) > 0$ such that*

(i) $\int_{t^0 - aA\rho^2}^{t^0} \operatorname{meas} A_{\mu - H/2^{s+1}, \rho_1}(t)\, dt \le \theta_1 \xi \rho_1^{n+1}$ *provided* $k \ge \mu/2$, $H = \mu - k > 0$, *and* $\operatorname{meas} A_{k, \rho_1}(t^0 - a\xi\rho^2) \le (\kappa_n/2)\rho_1^n$;

(ii) $\omega \le 2^{s+2}\rho^\varepsilon$ *or* $\int_{t^0 - a\eta_{k_{s+2}}\rho^2}^{t_0} \operatorname{meas} B_{k_{s+1}, \rho_1}(t)\, dt \le \theta_1 \eta_{k_{s+2}} \rho_1^{n+2}$, *provided* $\max_{[t^0 - a\eta\rho^2, t^0 - a\xi\rho^2]} \operatorname{meas} B_{\tilde{\mu}+\omega/2, \rho_1}(t) \le (\kappa_n/2)\rho_1^n$. *Here* $k_s = \tilde{\mu} + \omega/2^s$ *and* $\eta_{k_s} = (1/m)\Phi'(k_s)$.

PROOF. We treat (i) first. From the assumptions and Lemma 1 we have

$$\operatorname{meas}\{K(\rho_1)\backslash A_{k+\beta H, \rho_1}(t)\} \le b\kappa_n \rho_1^n, \qquad t \in [t^0 - a\xi\rho^2, t^0].$$

Choose r_0 with $1 - \beta \ge 2^{-r_0}$ and let $k_l = \mu - H/2^l$. Then

$$\operatorname{meas}\{K(\rho_1)\backslash A_{k_l, \rho_1}(t)\} \ge b\kappa_n \rho_1^n,$$

for $l \ge \gamma_0$ and $t \in [t^0 - a\xi\rho^2, t^0]$. By De Giorgi's lemma [49]

$$(k_{l+1} - k_l) \operatorname{meas}^{1-1/n} A_{k_{l+1}, \rho_1}(t) \le \frac{\beta \rho_1^n}{\operatorname{meas}\{K(\rho_1)\backslash A_{k_l, \rho_1}(t)\}} \int_{D_l(t)} |\nabla w|,$$

where $D_l(t) = A_{k_l, \rho_1}(t)\backslash A_{k_{l+1}, \rho_1}(t)$. Combining these two inequalities we have

$$\frac{H}{2^{l+1}} \operatorname{meas} A_{k_{l+1}, \rho_1}(t) \le \frac{\beta}{b\kappa_n^{1-1/n}} \rho_1 \left(\int_{D_l(t)} |\nabla w|^2 \right)^{1/2} (\operatorname{meas} D_l(t))^{1/2}.$$

Integrating this from $t^0 - a\xi\rho^2$ to t^0, as a result of the Schwarz inequality we have

$$\begin{aligned}&\left(\frac{H}{2^{l+1}}\right)^2 \left(\int_{t^0 - a\xi\rho^2}^{t^0} \operatorname{meas} A_{k_{l+1}, \rho_1}(t)\, dt \right)^2 \\ &\quad \le \left(\frac{\beta}{b\kappa_n^{1-1/n}} \right)^2 \rho_1^2 \int_{t^0 - a\xi\rho^2}^{t^0} \int_{A_{k_l, \rho_1}(t)} |\nabla w|^2 \cdot \int_{t^0 - a\xi\rho^2}^{t^0} \operatorname{meas} D_l(t).\end{aligned}$$

To estimate the term involving $|\nabla w|^2$ we take $\zeta = \zeta(x, \rho, \rho_1)$ in (1). An integration from $t^0 - a\xi\rho^2$ to t^0 yields

$$\frac{1}{2m\Lambda}\int_{t^0-a\xi\rho^2}^{t^0} e^{-\gamma\Phi'(k_l)t}\int_{A_{k_l,\rho_1}(t)}|\nabla w|^2$$
$$\leq e^{-\gamma\Phi'(k_l)(t^0-a\xi\rho^2)}\left[\int_{A_{k_l,\rho_1}(t^0-a\xi\rho^2)}\zeta^2\chi_{k_l}(w-k_l)\right.$$
$$\left.+\gamma a\xi\rho^2\frac{9}{(1-2^{-1/n})^2\rho^2}\left(\frac{H}{2^l}\right)^2\kappa_n\rho^n\right]$$
$$\leq e^{-\gamma\Phi'(k_\gamma)(t^0-a\xi\rho^2)}\left[\chi_{k_l}\left(\frac{H}{2^l}\right)+\frac{9a\gamma}{(1-2^{-1/n})^2}\frac{\xi H^2}{2^{2l}}\right]\kappa_n\rho^n.$$

Therefore

$$\int_{t^0-a\xi\rho^2}^{t^0}\int_{A_{k_l,\rho_l}(t)}|\nabla w|^2 \leq 2m\Lambda e^{\gamma\Phi'(k)a\xi\rho^2}\left[2m^3+\frac{9a\gamma}{(1-2^{-1/n})^2}\right]\frac{\xi H^2}{2^{2l}}\kappa_n\rho^n.$$

Substituting this estimate into the above inequality we get

$$\left(\int_{t^0-a\xi\rho^2}^{t^0}\operatorname{meas} A_{k_{l+1},\rho_1}(t)\right)^2 \leq C_1\xi\rho^{n+2}\int_{t^0-a\xi\rho^2}^{t^0}\operatorname{meas} D_l(t),$$

where $C_1 = C_1(n, m, M, \gamma, T)$. Summing over $l = r_0, r_0+1, \ldots, s$ we have

$$(s-r_0+1)\left(\int_{t^0-a\xi\rho^2}^{t^0}\operatorname{meas} A_{k_{s+1},\rho_1}(t)\right)^2 \leq C_1\kappa_n a\xi^2\rho^{2n+4}.$$

Choosing s so that $(C_1\kappa_n a/(s-r_0+1))^{1/2} < \theta_1/2^{1+2/n}$ we conclude

$$\int_{t^0-a\xi\rho^2}^{t^0}\operatorname{meas} A_{\mu-H/2^{s+1},\rho_1}(t) \leq \theta_1\xi\rho_1^{n+2}.$$

Next we establish (ii). When $\omega > 2^{s+2}\rho^\varepsilon$, take $k = \widetilde{\mu}+\omega/2$ (so $H = k-\widetilde{\mu} = \omega/2 \geq \rho^\varepsilon$) in Lemma 1. We have

$$\operatorname{meas}\{K(\rho_1)\backslash B_{\widetilde{\mu}+\omega(1-\beta)/2,\rho_1}(t)\} \geq b\kappa_n\rho_1^n, \qquad t\in[t^0-a\eta\rho^2, t^0].$$

Choose r_0 so that $2^{-r_0} \geq (1-\beta)/2$. Then for $l \geq r_0$,

$$\operatorname{meas}\{K(\rho_1)\backslash B_{k_l,\rho_1}(t)\} \geq b\kappa_n\rho^n, \qquad t\in[t^0-a\eta\rho^2, t^0],$$

where $k_l = \widetilde{\mu}+\omega/2^l$. Arguing as before we get

$$\frac{\omega^2}{2^{2l+2}}\left(\int_{t^0-a\eta_{k_{s+2}}\rho^2}^{t^0}\operatorname{meas} B_{k_l,\rho_1}(t)\right)^2$$
$$\leq\left(\frac{\beta}{b\kappa_n^{1-1/n}}\right)^2\rho_1^2\int_{t^0-a\eta_{k_{s+2}}\rho^2}^{t^0}\int_{B_{k_l,\rho_1}(t)}|\nabla w|^2\cdot\int_{t^0-a\eta_{k_{s+2}}\rho^2}^{t^0}\operatorname{meas}\widetilde{D}_l(t)\,dt,$$

for $r_0 \leq l \leq s$. Here $\widetilde{D}_l(t) = B_{k_l,\rho_1}(t)\backslash B_{k_{l+1},\rho_1}(t)$.

Letting $\zeta = \zeta(x, \rho, \rho_1)$ in (2), after an integration in t, we have

$$\begin{aligned}\frac{1}{2m\Lambda}\int_{t^0-a\eta_{k_{s+1}}\rho^2}^{t^0} e^{-\gamma\Phi'(k_l)t}\int_{B_{k_l,\rho_1}(t)}|\nabla w|^2 \\ \le e^{-\gamma\Phi'(k_l)(t^0-a\eta k_{s+2}\rho^2)}\Bigg[\int_{B_{k_l,\rho}(t^0-a\eta_{k_{s+2}}\rho^2)}\zeta^2\widetilde{\chi}_{k_l}(k_l-w) \\ +\gamma a\eta_{k_{s+2}}\rho^2\frac{9}{(1-2^{-1/n})^2\rho^2}\left(\frac{\omega}{2^{l+1}}\right)^2\kappa_n\rho^n \\ +\gamma a\eta_{k_{s+2}}\rho^2\kappa_n\rho^n\left(\frac{\omega}{2^{l+1}\rho^\varepsilon}\right)^2\Bigg]\end{aligned}$$

$(\because\ \omega \ge 2^{l+1}\rho^\varepsilon)$. Also notice that

$$\begin{aligned}\widetilde{\chi}_k(k_l-w) &\le m^2\Phi'(k_l)(k_l-w)^2 \\ &\le m^3\Phi'(k_{s+2})(k_l-\widetilde{\mu})^2 = m^4\eta_{k_{s+2}}\left(\frac{\omega}{2^l}\right)^2.\end{aligned}$$

Now we can deduce (ii) along the same lines as for (i). □

LEMMA 2′ (near-boundary estimate). *Let*

$$\omega_1 = \operatorname{osc}\{w\colon Q_\rho \cap \Gamma\} \le L\rho^{\varepsilon_1},$$

where L, ε_1 *are positive constants independent of* ρ. (*This condition can be deduced from* $\operatorname{osc}\{u\colon Q_\rho\cap\Gamma\} \le K\rho^{\varepsilon_1}$.) *Then when*

$$\operatorname{meas}\{K(\rho_1)\backslash K(\rho_1)\cap\Omega\} \ge b_1\rho_1^n \qquad (b_1 > 0),$$

for each $\theta_1 > 0$, *there exists* $s = s(\theta_1) > 0$ *such that one of the following inequalities holds*:

(i) $\omega = \operatorname{osc}\{w\colon Q_\rho \cap Q\} \le 2^{s+2}\rho^\varepsilon$ $(0 \le \varepsilon \le \varepsilon_1)$;

(ii) $\int_{t^0-a\xi\rho^2}^{t^0} \operatorname{meas} A_{\mu-\omega/2^{s+1},\rho_1}(t)\,dt \le \theta_1\xi\rho_1^{n+2}$;

(iii) $\int_{t^0-a\eta_{k_{s+2}}\rho^2}^{t^0} \operatorname{meas} B_{\widetilde{\mu}+\omega/2^{s+1},\rho_1}(t)\,dt \le \theta_1\eta_{k_{s+2}}\rho_1^{n+2}$ $(k_s = \widetilde{\mu}+\omega/2^s)$.

If $t^0 - a\xi\rho^2 \le 0$, *then* (ii) *and* (iii) *should be replaced by*

(ii)′ $\operatorname{meas} A_{\mu-\omega/4,\rho_1}(0) = 0$, $\int_0^{t^0} \operatorname{meas} A_{\mu-\omega/2^{s+1},\rho_1}(t)\,dt \le \theta_1 t^0\rho_1^n$; *and*

(iii)′ $\operatorname{meas} B_{\widetilde{\mu}+\omega/4,\rho_1}(0) = 0$, $\int_0^{t^0} \operatorname{meas} B_{\widetilde{\mu}+\omega/2^{s+1},\rho_1}(t)\,dt \le \theta_1 t^0\rho_1^n$.

PROOF. Fix $r_0 \ge 2$ which satisfies $2^{-r_0} \le 1-\beta$ and $2^{r_0} \ge 4L$. If $\omega \ge 2^{s+2}\rho^2$, $\omega \ge 2^{r_0}\rho^\varepsilon \ge 4L\rho^\varepsilon \ge 4\omega_1$ (for $s \ge r_0$). Therefore, $w(Q_\rho\cap\Gamma)$ intersects either $[\widetilde{\mu}, \widetilde{\mu}+\omega/4]$ or $[\mu-\omega/4, \mu]$, but not both at the same time. Assume it is disjoint from the latter. Let $k = \mu - \omega/2^{r_0}$. Then

$k \geq \max\{\mu/2, \sup_{Q_\rho \cap \Gamma} w(x, t)\}$ and $\operatorname{meas}\{K(\rho_1)\backslash A_{k, \rho_1(t)}\} \geq b_1 \rho_1^n$, for $t \in [t^0 - a\xi\rho^2, t^0]$. Now we can deduce (ii) as in Lemma 2.

If $t^0 - a\xi\rho^2 \leq 0$, $w(Q_\rho \cap \Gamma)$ is disjoint from $[\mu - \omega/4, \mu]$. As a result $\operatorname{meas} A_{\mu-\omega/4, \rho_1}(0) = 0$. The other inequality in (ii)$'$ can be established as in Lemma 2, only now the range of integration has to be changed from $a\xi\rho^2$ to t^0.

Similarly we obtain (iii) and (iii)$'$. The proof of the lemma is complete. □

LEMMA 3. *For each* $\theta_2 > 0$, *there exists* $\theta_1 > 0$ *such that*

(i) $\operatorname{meas} A_{k+H/2, \rho_2}(t) \leq \theta_2 \rho_2^n$, $t \in [t^0 - (a/4)\xi\rho^2, t^0]$, *for* $k \geq \max\{\mu/2, \sup_{Q_\rho \cap \Gamma} w(x, t)\}$, $H = \mu - k > 0$, *and* $\int_{t^0 - a\xi\rho^2}^{t^0} \operatorname{meas} A_{k, \rho_1}(t)\, dt \leq \theta_1 \xi \rho_1^{n+2}$. *If further* $\operatorname{meas} A_{k, \rho_1}(t^0 - a\xi\rho^2) = 0$, *this estimate also holds for* $t \in [t^0 - a\xi\rho^2, t^0]$.

(ii) $\operatorname{meas} B_{k-H/2, \rho_2}(t) \leq \theta_2 \rho_2^n$, $t \in [t^0 - (a/4)\eta_{k-H/2}\rho^2, t^0]$ *for* $k \leq \inf_{Q_\rho \cap \Gamma} w(x, t)$, $H = k - \tilde{\mu} \geq \rho^\varepsilon$, *and* $\int_{t^0 - a\eta_{k-H/2}\rho^2}^{t^0} \operatorname{meas} B_{k, \rho_1}(t)\, dt \leq \theta_1 \eta_{k-H/2} \rho_1^{n+2}$. *If further* $\operatorname{meas} B_{k, \rho_1}(t^0 - a\eta_{k-H/2}\rho^2) = 0$, *this estimate also holds for* $t \in [t^0 - a\eta_{k-H/2}\rho^2, t^0]$.

PROOF. We only prove (ii) as a typical case. Take $\zeta = \zeta(x, \rho_1, \rho_2)$ in (2) and integrate it from τ to t, where $t^0 - a\eta_{k-H/2}\rho^2 \leq \tau < t \leq t^0$. We have

$$
\begin{aligned}
&e^{-\gamma\Phi'(k)t} \tilde{\chi}_k(H/2) \operatorname{meas} B_{k-H/2, \rho_2}(t) \\
&\leq e^{-\gamma\Phi'(k)\tau} \left\{ \tilde{\chi}_k(H) \operatorname{meas} B_{k, \rho_1}(\tau) \right. \\
&\qquad \left. + \gamma \left[\frac{H^2}{(\rho_1 - \rho_2)^2} + 1 \right] \int_\tau^t \operatorname{meas} B_{k, \rho_1}(t)\, dt \right\}.
\end{aligned}
$$

From the assumptions in (ii) we deduce

$$
\int_{t^0 - a\eta_{k-H/2}\rho^2}^{t^0 - (a/4)\eta_{k-H/2}\rho^2} \operatorname{meas} B_{k, \rho_1}(t)\, dt \leq \theta_1 \eta_{k-H/2} \rho_1^{n+2}.
$$

So we can find $\tau \in [t^0 - a\eta_{k-H/2}\rho^2, t^0 - (a/4)\eta_{k-H/2}, \rho^2]$ such that

$$
\operatorname{meas} B_{k, \rho_1}(\tau) \leq \frac{4}{3a} \theta_1 \rho_1^n.
$$

Substituting this τ into the inequality above yields, for $t \in [t^0 - (a/4)\eta_{k-H/2}\rho^2, t^0]$,

$$\begin{aligned}\operatorname{meas} B_{k-H/2,\rho_2}(t) &\\ \le e^{\gamma\Phi'(k)a\eta_{k-H/2}\rho^2} &\left\{\frac{\widetilde{\chi}_k(H)}{\widetilde{\chi}_k(H/2)}\frac{4}{3a}\right.\\ &\left.+\gamma\frac{\eta_{k-H/2}}{\widetilde{\chi}_k(H/2)}\left[\frac{9H^2}{(1-2^{-1/n})^2}+\rho^2\right]\right\}\theta_1\rho_1^n.\end{aligned}$$

Since

$$\frac{\widetilde{\chi}_k(H)}{\widetilde{\chi}_k(H/2)} \le 1+16m^3, \quad \frac{\eta_{k-H/2}H^2}{\widetilde{\chi}_k(H/2)} \le 8m, \quad \Phi'(k)\eta_{k-H/2}\rho^2 \le K,$$

and $H \le \rho^\varepsilon$, the conclusion of (ii) holds for sufficiently small θ_1.

When $\operatorname{meas} B_{k,\rho_1}(t^0 - a\eta_{k-H/2}\rho^2) = 0$, the other conclusion of (ii) follows by taking $\tau = t^0 - a\eta_{k-H/2}\rho^2$. □

LEMMA 4. *There exists $\theta_2 > 0$ such that*

(i) $\operatorname{meas} A_{k+H/2,\rho_3}(t) = 0$, $t \in [t^0 - (a/4)\xi\rho^2, t^0]$, *for* $k \ge \max\{\mu/2, \sup_{Q_\rho\cap\Gamma} w(x,t)\}$, $H = \mu - k > 0$, *and* $\max_{[t^0 - a\xi\rho^2, t^0]} \operatorname{meas} A_{k,\rho_2}(t) \le \theta_2\rho_2^n$. *If further* $\operatorname{meas} A_{k,\rho_2}(t^0 - a\xi\rho^2) = 0$, *the conclusion continues to hold for* $t \in [t^0 - a\xi\rho^2, t^0]$.

(ii) $\operatorname{meas} B_{k-H/2,\rho_3}(t) = 0$, $t \in [t^0 - (a/4)\eta_k\rho^2, t^0]$, *for* $k \le \inf_{Q_\rho\cap\Gamma} w(x,t)$, $H = k - \widetilde{\mu} \ge \rho^\varepsilon$, *and* $\max_{[t^0 - a\eta_k\rho^2, t^0]} \operatorname{meas} B_{k,\rho_2}(t) \le \theta_2\rho_2^n$. *If further* $\operatorname{meas} B_{k,\rho_2}(t^0 - a\eta_k\rho^2) = 0$, *the conclusion continues to hold for* $t \in [t^0 - a\eta_k\rho^2, t^0]$.

PROOF. As a sample we establish (ii). We use the notation

$$\begin{aligned}&k_h = k - \frac{H}{2} + \frac{H}{2^{h+1}}, \qquad t_h = t^0 - \frac{a}{4}\eta_k\rho^2 - \frac{3a}{2^{h+2}}\eta_k\rho^2,\\ &\rho_h = \rho_3 + \frac{\rho_2 - \rho_3}{2^h}, \qquad \mu_h = \max_{[t_h, t^0]} \operatorname{meas} B_{k_h,\rho_h}(t),\\ &\zeta_h(x) = \zeta(x, \rho_h, \rho_{h+1}),\\ &I_h(t) = e^{-\gamma\Phi'(k_h)(t - t^0 + a\eta_k\rho^2)} \int_{B_{k_h,\rho_h}(t)} \widetilde{\chi}_{k_h}(k_h - w)\zeta_h^2\,dx.\end{aligned}$$

Since $k_h \ge h/2$, $\Phi'(k_h) \le 2m\Phi'(k) = 2m^2\eta_k$. Moreover, from the property of $\widetilde{\chi}_k$ we have

$$\widetilde{\chi}_k(k_h - w) \le m^2\Phi'(k_h)(k_h - w)^2 \le 2m^4\eta_k(k_h - w)^2.$$

Therefore, for $t \in [t^0 - a\eta_k\rho^2, t^0]$ we have:

$$\begin{aligned}
I_h(t) &\le \gamma_1\eta_k \int_{B_{k_h,\rho_h}(t)} (k_h - w)^2\zeta_h^2 \\
&\le \gamma_1\eta_k \left(\int_{B_{k_h,\rho_h}(t)} 1\right)^{2/n} \left\{\int_{B_{k_h,\rho_h}(t)} \left[(k_h - w)^2\zeta_h^2\right]^{n/(n-2)}\right\}^{(n-2)/n} \\
&\le \gamma_2\eta_k (\operatorname{meas} B_{k_h,\rho_h}(t))^{2/n} \int_{B_{k_h,\rho_h}(t)} |\nabla[(k_h - w)\zeta_h]|^2 \\
&\le \gamma_3\eta_k\mu_h^{2/n}\left[\int_{B_{k_h,\rho_h}(t)} |\nabla w|^2\zeta_h^2 + \frac{H^2}{(\rho_h - \rho_{h+1})^2}\mu_h\right],
\end{aligned} \tag{4}$$

where $\gamma_1, \gamma_2, \gamma_3$ are constants. From (2) we have

$$\begin{aligned}
I_h'(t) &+ \frac{1}{2m\Lambda}e^{-\gamma\Phi'(k_h)(t-t^0-a\eta_k\rho^2)} \int_{B_{k_h,\rho_h}(t)} |\nabla w|^2\zeta_h^2 \\
&\le \gamma\left[\frac{H^2}{(\rho_h - \rho_{h+1})^2} + 1\right]\mu_h, \quad \forall t \in [t_{h+1}, t^0].
\end{aligned} \tag{5}$$

The following three cases are possible:

(i) $I_h'(t) \ge 0$. From (5) we have

$$\begin{aligned}
\int_{B_{k_h,\rho_h}(t)} |\nabla w|^2\zeta_h^2 &\le 2m\Lambda\gamma e^{\gamma\Phi'(k_h)a\eta_k\rho^2}\left[\frac{H^2}{(\rho_h - \rho_{h+1})^2} + 1\right]\mu_h \\
&\le 2m\Lambda\gamma e^{\gamma C_1}\left[\frac{H^2}{(\rho_h - \rho_{h+1})^2} + 1\right]\mu_h.
\end{aligned}$$

Putting this into (4) gives

$$I_h(t) \le \gamma_3\eta_k\mu_h^{1+2/h}\left[(2m\Lambda\gamma e^{\gamma C_1} + 1)\frac{H^2}{(\rho_h - \rho_{h+1})^2} + 2m\Lambda\gamma e^{\gamma C_1}\right]. \tag{6}$$

(ii) $I_h'(t) < 0$ and there is $\tau \in [t_h, t]$ with $I_h'(\tau) = 0$. Fix τ such that $I_h'(s) < 0$ for $\tau < s \le t$. Since $I_h(t) < I_h(\tau)$ and (6) holds for $I_h(\tau)$, (6) also holds for $I_h(t)$.

(iii) $I_h'(\tau) < 0$ for all $\tau \in [t_h, t]$. From (5) we have

$$\begin{aligned}
\frac{1}{2m\Lambda}\int_{t_h}^{t} e^{-\gamma\Phi'(k_h)(t-t^0-a\eta_k\rho^2)} &\int_{B_{k_h,\rho_h}(t)} |\nabla w|^2\zeta_h^2 \\
&\le I_h(t_h) + \gamma(t - t_h)\left[\frac{H^2}{(\rho_h - \rho_{h+1})^2} + 1\right]\mu_h,
\end{aligned}$$

or,

$$\int_{t_h}^{t}\int_{B_{k_h,\rho_h}(t)} |\nabla w|^2\zeta_h^2 \le 2m\Lambda e^{\gamma C_1}\left\{I_h(t_h) + \gamma(t - t_h)\left[\frac{H^2}{(\rho_h - \rho_{h+1})^2} + 1\right]\mu_h\right\}.$$

Integrate (4) from t_h to t and use this inequality. We get

$$\int_{t_h}^t I_h(\tau)\,d\tau \le \gamma_3\eta_k\mu_h^{2/n}\Big[2m\Lambda e^{\gamma C_1}I_h(t_h)+(2m\Lambda\gamma e^{\gamma C_1}+1)\times(t-t_h)\frac{H^2}{(\rho_h-\rho_{h+1})^2}\mu_h+2m\Lambda e^{\gamma C_1}(t-t_h)\mu_h\Big].$$

We estimate $I_h(t_h)$ in the right-hand side as follows:

$$I_h(t_h)\le\gamma_1\eta_k\mu_h\left(k-\frac{H}{2}+\frac{H}{2^{h+1}}-\widetilde{\mu}\right)^2 = \gamma\eta_k\mu_h\left(\frac{1}{2}+\frac{1}{2^{h+1}}\right)^2H^2\le\gamma_1\eta_k\mu_hH^2.$$

As $I_h(\tau)$ is decreasing in $[t_h, t]$, $\int_{t_h}^t I_h(\tau)\ge I_h(t)(t-t_h)$. Hence

$$I_h(t)\le\gamma_3\eta_k\mu_h^{1+2/n}\Big[2m\Lambda\gamma_1e^{\gamma C_1}\frac{\eta_kH^2}{t-t_h}+(2m\Lambda\gamma e^{\gamma C_1}+1)\times\frac{H^2}{(\rho_h-\rho_{h+1})^2}+2m\Lambda\gamma e^{\gamma C_1}\Big].$$

In fact this estimate is valid in all cases. For $t\in[t_{h+1}, t^0]$ we deduce from it

$$I_h(t)\le\gamma_3\eta_k\mu_h^{1+2/n}\Big[2m\Lambda\gamma_1e^{\gamma C_1}\frac{\eta_kH^2}{t_{h+1}-t_h}+(2m\Lambda\gamma e^{\gamma C_1}+1)\times\frac{H^2}{(\rho_h-\rho_{h+1})^2}+2m\Lambda\gamma e^{\gamma C_1}\Big].$$

On the other hand, from the definition of $I_h(t)$ we have

$$I_h(t)\ge e^{-\gamma C_1}\widetilde{\chi}_{k_h}(k_h-k_{h+1})\operatorname{meas}B_{k_{h+1},\rho_{h+1}}(t) \ge\frac{e^{-\gamma C_1}}{4m}\eta_k(k_h-k_{h+1})^2\operatorname{meas}B_{k_{h+1},\rho_{h+1}}(t).$$

Notice we have used

$$\frac{\widetilde{\chi}(k_h-k_{h+1})}{(k_h-k_{h+1})}\ge\frac{1}{2m}\Phi'(k_h)\ge\frac{1}{4m^2}\Phi'(k)=\frac{\eta_k}{4m}.$$

Combining these two inequalities we obtain

$$\mu_{h+1}\le 4m\gamma_3e^{\gamma C_1}\mu_h^{1+2/n}2^{2(h+1)}\Big[2m\Lambda\gamma_1e^{\gamma C_1}\frac{\eta_k}{t_{h+1}-t_h}+(2m\Lambda\gamma e^{\gamma C_1}+1)\frac{1}{(\rho_h-\rho_{h+1})^2}+\frac{2m\Lambda\gamma e^{\gamma C_1}}{H^2}\Big].$$

In view of the definitions of t_h and ρ_h and the fact $H \geq \rho^\varepsilon$ this estimate becomes $\mu_{h+1} \leq (C_2/\rho^2)2^{4h}\mu_h^{1+2/n}$, where $C_2 = C_2(n, m, \gamma, M, T)$. Writing $y_h = \mu_h/\rho^n$, we have $y_{h+1} \leq C_2 2^{4h} y_h^{1+2/n}$. From the assumption

$$y_0 = \frac{\mu_0}{\rho^n} = \frac{1}{\rho^n} \max_{[t^0 - a\eta_k\rho^2, t^0]} \operatorname{meas} B_{k, \rho_2}(t) \leq \theta_2.$$

For sufficiently small θ_2, we can use induction to show $y_h \leq \theta_2 2^{-2nh}$ ($h = 0, 1, 2, \ldots$). This is because

$$\begin{aligned} y_{h+1} &\leq C_2 2^{4h} y_h^{1+2/n} \leq C_2 2^{4h} (\theta_2 2^{-2nh})^{1+2/n} \\ &= C_2 \theta_2^{1+2/n} 2^{-2nh} \leq \theta_2 2^{-2n(h+1)}. \end{aligned}$$

Thus it suffices to choose θ_2 satisfying $C_2\theta_2^{2/n} \leq 2^{-2n}$.

Now (ii) follows after letting $h \to \infty$. When $\operatorname{meas} B_{k, \rho}(t^0 - a\eta_k\rho^2) = 0$, we take $t_h = t^0 - a\eta_k\rho^2$ ($h = 0, 1, 2, \ldots$). Under this circumstance (iii) cannot happen. So the other conclusion holds. □

LEMMA 5. *Let $Q_\rho \subseteq Q$. There exists $s > 0$ such that*

(i) $\operatorname{meas} A_{\mu - H/2^{s+3}, \rho_3}(t) = 0$, $t \in [t^0 - (a/16)\xi\rho^2, t^0]$, *for* $k \geq \mu/2$, $H = \mu - k > 0$, *and* $\operatorname{meas} A_{k, \rho_1}(t^0 - a\xi\rho^2) \leq (\kappa_n/2)\rho_1^n$.

(ii) $\omega \leq 2^{s+2}\rho^\varepsilon$ *or* $\operatorname{osc}\{w, \widetilde{Q}_{\rho/4}\} \leq (1 - 1/2^{s+3}) \operatorname{osc}\{w, Q_\rho\}$, *where*

$$\widetilde{Q}_{\rho/4} = \{(x, t): x \in K(\rho/4), t^0 - a\xi(\rho/4)^2 < t < t^0\},$$

for $\max_{[t^0 - a\eta\rho^2, t^0 - a\xi\rho^2]} \operatorname{meas} B_{\widetilde{\mu} + \omega/2, \rho_1}(t) \leq \frac{\kappa_n}{2}\rho_1^n$.

PROOF. From Lemma 4 we determine θ_2. Using it in Lemma 3 we obtain θ_1. Then by Lemma 2 we obtain the desired s for (i). When $\omega > 2^{s+2}\rho^\varepsilon$,

$$\operatorname{meas} B_{\widetilde{\mu} + \omega/2^{s+3}, \rho_3}(t) = 0, \qquad t \in [t^0 - a\eta_{k_{s+1}}(\rho/4)^2, t^0],$$

where $\eta_{k_s} = (1/m)\Phi'(k_s)$, $k_s = \widetilde{\mu} + \omega/2^s$. As $\rho_3 \geq \rho/4$ and $\eta_{k_{s+2}} \geq \xi$,

$$\operatorname{osc}\{w, \widetilde{Q}_{\rho/4}\} \leq \mu - (\widetilde{\mu} + \omega/2^{s+3}) = (1 - 1/2^{s+3})\omega.$$

LEMMA 5′. *Suppose $\partial\Omega$ satisfies an exterior cone condition*

$$\operatorname{meas}\{K(\rho) \cap \Omega\} \leq (1 - \theta_0) \operatorname{meas} K(\rho) \qquad (\rho \leq a_0, 0 < \theta_0 < 1),$$

and $w \in C^{s_1, s_{1/2}}(\Gamma)$. If $\widetilde{Q}_{\rho/4} \cap \Gamma \neq \varnothing$, then for each $\varepsilon \in (0, \varepsilon_1)$, there is $s > 0$ such that either

$$\operatorname{osc}\{w, \widetilde{Q}_{\rho/4} \cap Q\} \leq 2^{s+2}\rho^\varepsilon,$$

or

$$\operatorname{osc}\{w, \widetilde{Q}_{\rho/4} \cap Q\} \le (1 - 1/2^{s+3}) \operatorname{osc}\{w, Q_\rho \cap Q\}.$$

PROOF. From the exterior cone condition and the fact $\widetilde{Q}_{\rho/4} \cap \Gamma \neq \varnothing$ we have

$$\operatorname{meas}\{K(\rho_1)\backslash(K(\rho_1) \cap \Omega)\} \ge b_1 \rho_1^n \quad \text{or} \quad t^0 - a\xi\rho^2 \le 0.$$

So Lemma 2′ is applicable.

In case (i) $\operatorname{osc}\{w, \widetilde{Q}_{\rho/4} \cap Q\} \le 2^{s+2}\rho^\varepsilon$ is readily deduced from $\operatorname{osc}\{w, Q_\rho \cap Q\} \le 2^{s+2}\rho^\varepsilon$ $(0 < \varepsilon \le \varepsilon_1)$.

In case (ii) we use Lemma 3 and Lemma 4 to get

$$\int_{t^0 - a\xi\rho^2}^{t^0} \operatorname{meas} A_{\mu-\omega/2^{s+1}}(t)\, dt \le \theta_1 \xi \rho_1^{n+2}.$$

This implies $\operatorname{meas} A_{\mu-\omega/2^{s+3}, \rho_3}(t) = 0$, $t \in [t^0 - (a/16)\xi\rho^2, t^0]$, and hence

$$\operatorname{osc}\{w, \widetilde{Q}_{\rho/4} \cap Q\} \le (1 - 1/2^{s+3}) \operatorname{osc}\{w, Q_\rho \cap Q\}.$$

The desired conclusion can be drawn in the cases (iii), (ii)′, and (iii)′ in a similar manner. □

One more lemma on the Hölder condition for the solution of a nondegenerate parabolic equation is needed before proving a Hölder condition for solutions in the degenerate case. Although our equation is nondegenerate on $u \ge M_1 = e^{-\Lambda T} \min\{\inf_\Omega u_0, \inf_{\partial\Omega\times[0,T]} \psi\}$, it is of no use to us as we need a Hölder estimate independent of M_1. Instead we shall use

LEMMA 6. *Let ρ_0 satisfy $0 < \rho_0 < 1$, $mM^{1-1/m}\Phi'(M)\rho_0 < 1$ and let $Q_{\rho_0} = \{(x, t)\colon |x - x^0| < \rho_0, t^0 - a\Phi'(\rho_0^\varepsilon)\rho_0^2 < t \le t^0\}$ be contained in Q. Suppose u is a classical solution in Q_{ρ_0} satisfying $0 < u \le M$. Write $w(x, t) = \int_0^{u(x,t)} \nu(s)\, ds$, $\mu_0 = \sup_{Q_{\rho_0}} w$, $\widetilde{\mu}_0 = \inf_{Q_{\rho_0}} w$, and $\omega_0 = \mu_0 - \widetilde{\mu}_0$. Suppose $\omega_0 \le \widehat{C}\rho_0^\varepsilon$ and $\mu_0 \ge 2\widehat{C}\rho_0^\varepsilon$ for some constants ε, $\widehat{C}$ which satisfy $0 < \varepsilon = \varepsilon(n, m, \Lambda, M, T) \le \frac{1}{2}$ and $\widehat{C} = \widehat{C}(n, m, \Lambda, M, T) \ge 1$. Then we have*

$$|w(x, t) - w(x^0, t^0)| \le C_1\widehat{C}(|x - x^0|^2 + |t - t^0|)^{\varepsilon_0/2},$$

where C_1, ε_0 only depend on n, m, Λ, T, M, and $\varepsilon_0 \le \varepsilon/4$.

PROOF. Map Q_{ρ_0} to

$$Q_1' = \left\{(x', t')\colon |x'| < 1, \frac{-\Phi'(\rho_0^\varepsilon)}{\Phi'(\mu_0)} < t' \le 0\right\}$$

by the change of variables $x' = (x - x^0)/\rho_0$, $t' = (t - t^0)/[a\rho_0^2\Phi'(\mu_0)]$. The function $v(x', t') = u(x, t)$ satisfies the equation

$$\frac{\partial v}{\partial t'} = \frac{\partial}{\partial x_i'}\left(A_{ij}(x', t')\frac{\partial v}{\partial x_j'}\right) + B_i(x', t')\frac{\partial v}{\partial x_i'} + C(x', t')v$$

in Q_1', where

$$A_{ij}(x', t') = a\Phi'(\mu_0)a_{ij}(x, t, v),$$
$$B_i(x', t') = a\Phi'(\mu_0)\rho_0 b_i(x, t, v),$$
$$C(x', t') = a\Phi'(\mu_0)\rho_0^2 c(x, t, v).$$

By our assumption $\mu_0 \geq w \geq \widetilde{\mu}_0 \geq \mu_0/2$. Hence

$$A_{ij}\xi_i\xi_j \geq a\Phi'(\mu_0)\frac{1}{\Lambda}\nu(v)|\xi|^2 = \frac{a\Phi'(\mu_0)}{\Lambda\Phi'(w)}|\xi|^2 \geq \frac{a}{\Lambda m 2^{1-1/m}}|\xi|^2,$$

and

$$A_{ij}\xi_i\xi_j \leq a\Phi'(\mu_0)\Lambda\nu(v)|\xi|^2 \leq a\Lambda m|\xi|^2.$$

Also B_i and C are bounded.

From the interior estimate for nondegenerate solutions we know there are $C_0 > 0$ and $\alpha \in (0, 1)$ such that

$$|v(x', t') - v(0, 0)| \leq C_0(|x'|^\alpha + |t'|^{\alpha/2}),$$

for $(x', t') \in Q_{1/2} = \{(x', t'): |x'| < 1/2, -1/2m < t' \leq 0\}$, where C_0, α only depend on n, m, M, Λ, T. Therefore

$$|u(x, t) - u(x^0, t^0)| \leq C_0\left\{\frac{|x - x^0|^\alpha}{\rho_0^\alpha} + \frac{|t - t^0|^{\alpha/2}}{(a\rho_0^2\Phi'(\mu_0))^{\alpha/2}}\right\}$$

in $\{(x, t): |x - x^0| < \rho_0/2, t^0 - (a/2m)\Phi'(\mu_0)\rho_0^2 < t \leq t^0\}$. In the domain $\Sigma = \{(x, t): |x - x^0| < \rho_0^2/4, (t_0 - (a/2m)\Phi'(\mu_0)\rho_0^2)^2 < t \leq t^0\}$ we have

$$\begin{aligned}|u(x, t) - u(x^0, t^0)| &\leq C_0(|x - x^0|^{\alpha/2} + |t - t^0|^{\alpha/4}) \\ &\leq 2C_0(|x - x^0|^2 + |t - t^0|)^{\alpha/4}.\end{aligned}$$

For $(x, t) \in Q_{\rho_0}\backslash\Sigma$,

$$\begin{aligned}(|x - x^0|^2 + |t - t^0|)^{1/4} &\geq \min\left\{\frac{\rho_0}{2}, \left(\frac{a}{2m}\Phi'(\mu_0)\rho_0^2\right)^{1/2}\right\} \\ &\geq \min\left\{\frac{1}{2}, \left(\frac{a}{2m^2}\Phi'(M)\right)^{1/2}\right\}\rho_0 \\ &\geq \min\left\{\frac{1}{2}, \left(\frac{a}{2m^2}\Phi'(M)\right)^{1/2}\right\}(\widehat{C}^{-1}\omega)^{1/\varepsilon}.\end{aligned}$$

Notice we have used $\omega \leq \omega_0 \leq \widehat{C}\rho_0^\varepsilon$ in the last line. From this we deduce

$$\omega \leq \max\left\{2, \left(\frac{2m^2}{a\Phi'(M)}\right)^{1/2}\right\}^\varepsilon \widehat{C}(|x - x^0|^2 + |t - t^0|)^{\varepsilon/4}$$

in $Q_{\rho_0}\backslash\Sigma$. Now the lemma follows by taking $\varepsilon_0 = (1/4)\min\{\alpha, \varepsilon\}$ and $C_1 = \max\{2C_0, 2^\varepsilon, (2m^2/a\Phi'(M))^{\varepsilon/2}\}$. □

THEOREM 7 (interior estimate). *Suppose that the coefficients of the equation* $u_t = (a_{ij}u_{x_j})_{x_i} + b_i u_{x_i} + cu$ *satisfy*

$$\frac{1}{\Lambda}\nu(|r|)|\xi|^2 \le a_{ij}(x, t, r)\xi_i\xi_j \le \Lambda\nu(|r|)|\xi|^2, \quad \forall\xi \in \mathbb{R}^n,$$

$$|b_i(x, t, r)| \le \Lambda, \qquad \left|\frac{\partial b_i}{\partial x_i}(x, t, r)\right| \le \Lambda, \qquad -\Lambda \le c(x, t, r) \le 0,$$

for $(x, t, r) \in \overline{Q} \times \mathbb{R}$. *Here* ν *satisfies* $\nu(0) = 0$, $\nu(r) > 0$ $(r > 0)$, *and* $1 \le r\nu(r)/\int_0^r \nu(s)\,ds \le m$ *for* $r > 0$. *Let* u *be a solution of this equation in* $C^{2,1}(Q) \cap C(\overline{Q})$ *with* $0 < u \le M$. *Then* $w = \int_0^u \nu(s)\,ds$ *satisfies*

$$|w(x, t) - w(x^0, t^0)| \le \frac{C}{\rho_0}(|x - x^0|^2 + |t - t^0|)^{\alpha/2}, \qquad (x, t) \in Q^*_{\rho_0},$$

where C *and* α *only depend on* n, m, Λ, M, *and* T. *Here* $\rho_0 = \min\{1, 1/mM^{1-1/m}\Phi'(M),\ d(x^0, \partial\Omega),\ (mt^0/a\Phi'(1))^{1/2}\}$ *and* $Q^*_{\rho_0} = \{(x, t)$: $|x - x^0| < \rho_0,\ t^0 - (a/m)\Phi'(1)\rho_0^2 < t \le t^0\}$.

PROOF. Let $\widehat{C} = \max\{(2^{s+8}m^3)^m,\ (4/\rho_0)\int_0^M \nu(s)\,ds\}$ and $\tau = 64m^3\widehat{C}^{1-1/m}$, where s is determined from Lemma 5. Fix ε with $0 < \varepsilon < 1/2$ and $(1/\tau)^\varepsilon \ge 1 - 2^{-1-\tau(s+3)/16}$. Denote

$$\begin{aligned}
&\rho_l = \rho_0/\tau^l \qquad (l = 0, 1, 2, \ldots),\\
&Q_l = \{(x, t)\colon |x - x^0| < \rho_l,\ t^0 - a\Phi'(\rho_l^\varepsilon)\rho_l^2 < t \le t^0\},\\
&\mu_l = \sup_{Q_l} w, \qquad \widetilde{\mu}_l = \inf_{Q_l} w, \qquad \omega_l = \mu_l - \widetilde{\mu}_l,\\
&l^* = \min_{l \ge 0}\{l\colon \mu_l \ge 2\widehat{C}\rho_l^\varepsilon\}.
\end{aligned}$$

We observe that since $\inf_Q u \ge M_1$, $\mu_l \ge M_1 \ge 2\widehat{C}\rho_l^\varepsilon$ for large l and so l^* is finite. However, our Hölder estimate should be independent of l^*.

Since

$$\omega_0 \le \int_0^M \nu(s)\,ds \le \frac{\widehat{C}}{4}\rho_0 \le \widehat{C}\rho_0^\varepsilon,$$

$\mu_0 \ge 2\widehat{C}\rho_0^\varepsilon$ holds provided $l^* = 0$. The theorem then follows from Lemma 6. So we may assume $l^* > 0$ and show that $\omega_l \le \widehat{C}\rho_l^\varepsilon$ $(l \le l^*)$ holds.

We use induction. Suppose this inequality is valid for l $(l < l^*)$. We shall establish it for $l + 1$. In case $\mu_l \le 2\rho_l^\varepsilon$, $\omega_{l+1} \le \mu_l \le \widehat{C}\rho_{l+1}^\varepsilon$. Thus we may assume $\mu_l \ge 2\rho_l^\varepsilon$, i.e., $\omega_l \le \widehat{C}\rho_l^\varepsilon$, $2\rho_l^\varepsilon \le \mu_l \le 2\widehat{C}\rho_l^\varepsilon$. Under this circumstance we can use Lemma 5, for Lemmas 1–5 are valid under $\mu \ge 2\rho^\varepsilon$.

We distinguish two cases:

CASE I. $\sup_{[t^0 - a\eta\rho_l^2,\, t^0 - a\xi\rho_l^2]} \operatorname{meas} B_{\widetilde{\mu}_l + \omega_l/2,\, \rho_{1l}}(t) \le (\kappa_n/2)\rho_{1l}^n$, where $\eta = \Phi'(\rho_l^\varepsilon)$, $\xi = (1/m^2)\Phi'(\mu_l)$, $\rho_{il} = [1 - (i/3)(1 - 2^{-1/n})]\rho_l$ $(i = 1, 2, 3)$.

From Lemma 5 we know either $\omega_l \le 2^{s+2}\rho_l^\varepsilon$, from which we get

$$\omega_{l+1} \le \omega_l \le 2^{s+2}\rho_l^\varepsilon \le 2^{s+2}\tau\rho_{l+1}^\varepsilon = 2^{s+2}64m^3\widehat{C}^{1-1/m}\rho_{l+1}^\varepsilon \le \widehat{C}\rho_{l+1}^\varepsilon,$$

or

$$\operatorname{osc}\left\{w, \frac{\widetilde{Q}_{\rho_l}}{4}\right\} \le \left(1-\frac{1}{2^{s+3}}\right)\operatorname{osc}\{w, Q_{\rho_l}\} \le \left(\frac{1}{\tau}\right)^\varepsilon \widehat{C}\rho_l^\varepsilon = \widehat{C}\rho_{l+1}^\varepsilon.$$

As $\rho_{l+1} = \rho_l/\tau < \rho_l/4$ and

$$\begin{aligned}\frac{\Phi'(\mu_l)(\rho_l/4)^2/m^2}{\Phi'(\rho_{l+1}^\varepsilon)\rho_{l+1}^2} &\ge \frac{\tau^2}{16m^3}\left(\frac{\rho_{l+1}^\varepsilon}{\mu_l}\right)^{1-1/m} \ge \frac{\tau^2}{16m^3}(2\widehat{C}\tau^\varepsilon)^{1/m-1} \\ &\ge \frac{\tau^{1+1/m}}{32m^3\widehat{C}^{1-1/m}} \ge 2,\end{aligned}$$

$$\begin{aligned}Q_{l+1} &= \{(x, t)\colon |x-x^0| < \rho_{l+1}, t^0 - a\Phi'(\rho_{l+1}^2)\rho_{l+1}^2 < t \le t^0\} \\ &\subseteq \widetilde{Q}_{\rho_{l/4}} = \left\{(x, t)\colon x \in K\left(\frac{\rho_l}{4}\right), t^0 - \frac{a}{m^2}\Phi'(\mu_l)\left(\frac{\rho_l}{4}\right)^2 < t \le t^0\right\}.\end{aligned}$$

Therefore $\omega_{l+1} \le \widehat{C}\rho_{l+1}^\varepsilon$.

CASE II. There exists $\bar{t} \in [t^0 - a\eta\rho_l^2, t^0 - a\xi\rho_l^2]$ such that

$$\operatorname{meas} A_{\mu_l-\omega_l/2, \rho_{1l}}(\bar{t}) \le \frac{\kappa_n}{2}\rho_{1l}^n.$$

We divide $[\bar{t}, t^0]$ equally into N subintervals so that $(a/2)\xi\rho_l^2 \le (t^0-\bar{t})/N \le a\xi\rho_l^2$. Then

$$\begin{aligned}N &\le \frac{a\eta\rho_l^2}{a\xi\rho_l^2/2} \le 2m^2\frac{\Phi'(\rho_l^\varepsilon)}{\Phi'(\mu_l)} \le 2m^3(2\widehat{C})^{1-1/m} \\ &\le 4m^3\widehat{C}^{1-1/m} = \frac{\tau}{16}.\end{aligned}$$

Writing $t_p = \bar{t} + ((p-1)/N)(t^0-\bar{t})$ $(p = 1, \dots, N)$ and $t_{N+1} = t^0$, and applying Lemma 5 to the interval $[t_1, t_2]$ we have

$$\operatorname{meas} A_{\mu-\omega/2^{s+4}, \rho_3}(t_2) = 0 \quad \text{(the subscript } l \text{ is omitted)}$$

and

$$\operatorname{meas} A_{\mu-\omega/2^{s+4}, \rho_1}(t_2) \le \operatorname{meas} A_{\mu-\omega/2^{s+4}, \rho_3}(t_2) + \kappa_n(\rho_1^n - \rho_3^n) \le \frac{\kappa_n}{2}\rho_1^n.$$

Repeating this finitely many times we arrive at

$$\begin{aligned}&\operatorname{meas} A_{\mu-\omega/2^{[(s+3)(N+1)+1]}, \rho_1}(t_N) \le \frac{\kappa_n}{2}\rho_1^n, \\ &\operatorname{meas} A_{\mu-\omega/2^{[(s+3)N+1]}, \rho_3}(t) = 0, \qquad t \in [t^0 - (a/32)\xi\rho^2, t^0].\end{aligned}$$

Since $N \le \tau/16$,

$$\omega_{l+1} \le \left[1 - \frac{1}{2^{(s+3)N+1}}\right]\omega_l \le \left(\frac{1}{\tau}\right)^{\varepsilon} \widehat{C}\rho_l^{\varepsilon} = \widehat{C}\rho_{l+1}^{\varepsilon}.$$

By induction we have shown $\omega_{l}^{*} \le \widehat{C}\rho_{l^*}^{\varepsilon}$. From the definition of l^*, $\mu_{l^*} \ge 2\widehat{C}\rho_{l^*}^{\varepsilon}$. It follows from Lemma 6 that the theorem holds for $(x, t) \in Q_{l^*}$.

For $(x, t) \in Q_l \backslash Q_{l+1}$ $(0 \le l < l^*)$ we deduce from $\omega_l \le \widehat{C}\rho_l^{\varepsilon}$ that

$$\begin{aligned}(|x - x^0|^2 + |t - t^0|)^{1/2} &\ge \min\{\rho_{l+1}, [a\Phi'(\rho_{l+1}^{\varepsilon})\rho_{l+1}^2]^{1/2}\} \\ &\ge \min\left\{1, \left(\frac{a}{m}\Phi'(M)\right)^{1/2}\right\}\rho_{l+1} \\ &\ge \min\left\{1, \left(\frac{a}{m}\Phi'(M)\right)^{1/2}\right\}\frac{1}{\tau}\left(\frac{\omega_l}{\widehat{C}}\right)^{1/\varepsilon}.\end{aligned}$$

Consequently

$$\begin{aligned}&|w(x, t) - w(x^0, t^0)| \\ &\quad \le \tau^{\varepsilon}\max\left\{1, \left[\frac{m}{a\Phi'(M)}\right]^{\varepsilon/2}\right\}\widehat{C}(|x - x^0|^2 + |t - t^0|)^{\varepsilon/2}. \quad \square\end{aligned}$$

THEOREM 8 (near-boundary estimate). *In addition to the hypotheses in Theorem 7, assume that $\partial\Omega$ satisfies an exterior cone condition and the solution satisfies $0 < u \le M$ and $u|_{\Gamma} \in C^{\varepsilon_1, \varepsilon_1/2}(\Gamma)$ $(\varepsilon_1 > 0)$. Then for all $(x^0, t^0) \in \overline{Q}$ and $(x, t) \in Q_{\rho_0}^{*} \cap Q$ where $\rho_0 = a_0$ is the constant in the exterior cone condition and $Q_{\rho_0}^{*} = \{(x, t)\colon |x - x^0| < \rho_0,\ t_0 - (a/m)\Phi'(1)\rho_0^2 < t \le t^0\}$,*

$$|w(x, t) - w(x^0, t^0)| \le C(|x - x^0|^2 + |t - t^0|)^{\alpha/2}.$$

Here C, α only depend on $n, m, \Lambda, M, T, \varepsilon_1$, $\|u\|_{C^{\varepsilon_1, \varepsilon_1/2}(\Gamma)}$, and θ_0, a_0 (appearing in the exterior cone condition).

PROOF. We take $\rho_0 = a_0$, $\varepsilon \le \varepsilon_1$ and determine $\widehat{C}, \tau, \varepsilon$ as in the previous theorem.

First consider $(x^0, t^0) \in Q$. When $Q_{\rho_0} \subseteq Q$, the Hölder estimate is already contained in Theorem 7. Therefore we may assume $Q_{\rho_0} \cap \Gamma \ne \varnothing$. Then for some $l_0 \ge 1$, $Q_{l_0 - 1} \cap \Gamma \ne \varnothing$ and $Q_{l_0} \cap \Gamma = \varnothing$. We shall prove

$$\omega_l \le \widehat{C}\rho_l^{\varepsilon} \qquad (l \le l_0 - 1)$$

by induction. Assuming this inequality is valid for l we establish it for $l+1$. If $\mu_l \le 2\rho_l^{\varepsilon}$, $\omega_{l+1} \le \mu_l \le \widehat{C}\rho_{l+1}^{\varepsilon}$. If $\mu_l > 2\rho_l^{\varepsilon}$, according to Lemma 5′, either $\operatorname{osc}\{w, \widetilde{Q}_{\rho_l/4} \cap Q\} \le 2^{s+2}\rho_l^{\varepsilon}$,

$$\begin{aligned}\omega_{l+1} &\le \operatorname{osc}\{w, \widetilde{Q}_{\rho_l/4} \cap Q\} \le 2^{s+2}\rho_l^{\varepsilon} \le 2^{s+2}\tau\rho_{l+1}^{\varepsilon} \\ &= 2^{s+2}64m^3\widehat{C}^{1-1/m}\rho_{l+1}^{\varepsilon} \le \widehat{C}\rho_{l+1}^{\varepsilon},\end{aligned}$$

or

$$\begin{aligned}\omega_{l+1} &\le \operatorname{osc}\{w, \widetilde{Q}_{\rho_l/4} \cap Q\} \le \left(1 - \frac{1}{2^{s+3}}\right) \operatorname{osc}\{w, Q_{\rho_l} \cap Q\} \\ &= \left(1 - \frac{1}{2^{s+3}}\right)\omega_l \le \left(\frac{1}{\tau}\right)^{\varepsilon} \widehat{C}\rho_l^{\varepsilon} = \widehat{C}\rho_{l+1}^{\varepsilon}.\end{aligned}$$

Hence no matter in which case, $\omega_l \le \widehat{C}\rho_l^{\varepsilon}$ implies $\omega_{l+1} \le \widehat{C}\rho_{l+1}^{\varepsilon}$. That is to say, we can deduce from $\omega_0 \le \widehat{C}\rho_0^{\varepsilon}$ that

$$\omega_{l_0} = \operatorname{osc}\{w, Q_{l_0} \cap Q\} \le \widehat{C}\rho_{l_0-1}^{\varepsilon} = \widetilde{C}\rho_{l_0}^{\varepsilon} \qquad (\widetilde{C} = \widehat{C}\tau^{\varepsilon}).$$

We may now determine l^* from $l^* = \min_{l\ge 0}\{l\colon \mu_l \ge 2\widehat{C}\rho_l^{\varepsilon}\}$ and complete the proof of Theorem 8 as we have done in the previous theorem.

Taking the limit we see that the Hölder estimate holds for all $(x^0, t^0) \in \overline{Q}$. □

A Hölder estimate for u can be obtained from the one for w. For, letting $w_1 \le w_2$, from $\Phi(w_1)/\Phi(w_2) \ge w_1/w_2$ we have

$$\frac{\Phi(w_2) - \Phi(w_1)}{\Phi(w_2)} \le \frac{w_2 - w_1}{w_2},$$

$$\Phi(w_2) - \Phi(w_1) \le \frac{\Phi(w_2)}{w_2}(w_2 - w_1).$$

On the other hand $1/mw \le \Phi'(w)/\Phi(w)$ implies $\Phi(w_2)/w_2 \le m\Phi'(w_2)$. Consequently

$$\begin{aligned}\Phi(w_2) - \Phi(w_1) &\le m\Phi'(w_2)(w_2 - w_1) \\ &\le m^2\Phi'(M)M^{1-1/m}\frac{w_2 - w_1}{w_2^{1-1/m}} \\ &\le m^2\Phi'(M)M^{1-(1/m)}(w_2 - w_1)^{1/m}.\end{aligned}$$

Since Φ is monotonic in w, we obtain from this inequality

$$|u_2(x, t) - u_1(x, t)| \le m^2\Phi'(M)M^{1-1/m}|w_2 - w_1|^{1/m}.$$

Clearly this estimate, which now holds for any w_1, w_2, shows that a Hölder estimate on w yields a similar estimate for u.

§2. Solvability for the first boundary value problem

Consider the first boundary value problem

$$\begin{cases} u_t = (a_{ij}u_{x_j})_{x_i} + b_i u_{x_i} + cu, \\ u|_{t=0} = u_0(x), \\ u|_{\partial\Omega\times[0,T]} = \varphi(s, t), \end{cases}$$

where u_0, ψ are nonnegative continuous functions satisfying the compatibility condition $u_0|_{\partial\Omega} = \psi|_{t=0}$. A nonnegative continuous function u in

$\overline{Q}$ is called a *generalized solution* of the first boundary value problem if it satisfies the initial and boundary values pointwisely and fulfills the equation in the following sense: For all $\varphi \in C^{2,1}(Q) \cap C^1(\overline{Q})$, $\varphi|_{t=T} = 0$ $(x \in \Omega)$, $\varphi|_{\partial\Omega\times[0,T]} = 0$,

$$\int_Q [u\varphi_t + A_{ij}(x, t, u)\varphi_{x_i x_j} - (A_i(x, t, u) + B_i(x, t, u))\varphi_{x_i} + (c(x, t, u) + B(x, t, u))\varphi]\,dx\,dt + \int_\Omega u_0(x)\varphi(x, 0)\,dx - \int_{\partial\Omega\times[0,T]} A_{ij}(s, t, \psi(s, t))\frac{\partial\varphi(s, t)}{\partial N}\cos(N, x_i)\cos(N, x_j)\,ds\,dt = 0,$$

where N is the unit outer normal at $\partial\Omega$ and

$$A_{ij}(x, t, r) = \int_0^r a_{ij}(x, t, s)\,ds, \qquad A_j(x, t, r) = \int_0^r \frac{\partial a_{ij}}{\partial x_i}(x, t, s)\,ds,$$

$$B_i(x, t, r) = \int_0^r b_i(x, t, s)\,ds, \qquad B(x, t, r) = \int_0^r \frac{\partial b_i}{\partial x_i}(x, t, s)\,ds.$$

We shall assume the coefficients of the equation and the domain Ω satisfy:

(i) a_{ij}, b_i, c, $\partial a_{ij}/\partial x_i$, and $\partial b_i/\partial x_i$ are in $C(\overline{Q}\times\mathbb{R})$.

(ii) There exist Λ and a function ν such that

$$\frac{1}{\Lambda}\nu(|r|)|\xi|^2 \le a_{ij}\xi_i\xi_j \le \Lambda\nu(|r|)|\xi|^2, \quad \forall\xi\in\mathbb{R}^n,$$

in $\overline{Q}\times\mathbb{R}$. Here $\nu \in C[0,\infty)$, $\nu(0) = 0$, $\nu(r) > 0$ for $r > 0$, and $\exists\delta > 0$, $m > 1$ such that

$$1 \le \frac{r\nu(r)}{\int_0^r \nu(s)\,ds} \le m \quad \text{for } 0 < r \le \delta.$$

(iii) $\partial\Omega \in C^{1+\beta_0}$, $\beta_0 > 0$.

THEOREM 9. *Under the above assumptions, for* $u_0 \in C^\beta(\overline{\Omega})$ *and* $\psi \in C^{\beta,\beta/2}(\partial\Omega\times[0,T])$, $\beta > 0$, *the first boundary value problem has a generalized solution* u. *Moreover,* u *is Hölder continuous in* $\overline{Q}$.

PROOF. Let

$$\overline{u}_{0k}(x) = \begin{cases} u_0(x), & u_0(x) > 1/k, \\ 1/k, & 0 \le u_0(x) \le 1/k, \end{cases}$$

and

$$\overline{\psi}_k(s, t) = \begin{cases} \psi(s, t), & \psi(s, t) > 1/k, \\ 1/k, & 0 \le \psi(s, t) \le 1/k. \end{cases}$$

We approximate $\overline{u}_{0k}$ by a smooth function u_{0k} so that $|\overline{u}_{0k} - u_{0k}| \le (1/k)\nu(1/k)$ in $\overline{\Omega}$. (This can be done by first extending $\overline{u}_{0k}$ to a larger domain containing $\overline{\Omega}$ while preserving the C^β-condition.) To approximate $\overline{\psi}_k$ we first extend $\overline{\psi}_k(s, t) - \overline{\psi}(s, 0)$ outside $\partial\Omega\times[0, T]$ so that it is odd in t and still satisfies the $C^{\beta,\beta/2}$-condition. Adding $u_{0k}(s)$ to the extension we obtain a smooth function ψ_k which satisfies $|\overline{\psi}_k - \psi_k| \le (2/k)\nu(1/k)$ on $\partial\Omega \times [0, T]$. We also extend $\overline{a}_{ij} = a_{ij}/\nu$, b_i, and c beyond $\overline{\Omega}$ so that they and $\partial a_{ij}/\partial x_i$, $\partial b_i/\partial x_i$ are continuous. Let $\overline{a}_{ij\varepsilon}$, $b_{i\varepsilon}$, c_ε be their smooth approximations.

The construction of a smooth approximation for ν is more delicate. Let

$$\alpha(r) = (r\nu(r)) \Big/ \int_0^r \nu(s)\,ds.$$

We have

$$\nu(r) = \int_0^r \nu(s)\,ds \frac{\alpha(r)}{r} \exp\left(\int_1^r \alpha(s)\frac{ds}{s}\right).$$

Also $1 \le \alpha(r) \le m$ and $\alpha \in C[0, \infty)$. Construct a smooth approximation of α, α_ε, such that

$$|\alpha_\varepsilon(r) - \alpha(r)| \le \frac{\varepsilon}{(\log r)^2 + 1}.$$

To achieve this first let $\beta(s) = \alpha(e^s) \in C(-\infty, \infty)$. For each $\varepsilon > 0$, we can find $\delta_{k\varepsilon}$ $(k \ge 1)$ such that

$$|\beta(s) - \beta(s^*)| \le \frac{\varepsilon}{k^2+1}, \quad \forall |s| \le k,\ |s - s^*| \le \delta_{k\varepsilon}.$$

Let $\sigma \in C^\infty[0, \infty)$ be a decreasing function with $\sigma(k-1) = \delta_{k\varepsilon}$ $(k \ge 1)$. Let $\beta_\varepsilon \in C^\infty(-\infty, \infty)$ be a mollifier of β of which radius is $\sigma(s)$, that is,

$$\beta_\varepsilon(s) = \frac{1}{K(s)} \int_{|\tau|\le\sigma(s)} \beta(s-\sigma) \exp\left[\left(\frac{\tau^2}{\sigma(s)^2} - 1\right)^{-1}\right] d\tau,$$

where

$$K(s) = \int_{|\tau|\le\sigma(s)} \exp\left[\left(\frac{\tau^2}{\sigma(s)^2} - 1\right)^{-1}\right] d\tau.$$

For $k - 1 \le |s| \le k$,

$$\begin{aligned} |\beta_\varepsilon(s) - \beta(s)| &= \left|\frac{1}{K(s)} \int_{|\tau|\le\sigma(s)} (\beta(s-\sigma) - \beta(s)) \exp\left[\left(\frac{\tau^2}{\sigma(s)^2} - 1\right)^{-1}\right] d\tau\right| \\ &\le \max_{|s^*-s|\le\sigma(s)} |\beta(s^*) - \beta(s)| \le \max_{\substack{|s|\le k, \\ |s^*-s|\le\delta_{k\varepsilon}}} |\beta(s^*) - \beta(s)| \\ &\le \frac{\varepsilon}{k^2+1} \le \frac{\varepsilon}{s^2+1}. \end{aligned}$$

So we may take $\alpha_\varepsilon(r) = \beta_\varepsilon(\log r)$. Let

$$\nu_\varepsilon(r) = \int_0^1 \nu(s)\,ds \frac{\alpha_\varepsilon(r)}{r} \exp\left(\int_1^r \alpha_\varepsilon(s)\frac{ds}{s}\right).$$

It is readily seen that $\nu_\varepsilon \in C[0, \infty)$, $\nu_\varepsilon(0) = 0$, $\nu_\varepsilon \in C^\infty(0, \infty)$,

$$1 \le \frac{r\nu_\varepsilon(r)}{\int_0^r \nu_\varepsilon(s)\,ds} \le m, \qquad \frac{1}{m}e^{-\pi/2} \le \frac{\nu_\varepsilon(r)}{\nu(r)} \le me^{\pi/2}.$$

Consider the first boundary value problem

$$\begin{cases} u_t = (a_{ij\varepsilon} u_{x_j})_{x_i} + b_{i\varepsilon} u_{x_i} + c_\varepsilon u, & a_{ij\varepsilon} = \overline{a}_{ij\varepsilon} \nu_\varepsilon, \\ u|_{t=0} = u_{0k}, \\ u|_{\partial\Omega\times[0,T]} = \psi_k. \end{cases}$$

By the maximum principle its solution u satisfies

$$\frac{e^{-2\Lambda T}}{k} \le u \le M = e^{2\Lambda T} \max\left\{ \sup_\Omega u_0, \sup_{\partial\Omega\times[0,T]} \psi \right\}.$$

Hence the problem is nondegenerate and its solution $u_{k\varepsilon}$ exists and belongs to $C^{2,1}(Q) \cap C(\overline{Q})$. (First replace the domain and initial-boundary data by their smooth approximations. Extract a subsequence of the sequence of approximate solutions which converges (up to second order) to a solution uniformly in each compact subset of Q. Finally use a barrier argument to show that this solution attains the initial and the boundary values.) By Theorems 7 and 8 we know that $u_{k\varepsilon} \in C^{\alpha,\alpha/2}(\overline{Q})$, where α and $\|u_{k\varepsilon}\|_{C^{\alpha,\alpha/2}}$ are independent of k and ε. From the estimates in Chapter I we also have $u_{k\varepsilon} \in C^{1+\alpha,1}(\overline{Q})$, where $\|u_{k\varepsilon}\|_{C^{1+\alpha,1}}$ is independent of ε. Extract a subsequence $\{u_{k\varepsilon_j}\}$ so that $u_{k\varepsilon_j} \to u_k$ in $\overline{Q}$. Then u_k assumes the initial-boundary values u_{0k} and ψ_k. Furthermore, it is a generalized solution (where u_0 and ψ are replaced by u_{0k} and ψ_k respectively) which satisfies $e^{-2\Lambda T}/k \le u_k \le M$, $u_k \in C^{\alpha,\alpha/2}(\overline{Q})$, $\|u_k\|_{C^{\alpha,\alpha/2}}$ is independent of k. Therefore $\{u_k\}$ is compact in $C(\overline{Q})$. One of its subsequences (still denoted by $\{u_k\}$) converges to u. Clearly u is a generalized solution which assumes the initial and boundary values u_0 and ψ. Also $u \in C^{\alpha,\alpha/2}(\overline{Q})$. □

§3. Uniqueness of the solution

Next we study the uniqueness of the generalized solutions. Suppose $\widetilde{u} \in C(\overline{Q})$ is another nonnegative generalized solution which takes the same initial and boundary values as u does. Then for $\varphi \in C^{2,1}(Q) \cap C^1(\overline{Q})$, $\varphi|_{t=T} = \varphi|_{\partial\Omega\times[0,T]} = 0$,

$$\int_Q (u_k - \widetilde{u})(\varphi_t + a_{ij}^{(k)} \varphi_{x_i x_j} + b_i^{(k)} \varphi_{x_i} + c^{(k)} \varphi) + \int_\Omega (u_{0k}(x) - u_0(x))\varphi(x, 0)$$
$$- \int_{\partial\Omega\times[0,T]} (\psi_k - \psi) a_{ij}^{(k)}(s, t) \frac{\partial \varphi}{\partial N} \cos(N, x_i)\cos(N, x_j)\, ds\, dt = 0, \quad (7)$$

where

$$a_{ij}^{(k)} = \frac{A_{ij}(x, t, u_k) - A_{ij}(x, t, \widetilde{u})}{u_k - \widetilde{u}} = \int_0^1 a_{ij}(x, t, \theta u_k + (1-\theta)\widetilde{u})\, d\theta,$$

$$b_i^{(k)} = -\int_0^1 \left(\frac{\partial a_{ij}}{\partial x_j} + b_i \right)(x, t, \theta u_k + (1-\theta)\widetilde{u}) d\theta,$$

and $c^{(k)}$ is given in a similar manner. Let $\widetilde{u}_\varepsilon$ be a smooth approximation of $\widetilde{u}$ satisfying $\widetilde{u} \le \widetilde{u}_\varepsilon \le \widetilde{u} + \varepsilon$, and let

$$a_{ij}^{(k,\varepsilon)} = \int_0^1 a_{ij\varepsilon}(x, t, \theta u_k + (1-\theta)\widetilde{u}_\varepsilon)\, d\theta,$$

and $b_i^{(k,\varepsilon)}$, $c^{(k,\varepsilon)}$ be similarly defined expressions. We solve the initial-boundary value problem

$$\begin{cases} \varphi_t + a_{ij}^{(k,\varepsilon)} \varphi_{x_i x_j} + b_i^{(k,\varepsilon)} \varphi_{x_i} + c^{(k,\varepsilon)} \varphi = U, \\ \varphi|_{t=T} = \varphi|_{\partial\Omega \times [0,T]} = 0, \end{cases}$$

where U is a smooth function in $\overline{Q}$. Since $u_k \in C^{2,1}(Q) \cap C^{1+\alpha,1}(\overline{Q})$, the coefficients of this equation belong to $C^{2,1}(Q) \cap C^{1+\alpha,1}(\overline{Q})$. By linear parabolic theory this problem admits a solution $\varphi^{(k,\varepsilon)}$ in $C^{2,1}(Q) \cap C^{1+\alpha,1}(\overline{Q})$. Plugging $\varphi = \varphi^{(k,\varepsilon)}$ into (7) yields

$$\begin{aligned} \int_Q (u - \widetilde{u})U = & \int_Q (u - u_k)U + \int_Q (u_k - \widetilde{u})[(a_{ij}^{(k,\varepsilon)} - a_{ij}^{(k)})\varphi_{x_i x_j} \\ & + (b_i^{(k,\varepsilon)} - b_i^{(k)})\varphi_{x_i} + (c^{(k,\varepsilon)} - c^{(k)})\varphi] - \int_\Omega (u_{0k} - u_0)\varphi(x, 0) \\ & + \int_{\partial\Omega \times [0,T]} (\psi_k - \psi) a_{ij}^{(k)}(s, t) \frac{\partial \varphi}{\partial N} \cos(N, x_i)\cos(N, x_j). \quad (8) \end{aligned}$$

Various terms in the right-hand side of (8) will be handled by the following three lemmas.

LEMMA 10. $|\varphi^{(k,\varepsilon)}| \le K = e^{2\Lambda T} \sup_{\overline{Q}} |U(x, t)|$.

PROOF. This is a direct consequence of the maximum principle. □

We shall need an inequality involving ν. First let us derive it. From

$$1 \le \frac{r\nu(r)}{\int_0^r \nu(s)\, ds} \le m \qquad (0 < r < \infty),$$

we have

$$\frac{r_2}{r_1} \le \frac{\int_0^{r_2} \nu(s)\, ds}{\int_0^{r_1} \nu(s) ds} \le \left(\frac{r_2}{r_1}\right)^m$$

and

$$\frac{1}{m} \le \frac{r_2 \nu(r_2)}{r_1 \nu(r_1)} \frac{\int_0^{r_1} \nu(s)\, ds}{\int_0^{r_2} \nu(s)\, ds} \le m, \qquad r_1 \le r_2.$$

Therefore $1/m \le \nu(r_2)/\nu(r_1) \le m(r_2/r_1)^{m-1}$, $r_1 \le r_2$.

LEMMA 11. *Suppose* $\partial\Omega \in C^2$. *Then* $|\partial\varphi^{(k,\varepsilon)}/\partial N| \nu(1/k) \le K$.

PROOF. Let $\varphi^{(k,\varepsilon)} = e^{-2\Lambda T}\widetilde{\varphi}$. Then $\widetilde{\varphi}$ satisfies the equation

$$\mathscr{L}(\widetilde{\varphi}) = \widetilde{\varphi}_t + a_{ij}^{(k,\varepsilon)} \widetilde{\varphi}_{x_i x_j} + b_i^{(k,\varepsilon)} \widetilde{\varphi}_{x_i} + (c^{(k,\varepsilon)} - 2\Lambda)\widetilde{\varphi} = U e^{2\Lambda t}.$$

For each $(x^0, t^0) \in \partial\Omega \times [0, T]$, we can pick $\overline{x}^0$ such that $\{x\colon |x - \overline{x}^0| \le R\} \cap \overline{\Omega} = \{x^0\}$, $R = |x^0 - \overline{x}^0|$, because $\partial\Omega$ is in C^2. Let

$$\Psi = 1 - \frac{R^q}{(|x - \overline{x}^0|^2 + (t - t^0)^2)^{q/2}}, \qquad q \text{ to be specified.}$$

We have

$$\begin{aligned}\mathscr{L}(\Psi) = &-q\frac{R^q}{(|x - \overline{x}^0|^2 + (t - t^0)^2)^{q/2+2}}\\ &\times \left[(q+2)a_{ij}^{(k,\varepsilon)}\frac{(x_i - \overline{x}_i^0)(x_j - \overline{x}_j^0)}{|x - \overline{x}^0|^2 + (t - t^0)^2}\right.\\ &\left.- a_{ii}^{(k,\varepsilon)} - (t - t^0) - b_i^{(k,\varepsilon)}((x_i - \overline{x}_i^0)\right] + (c^{(k,\varepsilon)} - 2\Lambda)\Psi.\end{aligned}$$

Taking $q = K/\nu(1/k)$, we have

$$\begin{aligned}&qa_{ij}^{(k,\varepsilon)}\frac{(x_i - \overline{x}_i^0)(x_j - \overline{x}_j^0)}{|x - \overline{x}^0|^2 + (t - t^0)^2}\\ &\ge \frac{q}{2\Lambda}\int_0^1 \nu_\varepsilon(\theta u_k + (1-\theta)\widetilde{u}_\varepsilon)\,d\theta \cdot \frac{R^2}{(R + \operatorname{diam}\Omega)^2 + T^2}\\ &\ge CK\int_0^1 \frac{\nu(\theta u_k + (1-\theta)\widetilde{u}_\varepsilon)}{\nu(1/k)}d\theta\\ &\ge \frac{CK}{2m}\frac{\nu(e^{-2\Lambda T}/2k)}{\nu(1/k)} \ge \frac{CK}{2m^2}\left(\frac{e^{-2\Lambda T}}{2}\right)^{m-1}.\end{aligned}$$

One can achieve

$$\begin{aligned}-\mathscr{L}(\Psi) &\ge C_1 q\left[\frac{R^2}{|(x - \overline{x}^0|^2 + (t - t^0)^2}\right]^{q/2} + \frac{\Lambda}{2}\Psi\\ &= C_1 q(1 - \Psi) + \frac{\Lambda}{2}\Psi \ge \min\left\{C_1 q, \frac{\Lambda}{2}\right\} \ge \frac{\Lambda}{2}\end{aligned}$$

provided K is large enough. As a result $\pm\widetilde{\varphi} + K_1\Psi$, where $K_1 = (3/\Lambda)e^{2\Lambda T}\sup_Q|U|$, cannot attain its nonpositive minimum in Q. Its minimum, which is zero, can only be taken on at (x^0, t^0). Therefore, at (x^0, t^0), $(\pm(\partial\widetilde{\varphi}/\partial N)) + K_1(\partial\Psi/\partial N)) \le 0$, that is

$$\pm e^{2\Lambda t^0}\frac{\partial\varphi^{(k,\varepsilon)}}{\partial N}(x^0, t^0) \le -K_1\frac{\partial\Psi}{\partial N}(x^0, t^0) \le K_2 q.$$

LEMMA 12. *Suppose $\partial\Omega \in C^2$. For sufficiently small ε*

$$\int_Q\left[(\varphi_{x_ix_j}^{(k,\varepsilon)})^2 + \sum_i(\varphi_{x_i}^{(k,\varepsilon)})^2\right] \le K(k).$$

In other words, the integral is bounded uniformly in ε.

Proof. From the equation satisfied by $\varphi = \varphi^{(k,\varepsilon)}$ we have

$$|\varphi_t + a_{ij}^{(k)}\varphi_{x_ix_j}| \le \omega(\varepsilon)\sum_{i,j}|\varphi_{x_ix_j}| + K\Big(\sum_i |\varphi_{x_i}| + 1\Big),$$

where $\omega(\varepsilon)$ is a decreasing function and $\lim_{\varepsilon\to 0}\omega(\varepsilon) = 0$. This can be seen from the fact that $\tilde{u}_\varepsilon$, $\tilde{a}_{ij\varepsilon}(x, t, r)\nu_\varepsilon(r)$ are continuous in ε. For a fixed $(x^0, t^0) \in \overline{Q}$, we can further write this inequality in the form

$$|\varphi_t + a_{ij}^{(k)}(x^0, t^0)\varphi_{x_ix_j}| \le [\overline{\omega}(|x - x^0| + |t - t^0|) + \omega(\varepsilon)]\sum_{i,j}|\varphi_{x_ix_j}| + K\Big(\sum_i |\varphi_{x_i}| + 1\Big), \tag{9}$$

where $\overline{\omega}$ has the same properties as ω. Notice this is possible because $a_{ij} \in C(\overline{Q})$, $\tilde{u} \in C(\overline{Q})$, and $u_k \in C^{\alpha,\alpha/2}(\overline{Q})$ (uniform in k).

After a rotation of axes we may assume $a_{ij}^{(k)}(x^0, t^0) = 0$ for $i \ne j$. Let ζ and η be cut-off functions so that

$$\zeta(x) = \begin{cases} 1, & |x - x^0| \le \frac{\delta}{2}, \\ 0, & |x - x^0| \ge \delta, \end{cases} \qquad |\nabla\zeta| \le \frac{2}{\delta},$$

and

$$\eta(t) = \begin{cases} 1, & |t - t^0| \le \frac{\delta}{2}, \\ 0, & |t - t^0| \ge \delta, \end{cases} \qquad |\eta'(t)| \le \frac{2}{\delta}.$$

It follows from (9) that

$$\begin{aligned}
&\int_Q \Big(\varphi_t^2 + \sum_{i,j} a_{ij}^{(k)}(x^0, t^0)a_{jj}^{(k)}(x^0, t^0)\varphi_{x_ix_j}^2\Big)\zeta^2\eta \\
&\qquad\qquad - \int_Q \sum_i a_{ii}^{(k)}(x^0, t^0)\varphi_{x_i}^2\zeta^2\eta\Big|_0^T \\
&+ \sum_{i,j} a_{ii}^{(k)}(x^0, t^0)a_{jj}^{(k)}(x^0, t^0) \\
&\quad \times \int_{\partial\Omega\times[0,T]} \varphi_{x_i}(\varphi_{x_ix_j}\cos(N, x_j) - \varphi_{x_jx_j}\cos(N, x_i))\zeta^2\eta \\
&+ \int_Q \Big[\sum_i a_{ii}^{(k)}(x^0, t^0)(-4\varphi_t\varphi_{x_i}\zeta\zeta_{x_i}\eta + \varphi_{x_i}^2\zeta^2\eta') \\
&\qquad + 2\sum_{i,j} a_{ii}^{(k)}(x^0, t^0)a_{jj}^{(k)}(x^0, t^0)(\varphi_{x_i}\varphi_{x_jx_j}\zeta\zeta_{x_i}\eta - \varphi_{x_i}\varphi_{x_ix_j}\zeta\zeta_{x_j}\eta\Big] \\
&\le 2n^2[\overline{\omega}(2\delta) + \omega(\varepsilon)]\int_Q \sum_{i,j}\varphi_{x_ix_j}^2\zeta^2\eta + K\int_Q \Big(\sum_i \varphi_{x_i}^2 + 1\Big)\zeta^2\eta.
\end{aligned}$$

From $\varphi(x, T) = 0$ we know $-\int_Q \varphi_{x_i}^2\zeta^2\eta|_0^T \ge 0$. Hence the second term in the left-hand side of the above inequality is nonnegative. As $u_k(x^0, t^0) \geq 1/ke^{-2\Lambda T}$, we have

$$a_{ii}^{(k)}(x^0, t^0) \geq \frac{1}{\Lambda}\int_0^1 \nu(\theta u_k(x^0, t^0) + (1-\theta)\widetilde{u}(x^0, t^0))\, d\theta$$
$$\geq \frac{1}{2m\Lambda}\nu\left(\frac{e^{-2\Lambda T}}{2k}\right).$$

Combining this with the previous inequality we obtain

$$\begin{aligned}
&\int_Q \left(\varphi_t^2 + \sum_{i,j}\varphi_{x_ix_j}^2\right)\zeta^2\eta \\
&\quad\leq K(k)\int_Q \left(\varphi_t^2 + \sum_{i,j} a_{ii}^{(k)}(x^0, t^0)a_{jj}^{(k)}(x^0, t^0)\varphi_{x_ix_j}^2\right)\zeta^2\eta \\
&\quad\leq K(k)(\overline{\omega}(2\delta) + \omega(\varepsilon) + \varepsilon_1)\int_Q \left(\varphi_t^2 + \sum_{i,j}\varphi_{x_ix_j}^2\right)\zeta^2\eta \\
&\qquad + K(k, \delta, \varepsilon_1)\int_{Q\cap\{|x-x^0|<\delta,\, |t-t^0|<\delta\}} \left(\sum_i \varphi_{x_i}^2 + 1\right) \\
&\qquad + K(k)\sum_{i,j} a_{ii}^{(k)}(x^0, t^0)a_{jj}^{(k)}(x^0, t^0) \\
&\qquad \times \int_{\partial\Omega\times[0,T]} \varphi_{x_i}(\varphi_{x_ix_j}\cos(N, x_i) - \varphi_{x_ix_j}\cos(N, x_j))\zeta^2\eta. \qquad (10)
\end{aligned}$$

Let $(s_1, s_2, \dots, s_{n-1})$ be a local coordinate of $\partial\Omega$. Denote

$$\begin{aligned}
\sigma_{ijl} &= \cos(N, x_i)\cos(N, x_j)(\cos(N, x_i)\cos(s_l, x_j) \\
&\qquad\qquad - \cos(N, x_j)\cos(s_l, x_i)], \\
\tau_{ij} &= \left(\frac{\partial\cos(N, x_j)}{\partial x_i} - \frac{\partial\cos(N, x_i)}{\partial x_j}\right)\cos(N, x_i).
\end{aligned}$$

We have

$$\begin{aligned}
&\int_{\partial\Omega} \varphi_{x_i}(\varphi_{x_ix_j}\cos(N, x_i) - \varphi_{x_ix_j}\cos(N, x_j))\zeta^2 \\
&\quad = \int_{\partial\Omega}\left(\sum_l \varphi_N\varphi_{Ns l}\sigma_{ijl} + \varphi_N^2\tau_{ij}\right)\zeta^2 \\
&\quad = \int_{\partial\Omega}\varphi_N^2\left[\left(\tau_{ij} - \frac{1}{2}\frac{\partial\sigma_{ijl}}{\partial s_l}\right)\zeta^2 - \sigma_{ijl}\zeta\zeta_{s_l}\right] \\
&\quad \leq \frac{K}{\delta}\int_\Omega \frac{\partial}{\partial N}(\varphi_N^2\zeta) = \frac{K}{\delta}\int_\Omega(2\varphi_N\varphi_{NN}\zeta + \varphi_N^2\zeta_N) \\
&\quad \leq \varepsilon_2\int_\Omega\sum_{i,j}\varphi_{x_ix_j}^2\zeta^2 + K(\varepsilon_2, \delta)\int_{\Omega\cap\{|x-x^0|<\delta\}}\sum_i\varphi_{x_i}^2.
\end{aligned}$$

Putting this into (10) and choosing $K(k)(\overline{\omega}(2\delta)+\omega(\varepsilon_0)+\varepsilon_1+\varepsilon_2)\le 1/2$, we have

$$\int_{Q\cap\{|x-x^0|<\delta/2,\,|t-t^0|<\delta/2\}}\Big(\varphi_t^2+\sum_{i,j}\varphi_{x_ix_j}^2\Big)\le K(k)\int_{Q\cap\{|x-x^0|<\delta,\,|t-t^0|<\delta\}}\Big(\sum_i\varphi_{x_i}^2+1\Big),$$

for $\varepsilon\le\varepsilon_0$. We can choose a finite number of (x^0, t^0)'s such that the union of the subsets $\{(x,t): |x-x^0|<\delta/2, |t-t^0|<\delta/2\}$ covers Q. Hence

$$\int_Q\Big(\varphi_t^2+\sum_{i,j}\varphi_{x_ix_j}^2\Big)\le K(k)\int_Q\Big(\sum_i\varphi_{x_i}^2+1\Big)\qquad(\varepsilon\le\varepsilon_0).$$

On the other hand,

$$\int_Q\sum_i\varphi_{x_i}^2=-\int_Q\varphi\Delta\varphi\le\varepsilon_3\int_Q\sum_{i,j}\varphi_{x_ix_j}^2+K(\varepsilon_3).$$

Taking $\varepsilon_3=1/2K(k)$ and substituting the resulting inequality into the previous one we obtain the desired estimate. □

THEOREM 13. *Suppose $\partial\Omega\in C^2$. Under the hypotheses in Theorem 9 the first boundary value problem has at most one generalized solution.*

PROOF. Substitute the estimates obtained in Lemmas 10, 11, and 12 into (8). We have

$$\begin{aligned}\left|\int_Q(u-\tilde u)U\right|&\le\sup_Q|U|\int_Q|u-u_k|+\sup_\Omega|\varphi(x,0)|\int_Q|u_{0k}-u_0|\\&\quad+K\Big\{\sup_Q\Big[\sum|a_{ij}^{(k,\varepsilon)}-a_{ij}^{(k)}|+\sum|b_i^{(k,\varepsilon)}-b_i^{(k)}|+|c^{(k,\varepsilon)}-c^{(k)}|\Big]\\&\qquad\qquad\times\int_Q\Big(\sum\varphi_{x_ix_j}^2+\sum\varphi_{x_i}^2-\varphi^2\Big)\Big\}^{1/2}\\&\quad+\int_{\partial\Omega\times[0,T]\cap\{\psi\le1/k\}}\Big(\frac1k+\frac1k\nu\Big(\frac1k\Big)\Big)\Lambda\int_0^1\nu\Big(\frac\theta k+\psi(1-\theta)\Big)d\theta\frac{K}{\nu(1/k)}\\&\quad+\int_{\partial\Omega\times[0,T]\cap\{\psi>1/k\}}\frac1k\nu\Big(\frac1k\Big)\frac{K}{\nu(1/k)}.\end{aligned}$$

Since $(1/\nu(1/k))\int_0^1\nu(\theta/k+\psi(1-\theta))\,d\theta\le m$ for $\psi\le1/k$, by first letting $\varepsilon\to0$ and then $k\to\infty$ in this inequality we conclude $\int_Q(u-\tilde u)U=0$. Since U is arbitrary, u is equal to $\tilde u$ in $\overline{Q}$. We have finished the proof of this theorem. □

§4. The initial value problem

Consider the initial value problem

$$\begin{cases} u_t = (a_{ij}(x, t, u)u_{x_j})_{x_i} + b_i(x, t, u)u_{x_i} + c(x, t, u)u, \\ (x, t) \in \mathbb{R}^n \times (0, T], \\ u|_{t=0} = u_0(x), \qquad x \in \mathbb{R}^n, \end{cases}$$

where the regularity and degeneracy of the coefficients are the same as in the first boundary value problem. We also assume $0 \le u_0(x) < M$. We would like to find a nonnegative, bounded solution for this Cauchy problem.

Let

$$u_0^k(x) = \begin{cases} u_0(x), & u_0 \ge 1/k, \\ 1/k, & u_0 < 1/k, \end{cases} \qquad |x| \le k,$$

and

$$u\Big|_{\substack{|x|=k \\ 0\le t\le T}} = \min_{|x|=k}\{u_0^k(x), 1/k\}.$$

Denote the solution of this first boundary value problem by $u^k(x)$. Since u^k is a solution for a nondegenerate problem,

$$u^k \in C^{2,1}(\{|x| < k\} \times \{0 < t \le T\}) \cap C(\{|x| < k\} \times \{0 \le t \le T\}).$$

Let $u^k = e^{2\Lambda t + \eta(|x|^2+1)^{1/2}} v$, where $\eta > 0$ is a constant. Then v satisfies

$$v_t = (a_{ij}v_{x_j})_{x_i} + \left[b_i + \eta\frac{a_{ij}x_j}{(|x|^2+1)^{1/2}}\right]v_{x_i} + \left\{c - 2\Lambda + \eta\left[\frac{a_{ij}x_j}{(|x|^2+1)^{1/2}}\right]_{x_i} + \eta\frac{b_ix_i}{(|x|^2+1)^{1/2}} + \eta^2\frac{a_{ij}x_ix_j}{|x|^2+1}\right\}v.$$

From this equation we see that v cannot attain its positive maximum in $\{|x| < k\} \times \{0 < t \le T\}$ provided η is sufficiently small. From $u_0 \le M$ we conclude $v \le M$, or $u^k \le Me^{2\Lambda T + \eta(k^2+1)^{1/2}}$. Letting $\eta \downarrow 0$ we have $u^k \le Me^{2\Lambda T}$.

After showing that the u^k's are uniformly bounded, and have a uniform bound on their Hölder norms of the same exponent (when $u_0 \in C^\beta(-\infty, \infty)$), we can select a subsequence which converges to a solution of the Cauchy problem. The uniqueness of this solution can be proved along the same lines (actually it is simpler) as in the proof of uniqueness for the first boundary value problem.

Open problems. 1. Study the degenerate oblique derivative boundary value problem.

2. Generalize the results of this chapter to the degenerate equation

$$u_t = (a_i(x, t, u, \nabla u))_{x_i} + b(x, t, u, \nabla u).$$

3. Generalize the results of this chapter to the degenerate equation

$$u_t = a_{ij}(x, t, u, \nabla u)u_{x_i x_j} + b(x, t, u, \nabla u).$$

4. Study the same problem when there are two lines of degeneracy, for instance, $\nu(u) = u^{\alpha}(1-u)^{\beta}$.

5. Study initial or initial-boundary value problems for the equation

$$\alpha(x, t, u)u_t + (a_j(x, t, u, \nabla u))_{x_j} + b(x, t, u, \nabla u) = 0,$$

where $\alpha(x, t, 0) = 0$.

6. Study the asymptotic behavior $(t \to \infty)$ of the solutions for the quasi-linear degenerate parabolic equations.

7. Study analogous problems for the degenerate elliptic equations.

8. Does Harnack's inequality still hold when the equation is degenerate in the t-direction (i.e., $\alpha(x, t, 0) = 0$ as in (5))?

CHAPTER VI

The Speed of Propagation of the Solution of a Degenerate Quasilinear Parabolic Equation

In this chapter we investigate an important property of the degenerate quasilinear parabolic equation of the previous chapter. Namely, we would like to study the domain of dependence of its solution. Under more restrictions on the coefficients we shall show that the solution has a finite speed of propagation. Works on this topic include [29] and [30].

§1. An estimate on the domain of dependence

Let Ω be a bounded domain in $\mathbb{R}^n$ and $T > 0$. In $Q = \Omega \times (0, T]$ we consider the equation

$$u_t = (a_{ij}(x, t, u)u_j)_i + b_i(x, t, u)u_i + c(x, t, u)u. \tag{1}$$

For $0 \leq u < \infty$ and $\xi \in \mathbb{R}^n$, we assume

$$\frac{1}{\Lambda}\nu(|u|)|\xi|^2 \leq a_{ij}(x, t, u)\xi_i\xi_j \leq \Lambda\nu(|u|)|\xi|^2 \tag{2}$$

where Λ is a positive constant and ν satisfies

$$\nu(s) \in C[0, \infty); \qquad \nu(0) = 0; \qquad \nu(s) > 0 \quad \text{for } s > 0. \tag{3}$$

Moreover, let

$$w = \varphi(u) = \int_0^u \nu(s)\,ds \quad \text{and its inverse} \quad u = \Phi(w). \tag{4}$$

We assume there exist $\delta > 0$ and $m > 1$ such that for all w_1, w_2 with $0 < w_1 < w_2 < \delta$

$$\frac{1}{m}\left(\frac{w_1}{w_2}\right)^{1-1/m} \leq \frac{\Phi'(w_2)}{\Phi'(w_1)} \leq \lambda\left(\frac{w_1}{w_2}\right) \tag{5}$$

where $\lambda(\tau)$ is nondecreasing and $\lambda(\tau) \to 0$ as $\tau \to 0^+$. Finally a_{ij}, b_i, b_{ix_j} $(i, j = 1, \dots, n)$, c are assumed to be in $C(\overline{Q} \times [0, \infty))$ and, for $(x, t) \in \overline{Q}$ and $0 < u < \infty$,

$$\sum_1^{\eta}\left(\frac{1}{\nu(u)^{1/2}}|b_i(x, t, u) - b_i(x, t, 0)| + \left|\frac{\partial b_i(x, t, u)}{\partial x_i}\right|\right) + |c(x, t, u)| < \eta(u) \tag{6}$$

where $\eta(t)$ is a nondecreasing function.

We shall also consider the equation (1) in $Q_0 = \mathbb{R}^n \times (0, T]$. In this case Q in (2)–(6) should be replaced by Q_0.

DEFINITION 1. A function u in $\overline{Q}$ or $\overline{Q}_0$ is called a (nonnegative) *weak solution* of (1) if there exists a sequence of positive, bounded, classical solutions $\{u^k\}$ converging uniformly to u in $\overline{Q}$ or $\overline{Q}_0$.

Now let us assume (2)–(6) are fulfilled. Besides

$$0 < u^k \le M \quad \text{in } Q \text{ or } Q_0 . \tag{7}$$

Before we can derive some inequalities for these classical solutions, we need some preparations.

By suitably modifying $\lambda(w_1/w_2)$ and m we may assume

$$\frac{1}{m}\left(\frac{w_1}{w_2}\right)^{1-1/m} \le \frac{\Phi'(w_2)}{\Phi'(w_1)} \le \lambda\left(\frac{w_1}{w_2}\right) \le m, \tag{5'}$$

for all w_1, w_2, $0 < w_1 < w_2 \le \widetilde{M} = \varphi(Me^{\eta(M)T})$. For instance, we can replace λ by $\tilde{\lambda}(\theta) = \max\{\lambda(\theta), \max_{\delta\le w_2\le\widetilde{M}, \theta\le\theta_1\le 1} \Phi'(w_2)/\Phi'(\theta_1 w_2)\}$. $\tilde{\lambda}$ is nondecreasing and $\tilde{\lambda}(\theta) \to 0$ (as $\theta \to 0^+$). On the other hand, we may replace m by

$$\widetilde{m} = \max\left\{\tilde{\lambda}(1), \max\{m, 1\}, \max_{\delta\le w_1\le w_2\le\widetilde{M}} \frac{\Phi'(w_1)w_1^{1-1/m}}{\Phi'(w_2)w_2^{1-1/m}}\right\}.$$

By Cauchy's mean value theorem, for $0 < w_1 < w_2 \le \widetilde{M}$, there is $\theta \in (0, 1)$ such that $\Phi(w_2)/\Phi(w_1) = w_2\Phi'(\theta w_2)/w_1\Phi'(\theta w_1)$. Combining this with (5′) we obtain

$$\frac{1}{m}\left(\frac{w_2}{w_1}\right)^{1/m} \le \frac{\Phi(w_2)}{\Phi(w_1)} \le m\frac{w_2}{w_1}. \tag{8}$$

Therefore, for $0 < u_1 < u_2 \le M$ we have

$$\frac{1}{m}\frac{u_2}{u_1} \le \frac{\varphi(u_2)}{\varphi(u_1)} \le m^m\left(\frac{u_2}{u_1}\right)^m, \tag{9}$$

$$\frac{1}{m} \le \frac{\nu(u_2)}{\nu(u_1)} \le m^m\left(\frac{u_2}{u_1}\right)^{m-1}. \tag{10}$$

We define the upper $\mathscr{B}_2$ function class as follows: For $x^0 \in \Omega$ and $0 \le t \le T$ denote $A_{k,\rho}(t, w) \equiv A_{k,\rho}(t) = \{x \in K(\rho) \cap \Omega : w(x, t) > k\}$, where $K(\rho) = K(x^0, \rho) = \{x : |x - x^0| < \rho\}$. Let

$$\zeta(x) = \zeta(x, \rho_1, \rho_2) = \begin{cases} 1, & |x - x^0| \le \rho_2, \\ \dfrac{\rho_1 - |x - x^0|}{\rho_1 - \rho_2}, & \rho_2 < |x - x^0| < \rho_1, \\ 0, & |x - x^0| \ge \rho_1, \end{cases} \tag{11}$$

for $\rho_1 > \rho_2 > 0$.

DEFINITION 2. A function w in $\overline{Q}$ belongs to the *upper* $\mathscr{B}_2(Q, \sigma, \gamma, M, \Phi)$ if

(i) $w \in C^{2,1}(Q) \cap C(\overline{Q})$ and $0 < w \le M$ in $\overline{Q}$;

(ii) For each $\rho \in (0, 1)$, $x^0 \in \Omega$, $t \in [0, T]$ and $k \ge \max_{\overline{K(\rho)\cap\partial\Omega}} w$,

$$\frac{\partial}{\partial t}\left(e^{-\gamma t}\int_{A_{k,\rho}(t)} \zeta^2(x)\chi_k(w-k) + \sigma e^{-\gamma t}\int_{A_{k,\rho}(t)} \zeta^2(x)|\nabla w|^2\right) \\ \le \gamma e^{-\gamma t}\int_{A_{k,\rho}(t)} |\nabla\zeta|^2(w-k)^2, \tag{12}$$

where

$$\chi_k(s) = \int_0^s \Phi'(k+\tau)\tau\, d\tau \tag{13}$$

and Φ satisfies the properties

(a) $\Phi \in C^1[0, M]$, $\Phi(0) = 0$, and $\Phi(s) > 0$ for $s > 0$;

(b) $(5')$ holds for Φ when $0 < w_1 < w_2 \le M$.

LEMMA 1. *Let u be a positive classical solution of* (1), $0 < u \le M$ *in* $\overline{Q}$. *Suppose in addition to* (2)–(6), $(5')$,

$$b_i(x, t, 0) = 0 \qquad (1 \le i \le n). \tag{14}$$

Then $w(x,t) = \varphi(e^{-\eta(M)t}u(x,t))$ belongs to the upper $\mathscr{B}_2(Q, \frac{1}{2m\Lambda}, \gamma, \widetilde{M}, \Phi)$, where $\widetilde{M} = \varphi(M)$, $\gamma > 0$, and Φ is the inverse of φ.

PROOF. Substitute $u = e^{\eta(M)t}v$ in (1) and multiply (1) by $\zeta^2(w-k)^+$. By integrating the resulting equality over Ω we have

$$\frac{\partial}{\partial t}\int_{A_{k,\rho}(t)} \zeta^2\chi_k(w-k) + \int_{A_{k,\rho}(t)} \zeta^2 a_{ij}w_i v_j \\ = \int_{A_{k,\rho}(t)} -2\zeta(w-k)a_{ij}\zeta_i v_j \\ + \int_{A_{k,\rho}(t)} \zeta^2(w-k)[b_i v_i + (c - \eta(M))v].$$

Set $h_i(x, t, r) = \int_0^r b_i(x, t, e^{\eta(M)t}\Phi(k+\tau))\Phi'(k+\tau)\tau\, d\tau$. By (6), (10), and (14)

$$h_i^2 \le \int_0^r b_i^2\Phi'(k+\tau)\tau\, d\tau \int_0^r \Phi'(k+\tau)\tau\, d\tau \\ \le \frac{1}{2}\eta^2(M)m^m e^{(m-1)\eta(M)}r^2\chi_k(r),$$

and $|\partial h_i/\partial x_i| \le \eta(M)\chi_k(r)$ $(1 \le i \le n)$. Consequently

$$\int_{A_{k,\rho}(t)} \zeta^2(w-k)b_i u_i = \int_{A_{k,\rho}(t)} \zeta^2\frac{dh_i(x, t, w-k)}{dx_i} \\ - \int_{A_{k,\rho}(t)} \zeta^2\left[\int_0^{w-k} \frac{\partial b_i(x, t, e^{\eta(M)t}\Phi(k+\tau))}{\partial x_i}\Phi'(k+\tau)\tau\, d\tau\right].$$

On the other hand, from integration by parts we have

$$\frac{\partial}{\partial t}\int_{A_{k,\rho}(t)}\zeta^2\chi_k(w-k)+\frac{1}{m\Lambda}\int_{A_{k,\rho}(t)}\zeta^2|\nabla w|^2$$
$$\leq \gamma_1\left[\int_{A_{k,\rho}(t)}\zeta|\nabla\zeta|\,|\nabla w|(w-k)+\int_{A_{k,\rho}(t)}\zeta|\nabla\zeta|\chi_k^{1/2}(w-k)(w-k)\right.$$
$$\left.+\int_{A_{k,\rho}(t)}\zeta^2\chi_k(w-k)\right].$$

By the Schwarz inequality,

$$\frac{\partial}{\partial t}\int_{A_{k,\rho}(t)}\zeta^2\chi_k(w-k)+\frac{1}{2m\Lambda}\int_{A_{k,\rho}(t)}\zeta^2|\nabla w|^2$$
$$\leq \gamma\left[\int_{A_{k,\rho}(t)}|\nabla\zeta|^2(w-k)^2+\int_{A_{k,\rho}(t)}\zeta^2\chi_k(w-k)\right],$$

or

$$\frac{\partial}{\partial t}\left[e^{-\gamma t}\int_{A_{k,\rho}(t)}\zeta^2\chi_k(w-k)\right]+\frac{1}{2m\Lambda}e^{-\gamma t}\int_{A_{k,\rho}(t)}\zeta^2|\nabla w|^2$$
$$\leq \gamma e^{-\gamma t}\int_{A_{k,\rho}(t)}|\nabla\zeta|^2(w-k)^2.$$

So w belongs to the upper $\mathscr{B}_2(Q, 1/2m\Lambda, \gamma, \widetilde{M}, \Phi)$. □

LEMMA 2. *Let w be in upper $\mathscr{B}_2$. For each $\beta\in(0,1)$ there is $a>0$ which only depends on β and the parameters in $\mathscr{B}_2$ so that*

$$\operatorname{meas} A_{\mu-(1-\beta)^2H,\,\rho/4}(t)=0,\qquad t\in[0,t^0],$$

whenever

$$\mu\geq\max_{\substack{x\in\overline{K(\rho)}\cap\Omega,\\ 0\leq t\leq t^0}} w,\qquad t^0\leq a\Phi'(\mu)\rho^2, \tag{15}$$

$$k\geq\max\left\{\beta\mu,\ \max_{\overline{K(\rho)}\cap\Omega} w(x,0),\ \max_{\substack{\overline{K(\rho)}\cap\partial\Omega,\\ 0\leq t\leq t^0}} w\right\}, \tag{16}$$

and $H=\mu-k>0$.

PROOF. Integrate (12) from 0 to t after choosing $\zeta = \zeta(x, \rho, \rho/2)$. Using the fact $\operatorname{meas} A_{k,\rho}(t) = 0$ which follows from (16), we obtain

$$e^{-\gamma t}\int_{A_{k,\rho}(t)} \zeta^2 \chi_k(w-k) \le \frac{4\gamma H^2 t^0}{\rho^2}\kappa_n \rho^n \qquad (t \le t^0),$$

where κ_n is the volume of the unit ball. From this and (15) we deduce

$$\chi_k(\beta H)\operatorname{meas} A_{k+\beta H,\rho/2}(t) \le 4a\gamma e^{\gamma T} H^2 \Phi'(\mu)\kappa_n \rho^n. \tag{17}$$

By $(5')$:

$$\frac{\chi_k(\beta H)}{\Phi'(\mu)} = \int_0^{\beta H} \frac{\Phi'(k+\tau)}{\Phi'(\mu)}\tau\, d\tau \ge \frac{1}{2m}(\beta H)^2. \tag{18}$$

Combining this with (17) yields

$$\operatorname{meas} A_{k+\beta H,\rho/2}(t) \le \frac{aC_1}{\beta^2}\rho^n, \qquad \forall t \in [0, t^0]. \tag{19}$$

The constants C_1 and $C_2, C_3, \ldots$ below are positive constants depending only on the parameters in $\mathscr{B}_2$. Let

$$k_h = k + \beta(2-\beta)H - \frac{\beta(1-\beta)H}{2^h}, \qquad \rho_h = \left(\frac{1}{4} + \frac{1}{2^{h+2}}\right)\rho,$$
$$\mu_h = \max_{0\le t\le t^0} \operatorname{meas} A_{k_h,\rho_h}(t), \qquad \zeta_h(x) = \zeta(x, \rho_h, \rho_{h+1}),$$
$$I_h(t) = e^{-\gamma t}\int_{A_{k_h,\rho_h}(t)} \zeta_h^2(x)\chi_{k_h}(w-k_h).$$

It follows from (19) that

$$\mu_0 \le \frac{aC_1}{\beta^2}\rho^n. \tag{20}$$

Our aim is to show $\lim_{h\to\infty}\mu_h = 0$ for sufficiently small a. When $w \ge k_h$, $k_h \ge k + \beta H \ge \beta\mu$, using $(5')$ and (13) we get

$$\frac{\chi_{k_h}(w-k_h)}{\Phi'(\mu)} \le \int_0^{w-k_h} \frac{\Phi'(k_h+\tau)}{\Phi'(\mu)}\tau\, d\tau$$
$$\le m\int_0^{w-k_h}\left(\frac{\mu}{k_h+\tau}\right)^{1-1/m}\tau\, d\tau \le \frac{m}{2\beta^{1-1/m}}(w-k_h)^2.$$

Consequently

$$I_h(t) \le \frac{m}{2\beta^{1-1/m}}\Phi'(\mu)\int_{A_{k_h,\rho_h}(t)} \zeta_h^2 (w-k_h)^2. \tag{21}$$

By Sobolev's inequality

$$\int_{\Omega\cap\{u>0\}} u^2 \le C(n)(\operatorname{meas}\Omega\cap\{u>0\})^{2/n}\int_{\Omega\cap\{u>0\}} |\nabla u|^2,$$

for all $u \in \overset{\circ}{W}{}_2^1(\Omega)$. Putting it into the right-hand side of (21) yields

$$I_h(t) \le C_2 \frac{1}{\beta^{1-1/m}} \Phi'(\mu) \mu_h^{2/n} \left[\int_{A_{k_h,\rho_h}(t)} \zeta_h^2 |\nabla w|^2 + \frac{H^2}{(\rho_h - \rho_{h+1})^2} \mu_h \right]. \tag{22}$$

On the other hand, from inequality (12) we have

$$I_h'(t) + \sigma e^{-\gamma t} \int_{A_{k_h,\rho_h}(t)} \zeta_h^2 |\nabla w|^2 \le \frac{\gamma H^2}{(\rho_h - \rho_{h+1})^2} \mu_h. \tag{23}$$

We set $\tau = \sup\{s \in [0, t] : I_h'(s) \ge 0\}$. Since $I_h(0) = 0$ and $I_h'(0) \ge 0$, τ exists. Obviously $I_h'(\tau) \ge 0$ and $I_h(t) \le I_h(\tau)$. From (22) and (23) we thus deduce

$$I_h(t) \le I_h(\tau) \le \frac{C_2}{\beta^{1-1/m}} \Phi'(\mu) \left(\frac{\gamma}{\delta} e^{\gamma T} + 1 \right) \frac{H^2 \mu_h^{1+2/n}}{(\rho_h - \rho_{h+1})^2}. \tag{24}$$

By (18)

$$\begin{aligned} I_h(t) &\ge e^{-\gamma T} \chi_{k_h}(k_{h+1} - k_h) \operatorname{meas} A_{k_{h+1},\rho_{h+1}}(t) \\ &\ge \frac{1}{2m} e^{-\gamma T} \Phi'(\mu)(k_{h+1} - k_h)^2 \operatorname{meas} A_{k_{h+1},\rho_{h+1}}(t). \end{aligned}$$

Combining this with (24) we finally obtain

$$(k_{h+1} - k_h)^2 \mu_{h+1} \le \frac{2mC_2 e^{\gamma T}}{\beta^{1-1/m}} \left(\frac{\gamma}{\delta} e^{\gamma T} + 1 \right) \frac{H^2 \mu_h^{1+2/n}}{(\rho_h - \rho_{h+1})^2}.$$

Hence

$$\mu_{h+1} \le \frac{C_3}{\beta^{3-1/m}} \frac{2^{4h} \mu_h^{1+2/n}}{\rho^2}.$$

Denote $y_h = \mu_h / \rho^n$. This inequality becomes

$$y_{h+1} \le \frac{C_3}{\beta^{3-1/m}} 2^{4h} y_h^{1+2/n},$$

whereas from (20) $y_0 \le aC_1/\beta^2$. Now, as shown in the last chapter, this leads to

$$y_h \le \frac{aC_1}{\beta^2} 2^{-2nh} \qquad (h = 0, 1, 2, \dots)$$

provided

$$a \le \frac{\beta^2}{C_1} \left(\frac{\beta^{3-1/m}}{C_3 2^{2n}} \right)^{n/2}. \tag{25}$$

Letting $h \to \infty$, we obtain $\lim_{h\to\infty} y_h = \lim_{h\to\infty} \mu_h = 0$. □

Since $\lambda(\tau) \to 0$ $(\tau \to 0^+)$, we can fix $q_0 \in (0, 1)$ such that

$$\lambda(q_0) \le \frac{1}{16}. \tag{26}$$

Let

$$\beta = 1 - (1 - q_0)^{1/3}. \tag{27}$$

LEMMA 3. *Let w be in upper $\mathscr{B}_2$. There exists a constant b depending only on the parameters of $\mathscr{B}_2$ such that*

$$w(x^0, t^0) \le \frac{1}{\beta} \max\left\{ \max_{K(x^0, b(t^0)^{1/2})\cap\Omega} w(x, 0), \max_{\substack{K(x^0, b(t^0)^{1/2})\cap\partial\Omega \\ 0\le t\le t^0}} w(x, t) \right\}, \tag{28}$$

for all $(x^0, t^0) \in Q$.

PROOF. Let

$$\mu_0 = M, \qquad \mu_l = q_0\mu_{l-1}, \qquad k_l = \beta\mu_l, \qquad \rho_0 = \left(\frac{t^0}{a\Phi'(M)}\right)^{1/2},$$

$$\rho_l = \frac{\rho_{l-1}}{4}, \qquad Q_l = \{(x, t) \in Q : |x - x^0| < \rho_l, 0 < t \le t^0\}$$

$$(l = 1, 2, \dots),$$

$$M_0 = \max\left\{ \max_{K(\rho_0)\cap\Omega} w(x, 0), \max_{\substack{K(\rho_0)\cap\partial\Omega \\ 0\le t\le t^0}} w(x, t) \right\}, \; l^* = \sup\{l : k_l > M_0\}, \tag{29}$$

where q, q_0, and β are as given in (25), (26), and (27).

Now we use induction to show, for $l \le l^* + 1$,

$$\max_{\overline{Q}_l} w \le \mu_l, \qquad t^0 \le a\Phi'(\mu_l)\rho_l^2. \tag{30}$$

From our choice of μ_0 and ρ_0 (30) holds for $l = 0$. We assume it holds for $l \le l^*$ and show that it is also valid for $l + 1$.

From $l \le l^*$ we know $k_l > M_0$. Hence $k_l \ge \max\{\beta\mu_l, M_0\}$. Applying Lemma 2 to Q_l we have, for $0 \le t \le t^0$, $\operatorname{meas} A_{\mu_l-(1-\beta)^2H_l, \rho_l/4}(t) = 0$, where $H_l = \mu_l - k_l = (1 - \beta)\mu_l$. As a result

$$\max_{\overline{Q}_{l+1}} w \le \mu_l - (1 - \beta)^2 H_l = \mu_l - (1 - \beta)^3 \mu_l = q_0\mu_l = \mu_{l+1}.$$

By $(5')$, (25), and (30)

$$\begin{aligned} a\Phi'(\mu_{l+1})\rho_{l+1}^2 &= \frac{\Phi'(\mu_{l+1})}{\Phi'(\mu_l)} \cdot \frac{a}{16}\Phi'(\mu_l)\rho_l^2 \\ &\ge \frac{1}{\lambda(\mu_{l+1}/\mu_l)} \frac{t^0}{16} = \frac{1}{\lambda(q_0)} \frac{t^0}{16} \ge t^0. \end{aligned}$$

It follows by induction that (30) holds for $l \le l^*+1$. In particular, $\max_{\overline{Q}_{l^*+1}} w \le \mu_{l^*+1}$. From the definition of l^* in (29) we conclude $k_{l^*+1} \le M_0$. Therefore

$$w(x^0, t^0) \le \max_{\overline{Q}_{l^*+1}} w \le \mu_{l^*+1} = \frac{1}{\beta}k_{l^*+1} \le \frac{M_0}{\beta},$$

and (28) follows by taking $b = (a\Phi'(M))^{-1/2}$. □

THEOREM 4. *Let $u(x, t)$ be a (nonnegative) weak solution of equation* (1) *in Q. Suppose* (2)–(6) *and* (5′) *are satisfied. Then for all $(x^0, t^0) \in Q$, there are constants K, b which only depend on $n, m, \delta, T, M, \Lambda, \eta(s)$, $\lambda(s)$, and functions $b_1, \dots, b_n$ such that*

$$u(x^0, t^0) \le K \max\left\{ \sup_{K(x^0, b(t^0)^{1/2})\cap\Omega} u(x, 0), \sup_{\substack{K(x^0, b(t^0)^{1/2})\cap\partial\Omega \\ 0\le t\le t^0}} u(x, t) \right\}, \tag{31}$$

where $K(x^0, \rho)$ is a ball centered at x^0 with radius ρ.

When u is a nonnegative weak solution of (1) *in Q_0,* (31) *is replaced by a simpler inequality*:

$$u(x^0, t^0) \le K \sup_{K(x^0, b(t^0)^{1/2})} u(x, 0). \tag{31′}$$

PROOF. First we consider the special case where equality (14) also holds. From Definition 1 there are positive classical solutions u^l satisfying (7) which converge to u uniformly. By Lemma 1, the function $W^l(x, t) = \varphi(u^l(x, t)e^{-\eta(M)t})$ belongs to upper $\mathscr{B}_2$ with parameters independent of l. By virtue of Lemma 3 there exists a constant b independent of l such that

$$W^l(x^0, t^0) \le \frac{1}{\beta} \max\left\{ \sup_{K(x^0, b(t^0)^{1/2})\cap\Omega} W^l(x, 0), \sup_{\substack{K(x^0, b(t^0)^{1/2})\cap\partial\Omega \\ 0\le t\le t^0}} W^l(x, t) \right\},$$

for all $(x^0, t^0) \in Q$. Combining this with (9) we have

$$u^l(x^0, t^0) \le \frac{me^{\eta(M)T}}{\beta^{1/m}} \max\left\{ \sup_{K(x^0, b(t^0)^{1/2})\cap\Omega} u^l(x, 0), \sup_{\substack{K(x^0, b(t^0)^{1/2})\cap\partial\Omega \\ 0\le t\le t^0}} u^l(x, t) \right\}.$$

After letting $l \to \infty$ we see that the theorem is valid.

The general case can be reduced to this special case as follows.

Under a change of variables

$$\begin{cases} x_i = x_i(y, \tau), & 1 \le i \le n, \\ t = \tau, \end{cases}$$

(1) becomes

$$\begin{aligned} \frac{\partial u}{\partial \tau} &= \frac{\partial}{\partial y_k}\left(a_{ij} \frac{\partial u}{\partial y_l} \frac{\partial y_k}{\partial x_i} \frac{\partial y_l}{\partial x_j} \right) \\ &\quad + \left[b_i \frac{\partial y_k}{\partial x_i} - \frac{\partial y_k}{\partial t} - a_{ij} \frac{\partial y_l}{\partial x_j} \frac{\partial}{\partial y_k}\left(\frac{\partial y_k}{\partial x_i} \right) \right] \frac{\partial u}{\partial y_k} + cu. \end{aligned} \tag{32}$$

We pick $y_k(x_1, \dots, x_n, t)$ $(1 \le k \le n)$ to be n-many independent first integrals of

$$\frac{\partial \varphi}{\partial t} - b_i(x, t, 0)\frac{\partial \varphi}{\partial x_i} = 0. \tag{33}$$

In other words, y_k can be solved via the solution $x_i = x_i(y_1, \dots, y_n, t)$ of

$$\begin{cases} \dfrac{dx_i}{dt} = -b_i(x, t, 0) & (1 \le i \le n), \\ x_i|_{t=0} = y_i. \end{cases}$$

Thus $\partial x_i/\partial y_j$ $(1 \le i, j \le n)$ is bounded. From

$$\frac{\partial(x_1, \dots, x_n)}{\partial(y_1, \dots, y_n)} = \exp\left(-\int_0^t \frac{\partial b_i}{\partial x_i} dt\right),$$

we know that it is also bounded below by a positive number. Hence $\partial y_j/\partial x_i$ is also bounded. The eigenvalues of $(\partial y_j/\partial x_i)^*(\partial y_j/\partial x_i)$ have positive upper and lower bounds Σ, σ. Hence

$$\frac{\sigma}{\Lambda}\nu(u)|\xi|^2 \le a_{ij}\frac{\partial y_k}{\partial x_i}\frac{\partial y_l}{\partial x_j}\xi_k\xi_l \le \Lambda\Sigma\nu(u)|\xi|^2, \quad \forall \xi \in \mathbb{R}^n.$$

Finally from the continuous dependence on initial data for ODE's $(\partial/\partial y_k)(\partial y_k/\partial x_i)$ is bounded. Hence (14) holds for equation (32). Switching back to (x, t) for the estimate

$$u(y^0, \tau^0) \le K \max\left\{\sup_{K(y^0, b(\tau^0)^{1/2})\cap\Omega} u(y, 0), \sup_{K(y^0, b(\tau^0)^{1/2})\cap\partial\Omega} u(y, \tau)\right\},$$

we conclude the proof of the theorem after, if necessary, replacing b by a large number which only depends on the transformation between x and y and is uniform in t for $0 \le t \le T$. □

COROLLARY 5. *Let u be a weak solution of (1) in $\overline{Q}$ and (2)–(6) hold. For each t, $0 \le t \le T$, there exists R_t such that*

$$u(x, t) = 0, \qquad |x| \ge R_t,$$

provided $u(x, 0)$ has compact support.

REMARK. This corollary asserts that the speed of propagation of the solution is finite.

PROOF. Let $u(x, 0)$ be supported in the ball $K(0, R_0)$. We may take $R_t = R_0 + bt^{1/2}$, where b is determined from Theorem 4. □

§2. A lower estimate for the solution

Further properties of the solution can be obtained by introducing a lower $\mathscr{B}_2$ function class. For $x^0 \in \Omega$, $\rho > 0$, and $0 \le t \le T$, we write $B_{k,\rho}(t, w) \equiv B_{k,\rho}(t) = \{x \in K(\rho) \cap \Omega : w(x, t) < k\}$.

DEFINITION 3. A function w defined in $\overline{Q}$ belongs to the *lower* $\mathscr{B}_2(Q, \sigma, \gamma, M, \Phi)$ if

(i) $w \in C^{2,1}(Q) \cap C(\overline{Q})$ and $0 < w(x, t) \leq M$ in $\overline{Q}$.

(ii) For all $\rho \in (0, 1)$, $x^0 \in \Omega$, $t \in [0, T]$, and $0 < k \leq \inf_{K(\rho)\cap\partial\Omega} w(x, t)$,

$$\frac{\partial}{\partial t}\left[e^{-\gamma t}\int_{B_{k,\rho}(t)} \zeta^2(x)\tilde{\chi}_k(k-w)\right] + \sigma e^{-\tau t}\int_{B_{k,\rho}(t)} \zeta^2|\nabla w|^2 \leq \gamma e^{-\gamma t}\int_{B_{k,\rho}(t)} |\nabla\zeta|^2(k-w)^2,$$

where $\zeta(x) = \zeta(x, \rho, (1-\sigma)\rho)$, $\forall\sigma \in (0, 1)$ (see (11) for the definition of $\zeta(x, \rho_1, \rho_2)$) and $\tilde{\chi}_k(s) = \int_0^s \Phi'(k-\tau)\tau\, d\tau$. The assumption on Φ is the same as in Definition 2.

LEMMA 6. *Let w be in the lower $\mathscr{B}_2(Q, \sigma, \gamma, M, \Phi)$ satisfying*

$$\inf_{K(x^0, \rho)} w(x, 0) \geq k > 0.$$

Then there exists a constant a depending solely on the parameters in $\mathscr{B}_2$ such that

$$\inf_{K(x^0, \rho/4)} w(x, t) \geq \frac{k}{4}, \quad t \in [0, t^0], \qquad t^0 = \min\{a\Phi'(k)\rho^2, T\}.$$

PROOF. From the last chapter we know

$$\frac{1}{2m}s^2 \leq \frac{\tilde{\chi}_k(s)}{\Phi'(k)} \leq m^2 s^2, \qquad k \geq s > 0.$$

Based on this Lemma 6 can be proved more or less like Lemma 2. We replace β, H, and $\chi_k(s)$ therein by $1/2$, $-k/2$, and $\tilde{\chi}_k(s)$. □

LEMMA 7. *Let u be a (nonnegative) weak solution of* (1) *and* (2)–(6), (5′), (14) *hold. There exists a constant b which only depends on $\mathscr{B}_2$ such that for all (x^0, t^0) in Q,*

$$u(x^0, t^0) \geq \frac{e^{-\eta(M)t^0}}{4m}\min\left\{\tilde{\mu}_\rho, \Phi\left(\tilde{M}\lambda^{-1}\left(\frac{\rho^2}{b^2 t^0}\right)\right)\right\},$$

where $\tilde{\mu}_\rho = \inf_{K(x^0, \rho)} u(x, 0)$, $\tilde{M} = \varphi(M)$, and $\lambda^{-1}(s)$ is the inverse function of λ given in (5′).

PROOF. The theorem obviously holds when $\tilde{\mu}_\rho = 0$. When $\tilde{\mu}_\rho > 0$, as in the proof of Lemma 1 we have, for a positive classical solution u^l of (1), $W^l = \varphi(e^{\eta(M)t}u^l(x, t))$ belongs to the lower $\mathscr{B}_2(Q, 1/2m\Lambda, \gamma, \widetilde{M}, \Phi)$.

In the following we drop the superscript l. Take k so that $a\Phi'(k)\rho^2 \geq t^0$. From $(5')$ we know that the desired estimate follows provided

$$\lambda\left(\frac{k}{\widetilde{M}}\right) \leq \frac{a\rho^2\Phi'(\widetilde{M})}{t^0}, \quad \text{i.e.,} \quad k \leq \widetilde{M}\lambda^{-1}\left(\frac{\rho^2}{b^2t^0}\right),$$

where $b = (a\Phi'(\widetilde{M}))^{1/2}$. By Lemma 6

$$w(x^0, t^0) \geq \frac{1}{4}\min\left\{\varphi(\tilde{\mu}_\rho), \widetilde{M}\lambda^{-1}\left(\frac{\rho^2}{b^2t^0}\right)\right\},$$

if we take

$$k = \min\left\{\varphi(\tilde{\mu}_\rho), \widetilde{M}\lambda^{-1}\left(\frac{\rho^2}{b^2t^0}\right)\right\}.$$

Hence, it follows from (9) that

$$u(x^0, t^0) \geq \frac{e^{-\eta(M)t^0}}{4m}\min\left\{\tilde{\mu}_\rho, \Phi\left(\widetilde{M}\lambda^{-1}\left(\frac{\rho^2}{b^2t^0}\right)\right)\right\}.$$

By taking limit the theorem continues to hold for a nonnegative weak solution. □

THEOREM 8. *Let u be a (nonnegative) weak solution of* (1) *in Q and* (2)–(6), $(5')$ *hold. For all* $(x^0, t^0) \in Q$,

$$u(x^0, t^0) \geq K^* \inf_{K(x^0, b(t^0)^{1/2})} u(x, 0), \tag{34}$$

where K^*, b only depend on n, m, δ, T, M, Λ, $\eta(s)$, $\lambda(s)$, and the functions $b_1, \dots, b_n$.

PROOF. In case, in addition, (14) holds, by Lemma 7 (taking $\rho = b(t^0)^{1/2}$)

$$\begin{aligned} u(x^0, t^0) &\geq K_1 \min\left\{\inf_{K(x^0, b(t^0)^{1/2})} u(x, 0), K_2\right\} \\ &\geq K_1 \min\left\{1, \frac{K_2}{M}\right\} \inf_{K(x^0, b(t^0)^{1/2})} u(x, 0), \end{aligned}$$

where $K_2 = \Phi(\widetilde{M}\lambda^{-1}(1))$. So (34) holds. In the general case we use the transformation in Theorem 4 to get rid of (14). □

REMARK. Theorem 8 supplements Theorem 4. In any case the domain of dependence for a solution of a degenerate quasilinear parabolic equation has been well understood. But this is not the case for the domain of influence. The domain determined by Lemma 6 is unlikely the genuine domain of influence since it is too small. It remains an open question how to determine this domain.

CHAPTER VII

Aleksandrov and Bony Maximum Principles for Parabolic Equations

The Aleksandrov maximum principle and the density theorem are fundamental in the study of quasilinear and fully nonlinear elliptic and parabolic equations. In this chapter we shall discuss the former. The density theorem will be covered in the next chapter.

§1. Introduction

When applied to linear equations, the Aleksandrov maximum principle generalises the classical maximum principle where the coefficients of the equation now are assumed merely to be measurable.

Let Ω be a bounded domain in $\mathbb{R}^n$. Consider a solution u for the linear elliptic equation

$$\begin{cases} a_{ij}(x)u_{ij} = f(x), & x \in \Omega, \\ u|_{\partial\Omega} \geq 0, \end{cases} \tag{1}$$

where $a_{ij}\xi_i\xi_j \geq \lambda|\xi|^2$ $(\lambda > 0)$ and $f > 0$. We compare u with the convex solution z of the Monge-Ampère equation

$$\begin{cases} \det(z_{ij}) = \dfrac{1}{\det(a_{ij})}\left(\dfrac{f}{n}\right)^n, \\ z|_{\partial\Omega} = 0. \end{cases} \tag{2}$$

By virtue of the geometric-arithmetic mean inequality

$$a_{ij}z_{ij} = \operatorname{tr}(a_{ij}z_{jk}) \geq n\det{}^{1/n}(a_{ij}z_{jk}) = n(\det(a_{ij})\det(z_{ij}))^{1/n} = f.$$

Therefore

$$a_{ij}(u-z)_{ij} \leq 0 \quad \text{in } \Omega.$$

Since $u - z \geq 0$ on $\partial\Omega$ and $u - z$ cannot attain a negative minimum in $\overline{\Omega}$, $\min_{\overline{\Omega}} u \geq \min_{\overline{\Omega}} z$. Now an explicit estimate for $\min_{\overline{\Omega}} u$ follows from a corresponding one for z. This maximum principle was first discovered by Aleksandrov [31]. Aleksandrov's proof is indirect in the sense that it involves the existence of a generalized solution for the Monge-Ampère equation. Is it possible to derive the maximum principle avoiding (2)? This was done by Pucci [32]. From the solvability of (1) (this is a well-known fact when the coefficients have better regularity) he constructed the negative convex envelope of u

$$\check{u}(x) = \inf\left\{ \min\{0, \alpha_i u(x^i)\}: \forall\alpha_i,\ \alpha_i \geq 0,\ \sum \alpha_i = 1; \right.$$
$$\left. x^i \in \Omega \text{ such that } x = \alpha_i x^i . \right\}$$

Then the Aleksandrov maximum principle follows from an explicit estimate of $\check{u}$ over the convex domain Ω. For details see [6].

The situation for parabolic equations is similar. Let Ω be a bounded convex domain in $\mathbb{R}^n$ and $Q = \Omega \times (0, T]$. Consider a solution u of

$$\begin{cases} a_{ij}(x, t)u_{ij} - u_t = f(x, t), & (x, t) \in Q, \\ u|_{\partial^* Q} \geq 0, \end{cases} \tag{3}$$

where $\partial^* Q = \Omega \times \{t = 0\} \cup \partial\Omega \times [0, T]$ is the parabolic boundary of Q. Suppose that $a_{ij}\xi_i\xi_j \geq \lambda|\xi|^2$ for some $\lambda > 0$ and $f > 0$. We compare u with the solution z of the parabolic Monge-Ampère equation

$$\begin{cases} -z_t \det(z_{ij}) = \dfrac{1}{\det(a_{ij})}\left(\dfrac{f}{n+1}\right)^{n+1}, \\ z|_{\partial^* Q = 0}, \end{cases} \tag{4}$$

which belongs to $C^{2,1}(Q) \cap C(\overline{Q})$ and is convex in x and decreasing in t.

$$\begin{aligned} a_{ij}z_{ij} - z_t &= \operatorname{tr}\begin{pmatrix} -z_t & 0 \\ 0 & a_{ij}z_{jk} \end{pmatrix} \\ &\geq (n+1)\det{}^{1/(n+1)}\begin{pmatrix} -z_t & 0 \\ 0 & a_{ij}z_{jk} \end{pmatrix} \\ &= (n+1)[(-z_t)\det(a_{ij})\det(z_{ij})]^{1/(n+1)} = f. \end{aligned}$$

Therefore

$$\begin{cases} a_{ij}(u-z)_{ij} - (u-z)_t \leq 0 & \text{in } Q, \\ (u-z)|_{\partial^* Q} \geq 0. \end{cases}$$

By the maximum principle $\min_{\overline{Q}} u \geq \min_{\overline{Q}} z$. Once again an explicit estimate for $\min_{\overline{Q}} u$ follows from a corresponding one for $\min_{\overline{Q}} z$. This is Krylov's work [33] on the parabolic version of the Aleksandrov maximum principle. Notice that to begin with one needs to show that (4) admits a generalized solution. Proofs independent of (4) were given in [34] and [48]. The idea is similar to [32]. The proof in [48] is quite simple, with Sard's lemma used in an implicit way. The proof in [34] is very complicated in taking limits. However, since we shall use the details of this proof in Chapter 10, we shall follow the treatment in [34] and supplement it with a rigorous proof.

THEOREM 1. *Let Ω be a bounded convex domain with C^2-boundary. Let $z \in C^{2,1}(\overline{Q})$, $Q = \Omega \times (0, T)$, be convex in x, decreasing in t and vanish on $\partial^* Q$. Then for all $x^0 \in \Omega$*

$$(-z(x^0, T))^{n+1} \le \frac{(n+1)R^n}{\kappa_n} \int_Q -z_t \det(z_{ij})\, dx\, dt, \tag{5}$$

where R is the diameter of Q and κ_n is the volume for the n-dimensional ball.

PROOF. First we claim

$$\int_Q -z \det(z_{ij})\, dx \Big|_{t=T} = (n+1) \int_Q -z_t \det(z_{ij})\, dx\, dt. \tag{6}$$

For the moment we assume $z \in C^{3,1}(\overline{Q})$ and $\partial\Omega \in C^3$. Then we have

$$\frac{d}{dt} \int_\Omega z \det(z_{ij})\, dx - \int_\Omega z_t \det(z_{ij})\, dx = \int_\Omega z z_{ijt} \frac{\partial}{\partial z_{ij}} \det(z_{kl})\, dx.$$

Applying integration by parts in x_i, x_j to the right-hand side of this equality yields

$$\int_\Omega z_{ij} z_t \frac{\partial}{\partial z_{ij}} \det(z_{kl})\, dx = n \int_\Omega z_t \det(z_{kl})\, dx.$$

Notice that we have used the identity

$$\frac{\partial}{\partial x_j} \frac{\partial}{\partial z_{ij}} \det(z_{kl}) = 0, \qquad 1 \le i \le n, \tag{7}$$

and Euler's formula for homogeneous functions. Therefore

$$\frac{d}{dt} \int_\Omega z \det(z_{ij})\, dx = (n+1) \int_\Omega z_t \det(z_{ij})\, dx. \tag{8}$$

Clearly (6) follows from (8).

In general, we approximate $\partial\Omega$ and z by C^3-boundary and $C^{3,1}(\overline{Q})$ functions respectively. After obtaining (6) we can take limit and show that the theorem remains valid for $\partial\Omega \in C^2$ and $z \in C^{2,1}(\overline{Q})$.

We supply a proof of the identity (7).

LEMMA 2. *Let $F\colon \mathbb{R}^{n+1} \to \mathbb{R}^n$ be a smooth function. Set*

$$A_i = \det \left[\frac{\partial F}{\partial x_0}, \dots, \frac{\partial F}{\partial x_{i-1}}, \frac{\partial F}{\partial x_{i+1}}, \dots, \frac{\partial F}{\partial x_n} \right].$$

Then

$$\sum_{i=0}^{n} (-1)^i \frac{\partial A_i}{\partial x_i} = 0.$$

PROOF. Let $i < j$, $i, j \in \{0, 1, \dots, n\}$ and let

$$C_{ij} = \det \left[\frac{\partial^2 F}{\partial x_i \partial x_j}, \frac{\partial F}{\partial x_0}, \dots, \frac{\partial F}{\partial x_{i-1}}, \frac{\partial F}{\partial x_{i+1}}, \dots, \frac{\partial F}{\partial x_{j-1}}, \frac{\partial F}{\partial x_{j+1}}, \dots, \frac{\partial F}{\partial x_n} \right].$$

When $i > j$ we write $C_{ij} = C_{ji}$. From the properties of determinants we have

$$\frac{\partial A_i}{\partial x_i} = \sum_{j<i}(-1)^j C_{ij} + \sum_{j>i}(-1)^{j-1} C_{ij} = \sum_{j=0}^{n}(-1)^j \sigma(i, j) C_{ij},$$

where

$$\sigma(i, j) = \begin{cases} 1 & j < i, \\ 0 & j = i, \\ -1 & j > i. \end{cases}$$

Using $\sigma(i, j) = -\sigma(j, i)$ and $C_{ij} = C_{ji}$

$$\sum_{i=0}^{n}(-1)^i \frac{\partial A_i}{\partial x_i} = \sum_{i,j=0}^{n}(-1)^{i+j}\sigma(i, j) C_{ij} = \sum_{i,j=0}^{n}(-1)^{i+j}\sigma(j, i) C_{ji}$$
$$= (-1)\sum_{i,j=0}^{n}(-1)^{i+j}\sigma(i, j) C_{ij}.$$

This implies $\sum_{i=0}^{n}(-1)^i \partial A_i / \partial x_i = 0$. □

(7) follows from this lemma by replacing $n+1$ by n and taking $F = (\partial z/\partial x_1, \ldots, \partial z/\partial x_{i-1}, \partial z/\partial x_{i+1}, \ldots, \partial z/\partial x_n)$.

THE PROOF OF THEOREM 1 (CONTINUED). Let z_1, z_2 be two convex functions in $C^2(\overline{\Omega})$ satisfying $z_1|_{\partial\Omega} = z_2|_{\partial\Omega} = 0$ and $z_1 \geq z_2$ in $\overline{\Omega}$. From (8)

$$\frac{d}{dt}\int_\Omega [tz_1 + (1-t)z_2] \det([tz_1 + (1-t)z_2]_{ij})\, dx$$
$$= (n+1)\int_\Omega (z_1 - z_2)\det([tz_1 + (1-t)z_2]_{ij})\, dx \geq 0.$$

Integrating from 0 to 1 in t we get

$$\int_\Omega -z_1 \det(z_{1ij})\, dx \leq \int_\Omega -z_2 \det(z_{2ij})\, dx. \tag{9}$$

Let $z(x^0, T) = z^0 (x^0 \in \Omega)$ and let $w(x)$ be the surface of the cone with vertex (x^0, z^0) and base Ω. By convexity $w(x) \leq z(x, T)$ in $\overline{\Omega}$. We would like to choose $z_1 = w(x)$ and $z_2 = z(x, T)$ in (9). However, since w is not smooth at x^0 we need to modify w slightly. Let $w^\delta(x) = \int w(x-y) K_\delta(y)\, dy$ and $z^\delta(x, T) = \int z(x, T) K_\delta(y)\, dy$ where K_δ is a mollifier of radius δ. Then $z_1 = w^\delta$ and $z_2 = z^\delta(\cdot, T)$ are convex, $z_1 \leq z_2$, and $z_i|_{\partial\Omega_\delta} = O(\delta)$, $i = 1, 2$ where $\Omega_\delta = \{x \in \Omega\colon d(x, \partial\Omega) > \delta\}$. Following the derivation of (9)

$$\begin{aligned}\int_{\Omega_\delta} -z^\delta \det(z_{ij}^\delta)\,dx\bigg|_{t=T} + O(\delta) &\geq \int_{\Omega_\delta} -w^\delta \det(w_{ij}^\delta)\,dx \\ &\geq \int_{w(x)\leq -z^0+\eta} -w^\delta \det(w_{ij}^\delta)\,dx \\ &\geq (-z^0+\eta+O(\delta)) \int_{w(x)\leq -z^0+\eta} \det(w_{ij}^\delta)\,dx \\ &= (-z^0+\eta+O(\delta)) \int_{\omega^{\eta,\delta}} dp_1\cdots dp_n \\ &= (-z^0+\eta+O(\delta))\operatorname{meas}\omega^{\eta,\delta}, \qquad (10)\end{aligned}$$

where $\delta = o(\eta)$, $\eta = o(1)$, $p_i = \partial w^\delta/\partial x_i$ $(1 \leq i \leq n)$ and $\omega^{\eta,\delta}$ is the image of $\{x\colon w(x) \leq -z^0 + \eta\}$ under the mapping $p_i = p_i(x)$. Letting $\delta \to 0$ we have

$$\int_{\Omega} -z\det(z_{ij})\,dx\bigg|_{t=T} \geq (-z^0+\eta)\operatorname{meas}\omega^{\eta,0}.$$

Notice that $p = (p_1, \dots, p_n) \in \omega^{\eta,0}$ if and only if the tangent hyperplane $z = p(x-x^0)+z^0$ lies below $\Omega\times\{z=0\}$, i.e., $z^0+p(x-x^0)<0$, $\forall x\in\Omega$.

The diameter of Ω is R. We may assume $\Omega \subseteq B_R = \{x\colon |x| < R\}$. Then $\omega_R \subseteq \omega$ where $\omega_R = \{p\colon z^0+p(x-x^0)<0,\ \forall x\in B_R\}$. Geometrically ω_R is an ellipsoid. By an easy computation

$$\operatorname{meas}\omega^{\eta,0} \geq \operatorname{meas}\omega_R = \kappa_n \frac{R(-z^0)^n}{(R^2-|x^0|^2)^{(n+1)/2}} \geq \kappa_n \frac{(-z^0)^n}{R^n}. \qquad (11)$$

Combining (6), (10), and (11), and letting $\eta \to 0$ we obtain (5). □

§2. Some properties of convex functions

In this section we describe some properties of convex functions which are useful in later discussions. We shall always assume the domain Ω in which the convex function z is defined is convex. Otherwise we can utilize $u(x) = \inf\{\sum\lambda_i u(x^i)\colon \lambda_i \geq 0,\ \sum\lambda_i = 1,\ x^i\in\Omega,\ \sum\lambda_i x^i = x\}$ to extend u to the convex hull of Ω, that is, the smallest convex set containing Ω. We shall further assume that Ω is bounded in $\mathbb{R}^n$.

Let $z_\varepsilon(x) = \int z(x-y)K_\varepsilon(y)\,dy$ where K_ε is a mollifier supported in a ball of radius ε be a smooth regularization of z. It is easy to see that z_ε, which now is defined in a subdomain Ω_ε of Ω, is again convex. In this way a family of smooth convex functions is obtained. We need to study sequences of convex functions.

LEMMA 3. *Let $\{z_j\}_{j=1}^\infty$ be a sequence of convex functions in a bounded convex domain Ω. Suppose $z_j(x) \to z(x)$ $(j\to\infty,\ \forall x\in\Omega)$. Then in* *each compact subset of* Ω $\{z_j\}$ *is uniformly bounded, satisfying a uniform Lipschitz condition, and converges uniformly to* z.

REMARK. That z is convex follows from $z_j(x) \to z(x)$.

PROOF. First we claim $\sup\{|z_j(x)|: x \in \Omega, d(x, \partial\Omega) \geq a\}$ (a a constant) is bounded. For, if there is $\{x^j\}$ such that $z_j(x^j) \to \infty$ and $d(x^j, \partial\Omega) \geq a$, $j = 1, 2, \ldots$, let x^0 be a limit point of $\{x^j\}$. For simplicity we assume $x^j \to x^0$. Select $X^i \in \Omega$ $(0 \leq i \leq n)$ so that x^0 belongs to the convex hull of $X^0, \ldots, X^n$. We may also assume the x^j's are all contained in it. Let M be an upper bound of $z_j(X^i)$ $(0 \leq i \leq n, j \geq 1)$. We have $z_j(x^j) \leq \sum_i \lambda_{ij} z_j(X^i) \leq \sum \lambda_{ij} M = M$, where $x^j = \sum_{i=0}^n \lambda_{ij} X^i$ $(\sum_{i=0}^n \lambda_{ij} = 1)$.

On the other hand, we choose a point X^{n+1} from the convex hull of $\{x^0, X^1, \ldots, X^n\}$ which is so close to x^0 that $\lambda_0 \geq \frac{3}{4}$ in the representation

$$X^{n+1} = \lambda_0 x^0 + \sum_{i=1}^n \lambda_i X^i \qquad \left(\sum_{i=0}^n \lambda_i = 1\right).$$

Then we have $\lambda_{0j} \geq \frac{1}{2}$ in

$$X^{n+1} = \lambda_{0j} x^j + \sum \lambda_{ij} X^i \qquad \left(\sum_{i=0}^n \lambda_{ij} = 1\right).$$

Therefore $z_j(X^{n+1}) \leq \lambda_{0j} z_j(x^j) + \sum \lambda_{ij} z_j(X^i)$ and so

$$z_j(x^j) \geq \frac{1}{\lambda_{0j}}\left(z_j(X^{n+1}) - \sum_i \lambda_{ij} z_j(X^i)\right).$$

Notice that the right-hand side of this inequality is bounded. In view of the upper and lower estimations we have just derived, it is impossible to have $|z_j(x^j)| \to \infty$.

Next we observe that when $d(x^i, \partial\Omega) \geq 2a$ $(i = 1, 2)$ one can find x^3 and x^4 on the line segments extending $\overrightarrow{x^1x^2}$ and $\overrightarrow{x^2x^1}$ such that $|x^3 - x^2| = a$, $d(x^3, \partial\Omega) \geq a$, and $|x^4 - x^1| = a$, $d(x^4, \partial\Omega) \geq a$.

By convexity

$$\frac{z_j(x^1) - z_j(x^4)}{|x^1 - x^4|} \leq \frac{z_j(x^2) - z_j(x^1)}{|x^2 - x^1|} \leq \frac{z_j(x^3) - z_j(x^2)}{|x^3 - x^2|}.$$

Consequently

$$\sup_{\substack{d(x^i, \partial\Omega) \geq 2a \\ i=1,2}} \left|\frac{z_j(x^2) - z_j(x^1)}{x^2 - x^1}\right| \leq \frac{2}{a} \sup_{d(x, \partial\Omega) \geq a} |z_j(x)| = K_a.$$

So z_j satisfies a uniform Lipschitz condition in every compact subset. From this it follows that z_j converges uniformly in every compact subset. The proof of the lemma is complete. □

For a continuous convex function z in a bounded convex domain $\Omega \subseteq \mathbb{R}^n$, a supporting hyperplane through $(x^0, z(x^0))$, $x^0 \in \Omega$, $x_{n+1} - z(x^0) = \sum_{i=1}^n p_i(x_i - x_i^0) \equiv p(x-x^0)$, is a hyperplane satisfying $x_{n+1} \leq z(x)$, $\forall x \in \Omega$. The mapping defined by $x \mapsto p = (p_1, \dots, p_n)$ is called the *normal mapping*. We denote it by $\nu(x) = \nu(z, x)$.

The normal mapping has a unique image if and only if $z(x)$ has a unique tangent hyperplane everywhere. Otherwise the image is not necessarily unique, and the normal map becomes multivalued whenever there is a jump in the first derivatives. The inverse image of the normal map is not unique too. Different x's may correspond to the same p when the graph of z contains a straight line through x. Therefore ν should be regarded as a set-valued mapping. Nevertheless, the discussion below will make it clear that this set-valued mapping is a rather simple one.

Consider the set $\nu(M) \equiv \nu(z, M)$ for $M \subseteq \Omega$. Let z be a convex function in $C^2(\Omega)$ (hence $\nu(x)$ is unique). For a measurable set M the integral

$$\int_M \det(z_{ij})\, dx = \int_{\nu(M)} dp_1 \cdots dp_n = \operatorname{meas} \nu(M)$$

exists. In other words, $\nu(M)$ is Lebesgue measurable if M has the same property. Also the Riemann measure $\operatorname{meas}\nu(M)$ given in this formula is completely additive.

Most of these properties are sustained for a convex z in $C(\Omega)$. Specifically $\nu(M)$ is Lebesgue measurable when M is a Borel set. The Riemann measure $\operatorname{meas}\nu(M)$ still exists and can be calculated via a smooth approximation. Also it is completely additive. Let us establish these facts.

LEMMA 4. *Let z be a continuous convex function in a bounded convex domain Ω. The normal image of a closed set M ($\subseteq \Omega$), $\nu(M)$, is also closed.*

PROOF. Let $p^j \in \nu(M)$, $j = 1, 2, \dots$, $p^j \to p^0$. Each p^j has a preimage x^j in M. Since M is bounded, $\{x^j\}$ contains a subsequence that converges to some x^0. To simplify notation we assume $x^j \to x^0$. x^0 belongs to M as M is closed. Taking the limit in

$$z(x^j) + p^j(x - x^j) \leq z(x), \qquad \forall x \in \Omega$$

we obtain

$$z(x^0) + p^0(x - x^0) \leq z(x), \qquad \forall x \in \Omega.$$

Hence $p^0 \in \nu(M)$. $\nu(M)$ is closed. □

The image of an open set is not necessarily open. For instance, suppose that $z(\omega)$, where ω is an open subset of Ω, belongs to a hyperplane. Then $\nu(\omega)$ consists of a single point. Nevertheless, one can show that it is always measurable. This can be seen by taking the complement and using Lemma 4 (see Corollary 7 below). To show the complement of a measurable set is still measurable we need to consider another normal mapping of which total measure is finite.

Let x be a point on a bounded, closed, convex hypersurface K and let $(p_1, \dots, p_n)$ be the normal for a supporting hyperplane of K through x. Denote the total image of the map

$$x \to \left(\frac{p_1}{\sqrt{1+|p|^2}}, \dots, \frac{p_n}{\sqrt{1+|p|^2}}\right), \qquad |p|^2 = \sum_{i=1}^n p_i^2,$$

from K to the unit sphere S_n by $\mu(K)$. We have

$$\operatorname{meas} \mu(M) = \int_M \frac{\nu(dx)}{\sqrt{1+|p|^2}},$$

and

$$\operatorname{meas} \nu(M) = \int_M \sqrt{1+|p|^2}\,\mu(dx).$$

So the measurability of $\nu(M)$ and $\mu(M)$ are equivalent.

We can close off the hypersurface K by adding a cap K_+: $z = z_+(x)$ given by

$$z_+(x) = \sup\left\{\sum \alpha_i z(x^i) \colon \alpha_i \ge 0, \sum \alpha_i = 1, x^i \in \partial K, \sum \alpha_i x^i = x\right\}.$$

It is easy to show that K and K_+ together form a convex hypersurface without boundary. Denote it still by K. On one hand we have $\mu(K) \subseteq S_n$. On the other hand a hyperplane π translating along a fixed direction touches K twice, at one of which π becomes a supporting hyperplane. Hence $S_n \subseteq \mu(K)$. So $\operatorname{meas}\mu(K)$ is finite.

LEMMA 5. *Let N be the subset of K consisting of points whose supporting hyperplanes touch more than one point (and hence a line segment). Then $\mu(N)$ is of measure zero.*

PROOF. The lemma is true for $n = 1$. For, there are finitely many line segments of length $\ge 1/k$ $(k = 1, 2, \dots)$ on K. So N is composed of at most countably many line segments. But then $\mu(N)$ is a countable set and $\operatorname{meas}\mu(N) = 0$.

Now we use induction on the dimension. Select $n+1$ many orthogonal directions $L_1, \dots, L_{n+1}$ in $\mathbb{R}^{n+1}$. Let us denote the image of the subset of N which is not on a line segment perpendicular to L_i by M_i $(\subseteq \mu(N))$, $i = 1, \dots, n$, and the image of the points in N which lie on line segments perpendicular to all L_i $(1 \le i \le n)$ (so they are parallel to L_{n+1}) by M_{n+1}. We have $\mu(N) = \bigcup_{i=1}^{n+1} M_i$.

The dimension of the union of all line segments on K parallel to L_{n+1} is at most $n-1$. Hence $\operatorname{meas} M_{n+1} = 0$.

For any $(q_0^0, \dots, q_{n+1}^0) \in M_1$, there is a supporting hyperplane

$$\pi: \sum_{i=1}^{n+1} q_i^0(x_i - x_i^0) = 0 \qquad (z(x) \equiv x_{n+1})$$

satisfying the following property: there exist two distinct $X^0 = (x_1^0, \dots, x_{n+1}^0)$ and $Y^0 = (y_1^0, \dots, y_{n+1}^0)$ in $\pi \cap K$ so that $\overline{X^0Y^0}$ is not perpendicular to L_1. To simplify the situation let us take $L_1 = (1, 0, \dots, 0)$. Then $x_1^0 \neq y_1^0$.

Project K, π, X^0, Y^0 along a fixed direction in the supporting hyperplane to obtain $\overline{K}$, $\overline{\pi}$, $\overline{X^0}$, $\overline{Y^0}$. Then $\overline{\pi}$ is a supporting hyperplane of $\partial\overline{K}$ and $\overline{X}^0$, $\overline{Y}^0$ are two distinct points in $\overline{\pi} \cap \partial\overline{K}$ (for $\overline{x}_1^0 = x_1^0 \neq y_1^0 = \overline{y}_1^0$). By the induction hypothesis $\operatorname{meas}(\operatorname{pr} M_1) = 0$, where pr stands for projection.

A direction of projection in the supporting hyperplane θ is perpendicular to L_1. Hence we can choose $(0, \cos\theta, \sin\theta, 0, \dots, 0)$ so that $q_2^0 \cos\theta + q_3^0 \sin\theta = 0$. When $q_2^{02} + q_3^{02} \neq 0$, θ is uniquely determined in $[0, \pi)$. When $q_2^0 = q_3^0 = 0$, θ is arbitrary. Therefore $\sup_{M_1^* \subseteq M_1} \operatorname{meas} M_1^* \leq \int_0^{2\pi} \operatorname{meas}(\operatorname{pr}_\theta M_1)\, d\theta = 0$ and so $\operatorname{meas} M_1 = 0$. Similarly $\operatorname{meas} M_i = 0$ $(2 \leq i \leq n)$. As $\sup_{M^* \subseteq \mu(N)} \operatorname{meas} M^* \leq \sum_{i=1}^{n+1} \operatorname{meas} M_i = 0$, we conclude $\operatorname{meas} \mu(N) = 0$. Notice that more precisely we have shown that the inner measure of $\mu(N)$ is zero. However, similarly one can show that its outer measure vanishes too. □

Lemma 6. *Let $\mu(M)$ be measurable, $M \subseteq K$. Then $\mu(K \backslash M)$ is also measurable.*

Proof. Since $\mu(M) \cup \mu(K\backslash M) = S_n$, $S_n \backslash \mu(M) \subseteq \mu(K\backslash M)$. Therefore

$$\begin{aligned} \mu(K\backslash M) &= \mu(K\backslash M) \cap [(S_n \backslash \mu(M)) \cup \mu(M)] \\ &= (S_n \backslash \mu(M)) \cup (\mu(K\backslash M) \cap \mu(M)). \end{aligned}$$

By Lemma 5 $\operatorname{meas}[\mu(K\backslash M) \cap \mu(M)] = 0$. Hence $\mu(K\backslash M)$ is measurable. □

Corollary 7. *$\mu(M)$ is measurable for open $M \subseteq K$.*

Proof. $\mu(K\backslash M)$ is measurable by Lemma 4, since $K\backslash M$ is closed. So by Lemma 5 $\mu(M)$ is measurable. □

Corollary 8. *$\mu(M)$ is measurable for any Borel subset M of K.*

Proof. From the definition of Borel sets it suffices to show $\mu(\bigcup_{j=1}^\infty M_j)$ and $\mu(\bigcap_{j=1}^\infty M_j)$ are measurable for measurable $\mu(M_j)$, $j = 1, 2, \dots$. But the former result follows from the identity $\mu(\bigcup_{j=1}^\infty M_j) = \bigcup_{j=1}^\infty \mu(M_j)$ whose right-hand side is measurable. For the latter $\mu(\bigcup_{j=1}^\infty (K\backslash M_j))$ is measurable since by Lemma 6 each $\mu(K\backslash M_j)$ is measurable. So using Lemma 6 once more we see that

$$\mu\left(K\setminus\bigcup_{j=1}^{\infty}(K\setminus M_j)\right)=\mu\left(\bigcap_{j=1}^{\infty}M_j\right)$$

is measurable too. □

COROLLARY 9 (ADDITIVITY OF THE MEASURE). *Let M_1 and M_2 be two disjoint Borel sets. Then*

$$\operatorname{meas}\mu(M_1\cup M_2)=\operatorname{meas}\mu(M_1)+\operatorname{meas}\mu(M_2).$$

PROOF. From $\mu(M_1\cup M_2)=\mu(M_1)\cup[\mu(M_2)\setminus(\mu(M_1)\cap\mu(M_2))]$ and the fact that $\operatorname{meas}(\mu(M_1)\cap\mu(M_2))=0$ (Lemma 5) we immediately deduce this corollary in view of Corollary 8. □

COROLLARY 10 (COMPLETE ADDITIVITY OF THE MEASURE). *Let M_j $(j=1,2,\dots)$ be mutually disjoint Borel sets. Then* $\operatorname{meas}\mu(\bigcup_{j=1}^{\infty}M_j)=\bigcup_{j=1}^{\infty}\mu(M_j)$.

PROOF. From Corollary 9

$$\operatorname{meas}\mu\left(\bigcup_{j=1}^{\infty}M_j\right)=\sum_{j=1}^{k}\operatorname{meas}\mu(M_j)+\operatorname{meas}\left(\bigcup_{j=k+1}^{\infty}M_j\right),\tag{12}$$

where $\mu(\bigcup_{j=k+1}^{\infty}M_j)$ is monotonic decreasing as k goes to ∞. Denote its limit by N. In case N is nonempty and it contains a point which has a unique preimage $p\in K$, there is j_0 such that $p\in M_{j_0}$. However, from $M_k\cap M_{j_0}=\varnothing$ $(k>j_0)$ one has $p\notin\bigcup_{j_0+1}^{\infty}M_j$ which is in contradiction to $\mu(p)\in N$. Hence every point in N has at least two preimages. By Lemma 5 $\operatorname{meas}N=0$. The lemma follows by letting $k\to\infty$ in (12). □

Lemmas 5, 6 and Corollaries 7–10 also hold for ν via the relation

$$\nu(M)=\int_M\sqrt{1+|p|^2}\mu(dx),$$

where now $\nu(M)$ $(M\subseteq\Omega)$ is the normal image of M for the convex function $z\in C(\Omega)$. Here Ω is a bounded convex domain in $\mathbb{R}^n$. Analogous results hold for unbounded domains by taking limits.

LEMMA 11. *Let $\{z^j\}$ be a sequence of convex functions satisfying $z^j(x)\to z(x)$ for all x in a convex domain Ω (Ω is not necessarily bounded). Suppose that M^j $(j=1,2,\dots)$ and M are bounded closed subsets of Ω and $M^j\to M$ $(j\to\infty)$. Then*

$$\limsup_{j\to\infty}\omega(z^j,M^j)\le\omega(z,M),\tag{13}$$

where $\omega(z,M)=\omega(M)=\operatorname{meas}\nu(M)$.

PROOF. Let G be a neighborhood of $\nu(M)$. It suffices to show that there is j_0 such that

$$\nu(z^j,M^j)\subseteq G,\quad\forall j\ge j_0.\tag{14}$$

(13) readily follows from (14) by passing to the limit. Were (14) not valid, one could find $x^j \in M^j$ such that there is a supporting hyperplane $x_{n+1} - z^j(x^j) = p^j(x - x^j)$ at $(x^j, z^j(x^j))$ whose slope $(p_1^j, \dots, p_n^j)$ does not belong to G. $\{x^j\}$ is bounded since $M^j \to M$. So there is a subsequence of $\{x^j\}$, which we denote by $\{x^j\}$ for simplicity, converging to some x_0 in M since M is closed. On the other hand, from the uniform boundedness of $\{p^j\}$ we can extract a subsequence (still denoted by $\{p^j\}$) converging to $p^0 \notin G$. Taking limit in $z^j(x) - z^j(x^j) \geq p^j(x - x^j)$ yields

$$z(x) - z(x^0) \geq p^0(x - x^0).$$

So $p^0 \in \nu(M)$—a contradiction to $p^0 \notin G$. So (14) must hold. □

LEMMA 12. *Let $\{z^j\}$ be a sequence of convex functions satisfying $z^j(x) \to z(x)$ in a convex domain Ω. Suppose G^j $(j = 1, 2, \dots)$ and G are bounded open subsets of Ω, $G^j \to G$, and $\omega(z, \partial G) = 0$. Then*

$$\lim_{j\to\infty} \omega(z^j, G^j) = \omega(z, G),$$

and

$$\lim_{j\to\infty} \omega(z^j, \partial G^j) = 0.$$

PROOF. As G^j and G are contained in a bounded region of Ω, we may take Ω to be bounded. Putting a cap to close off $K^j = \{(x, z^j(x)) \colon x \in \Omega\}$ to form a convex hypersurface (still denoted by K^j), μ^j maps K^j to S_n. Lemma 11 remains valid for this mapping. Take the complement in Lemma 11 and then return to ν. We obtain $\liminf_{j\to\infty} \omega(z^j, G^j) \geq \omega(z, G)$. On the other hand by Lemma 11 $\limsup_{j\to\infty} \omega(z^j, \overline{G}^j) \leq \omega(z, \overline{G})$. As $\omega(z, \partial G) = 0$,

$$\limsup_{j\to\infty} \omega(z^j, \overline{G}^j) \leq \omega(z, \overline{G}) = \omega(z, G) \leq \liminf_{j\to\infty} \omega(z^j, G^j).$$

The second conclusion thus follows. □

For any convex function z in a bounded convex domain Ω we may employ the usual construction to define the integral $\int_\Omega f\omega(z, dx)$, at least when near the boundary $\nu(z, \overline{\Omega}\backslash M)$ (closed $M \subseteq \Omega$) has no concentrated measure i.e., $\omega(z, \overline{\Omega}\backslash M) = 0$, or when $f \in \overset{\circ}{C}(\Omega)$ (i.e., $f \in C(\Omega)$ and has a compact support in Ω). In the latter case we may write the integral as $\int_{M_f} f(x)\omega(z, dx)$ where M_f is the support of f and $f|_{\partial M_f} = 0$.

LEMMA 13. *Let the convex sequence $z^j(x) \to z(x)$ be in Ω. Then $\omega(z^j)$ weakly converges to $\omega(z)$. In other words, for $f \in \overset{\circ}{C}(\Omega)$ or for $f \in C(\Omega)$ when $\nu(z^j, \overline{\Omega}\backslash M)$ $(M \subseteq \Omega$ closed$)$ has no concentrated measure, then*

$$\int_\Omega f\omega(z^j, dx) \to \int_\Omega f\omega(z, dx) \quad \text{as } j \to \infty.$$

PROOF. The number of x_i^0 satisfying $\omega(z, M_f \cap \{x_i = x_i^0\}) \geq 1/k$ does not exceed $k\omega(z, M_f)$. Hence there are at most countably many x_i^0 ($1 \leq i \leq n$) with $\omega(z, M_f \cap \{x_i = x_i^0\}) > 0$. Similarly, there are at most countably many x_i^0 with $\omega(z^j, M_f \cap \{x_i = x_i^0\}) > 0$ ($1 \leq i \leq n$, $j = 1, 2, \ldots$). The rectangles whose coordinates are not equal to these x_i^0's are dense in Ω. Since f is uniformly continuous, for each $\varepsilon > 0$ there corresponds $\delta > 0$ such that $|f(x) - f(\tilde{x})| \leq \varepsilon$ for $|x - \tilde{x}| \leq \delta$.

We can divide M_f into L_ε many such rectangles whose diameters do not exceed δ. Fix a point x^l ($1 \leq l \leq L_\varepsilon$) in each of these rectangles M^l.

By virtue of Lemma 12, $\omega(z^j, M^l) = \omega(z^j, \overline{M^l})$. Consequently,

$$\left|\int_\Omega f\omega(z^j, dx) - \int_\Omega f\omega(z, dx)\right|$$
$$\leq \sum_{l=1}^{L_\varepsilon} |f(x^l)|\,|\omega(z^j, M^l) - \omega(z, M^l)| + \varepsilon(\omega(z^j, M_f) + \omega(z, M_f)).$$

Notice that additivity of the integral has been used. By Lemmas 12 and 11 the right-hand side of this inequality $\leq 2\varepsilon \cdot \omega(z, M_f)$ as $j \to \infty$. Letting $\varepsilon \to 0$ we conclude the proof of this lemma. □

We also need to study the extension of convex functions.

LEMMA 14. *Let Ω be a bounded convex domain with piecewise C^1-boundary. Suppose z is a convex function satisfying a Lipschitz condition in $\overline{\Omega}$ and vanishing on $\partial\Omega$. Then z can be extended as a convex function in $\mathbb{R}^n$ so that $z > 0$ in $\mathbb{R}^n \setminus \overline{\Omega}$. Furthermore, this extension is Lipschitz continuous with Lipschitz constant a constant multiple of the original one.*

PROOF. Without loss of generality we may assume that the origin is contained in Ω. Let $d(0, \partial\Omega) = d_0$. Then $z(x)$ satisfies

$$|z(x) - z(\tilde{x})| \leq N|x - \tilde{x}|, \quad \forall x, \tilde{x} \in \overline{\Omega}.$$

Let Ox intersect $\partial\Omega$ at y. We claim that

$$z(x) = Nd_0\left(\frac{|x|}{|y|} - 1\right) \qquad (x \in \mathbb{R}^n \setminus \overline{\Omega})$$

is an admissible extension.

First we show that z is convex outside Ω (see Figure 7.1). We need to prove

$$\frac{CD}{OC} \leq \frac{AB}{OA}\frac{DF}{BF} + \frac{EF}{OE}\frac{BD}{BF}.$$

By the convexity of Ω the left-hand side of this inequality achieves its maximal value when C lies on the line segment AE. We claim

$$\frac{CD}{OC} = \frac{AB}{OA}\frac{DF}{BF} + \frac{EF}{OE}\frac{BD}{BF} \tag{15}$$

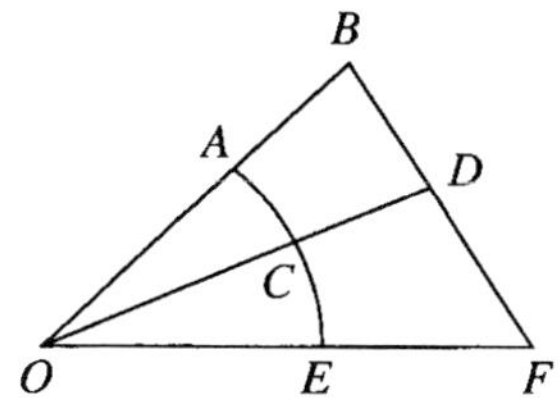

FIGURE 7.1

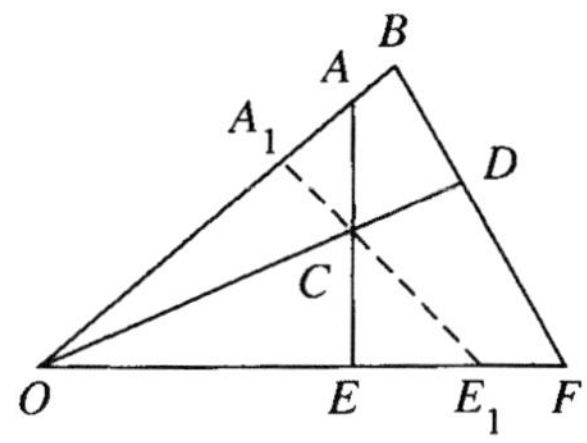

FIGURE 7.2

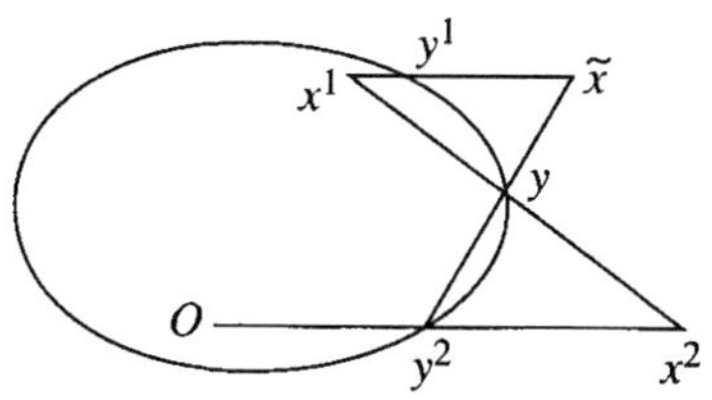

FIGURE 7.3

in this case. Adding 1 to both sides of this equality, we find that (15) is equivalent to

$$\frac{OD}{OC} = \frac{OB}{OA}\frac{DF}{BF} + \frac{OF}{OE}\frac{BD}{BF}.$$

We draw A_1CE_1 parallel to BDF (see Figure 7.2). Then $OB = OA_1 \cdot OD/OC$, $OF = OE_1 \cdot OD/OC$. Hence (15) is equivalent to

$$1 = \frac{OA_1}{OA}\frac{DF}{BF} + \frac{OE_1}{OE}\frac{BD}{BF},$$

or

$$\frac{A_1A}{OA}DF = \frac{EE_1}{OE}BD.$$

As $DF/BD = CE_1/A_1C$, (15) is finally equivalent to

$$\frac{AA_1}{OA}\frac{CE_1}{A_1C}\frac{ED}{E_1E} = -1. \tag{16}$$

(16) is nothing but the Menelaus theorem in trigonometry. Hence (15) holds.

Next we show z is convex across $\partial\Omega$ (see Figure 7.3). In other words, for $x^1 \in \Omega$, $x^2 \in \mathbb{R}^n \backslash \Omega$, $y \in \overline{x^1x^2} \cap \partial\Omega$, we want to show

$$\overline{yx^2}z(x^1) + \overline{x^1y}z(x^2) \geq 0. \tag{17}$$

Suppose $\overline{Ox^2}$ intersects $\partial\Omega$ at y^2. We draw $\overline{x^1y^1}$ parallel to $\overline{y^2x^2}$, $y^1 \in \partial\Omega$. The extended line segment of $\overline{x^1y^1}$ intersects the extended line segment of $\overline{y^2y}$ at $\tilde{x}$. We have

$$\begin{aligned}|z(x^1)| &= |z(x^1) - z(y^1)| \leq N\overline{x^1y^1} \leq N\overline{x^1\tilde{x}} \\ &= N\,\overline{y^2x^2}\,\overline{x^1y}\Big/\overline{yx^2} = \overline{x^1y}\Big/\overline{yx^2} \cdot \overline{Oy^2}z(x^2)/d_0 \\ &\leq \overline{x^1y}z(x^2)\Big/\overline{yx^2}.\end{aligned}$$

So (17) holds.

To conclude the proof we show the extension z satisfies a Lipschitz condition. It suffices to focus on a neighborhood of $(0, \dots, 0, x_n)$.

Locally $\partial\Omega$ can be described as a graph of the function $y_n = f(y_1, \dots, y_{n-1})$. Writing $|x|/|y| = r$, we have $r = x_1/y_1 = \cdots = x_{n-1}/y_{n-1} = x_n/y_n$. Thus

$$rf\left(\frac{x_1}{r}, \dots, \frac{x_{n-1}}{r}\right) - x_n \equiv g(x_1, \dots, x_n, r) = 0,$$

and

$$\frac{\partial g}{\partial r} = f\left(\frac{x_1}{r}, \dots, \frac{x_{n-1}}{r}\right) - \sum_{i=1}^{n-1} \frac{x_i}{r}\frac{\partial f}{\partial x_i} \approx f > 0.$$

By the implicit function theorem we may solve r as a piecewise C^1-function of $(x_1, \dots, x_n)$. From

$$\frac{\partial r}{\partial x_i} = -\frac{\partial g}{\partial x_i}\Big/\frac{\partial g}{\partial r} = \frac{\partial f}{\partial x_i}\Big/\left(f - \sum_{i=1}^{n-1}\frac{x_i}{r}\frac{\partial f}{\partial x_i}\right)$$

and

$$\frac{\partial r}{\partial x_n} = -1\Big/\left(f - \sum_{i=1}^{n-1}\frac{x_i}{r}\frac{\partial f}{\partial x_i}\right)$$

we see that the extension z satisfies a Lipschitz condition with Lipschitz constant a constant multiple of N. □

LEMMA 15. *Under the hypotheses of Lemma* 14, *there exist a sequence of convex functions* $\{z_k\}$ *in* $C^\infty(\mathbb{R}^n)$ *and a sequence of convex domains* $\{\Omega_k\}$, $\partial\Omega_k \in C^\infty$, $k = 1, 2, \dots$ *such that* $\Omega_{k+1} \subseteq \Omega_k$ $(\forall k = 1)$, $\Omega = \bigcap_{k=1}^\infty \Omega_k$, $z_k|_{\partial\Omega_k} = 0$, *and* $z_k(x) \to z(x)$ *as* $k \to \infty$, $\forall x \in \mathbb{R}^n$.

PROOF. Let z be the extension in $\mathbb{R}^n$ obtained by Lemma 14. We shall use induction to construct Ω_k and z_k. Suppose Ω_k and z_k have been determined and $\delta_k = d(\partial\Omega_k, \partial\Omega)$. We let z^{k+1} be a smooth regularization of z whose mollifier is supported in a ball of radius η_k. Then

$$z^{k+1}|_{\partial\Omega_k} \le Nd_0\frac{\eta_k}{\min_{\partial\Omega}|y|} \quad \text{and} \quad z^{k+1}|_{\partial\Omega_k} \ge Nd_0\frac{\delta_k - \eta_k}{\max_{\partial\Omega}|y|}.$$

Choose η_k so that

$$\left(1 + \frac{\max_{\partial\Omega}|y|}{\min_{\partial\Omega}|y|}\right)\eta_k < \delta_k,$$

and let $z_{k+1} = z^{k+1} - \sigma^{k+1}$, $\Omega_{k+1} = \{x\colon z_{k+1}(x) < 0\}$ where

$$\sigma^{k+1} = \frac{Nd_0}{2}\left(\frac{\eta_k}{\min_{\partial\Omega}|y|} + \frac{\delta_k - \eta_k}{\max_{\partial\Omega}|y|}\right).$$

It is easy to show that z_{k+1}, Ω_{k+1} satisfy our requirement. For example, to show that Ω_k is convex we observe that since z_k is convex, for all $x^0 \in \partial\Omega_k$, $p(x-x^0)+z \geq 0$. Letting $z=0$ we get $p(x-x^0) \geq 0$, that is to say, $\partial\Omega_k$ lies on one side of the supporting hyperplane. So it is convex.

To start the induction it is sufficient to take Ω_0 a ball containing Ω. □

§3. Convex envelopes

DEFINITION. For a function u in Ω or $\overline{\Omega}$, the function defined by

$$\check{u}(x) = \inf\left\{\sum \lambda_i u(x^i):\ x = \sum \lambda_i x^i,\ x^i \in \Omega \text{ or } \overline{\Omega},\ \lambda_i \geq 0,\ \sum \lambda_i = 1\right\}$$

is called the *lower convex envelope function* (or simply *lower convex envelope*) of u. It is defined in Ω_0, the convex hull of Ω (i.e., the smallest convex set containing Ω). The set $\{x:\ u(x) = \check{u}(x)\}$ is called the *lower contact set* of u.

LEMMA 16. *Let $u \in C(\overline{\Omega})$ where Ω is a bounded convex set. Then the lower convex envelope $\check{u}$ belongs to $C(\overline{\Omega})$.*

PROOF. By virtue of Lemma 3 $\check{u}$ is Lipschitz continuous in any compact subset of Ω. Hence it is continuous in Ω. □

Let Ω be a bounded convex domain. When $x^\sigma \in \Omega$ and $\check{u}(x^\sigma) < u(x^\sigma)$ the intersection of all supporting hyperplanes at $(x^\sigma, \check{u}(x^\sigma))$ is called the *supporting set* of $\check{u}$ at x^σ. The subset of Ω over which the graph of u belongs to this supporting set will be denoted by $D(x^\sigma)$.

LEMMA 17. *For $x^\sigma \in \Omega$ with $\check{u}(x^\sigma) < u(x^\sigma)$, there exist N, λ_i, and x^i $(0 \leq i \leq N)$ with $N \leq n$, $\lambda_i > 0$, $\sum_0^N \lambda_i = 1$, $x^i \in D(x^\sigma)$, $\sum_0^N \lambda_i x^i = x^\sigma$ such that $\check{u}(x^\sigma) = \sum_0^N \lambda_i u(x^i)$.*

PROOF. For $\varepsilon > 0$, $\exists M$, μ_i, X^i $(0 \leq i \leq M)$ with $\mu_i > 0$, $\sum_0^M \mu_i = 1$, $X^i \in \overline{\Omega}$, $\sum_0^M \mu_i X^i = x^\sigma$ such that $\sum_0^M \mu_i u(x^i) \leq \check{u}(x^\sigma) + \varepsilon$. When the supporting set consists of a single hyperplane π,

$$\sum_0^M \mu_i z_\pi(X^i) \leq \check{u}(x^\sigma) + \varepsilon, \tag{18}$$

where $z_\pi(x)$ is the function for π.

There must be $n+1$ many points $y^0, \dots, y^n$ in $D(x^\sigma)$ which satisfy $\sum_0^n \nu_i y^i = x^\sigma$ for some $\nu_0, \dots, \nu_n$, $\nu_i > 0$, $\sum \nu_i = 1$, otherwise the supporting hyperplane can be rotated and is not unique. Putting $u(y^i) = z_\pi(y^i)$ $(0 \leq i \leq M)$ and $\sum_0^n \nu_i z_\pi(y^i) = \sum_0^M \mu_i z_\pi(X^i)$ into (18) we have $\sum_0^n \nu_i u(y^i) \leq \check{u}(x^\sigma) + \varepsilon$. Here y^i and ν_i depend on ε. As $\varepsilon \to 0$, y^i and ν_i subconverge to x^i and λ_i respectively for $i = 0, 1, \dots, n$. Taking the limit we obtain $\sum_0^n \lambda_i u(x^i) \le \check{u}(x^\sigma)$. Combining with

$$\check{u}(x^\sigma) \le \sum_0^n \lambda_i \check{u}(x^i) \le \sum_0^n \lambda_i u(x^i)$$

the lemma follows.

In general, the supporting set π_* is the intersection of all supporting hyperplanes π. Denote the projection of π_* into x-space by $\mathbb{R}^n_*$ and write $\Omega \cap \mathbb{R}^n_*$ as Ω_*. Projecting (18) into $\mathbb{R}^n_*$ we get

$$\sum_0^M \mu_i z_{\pi_*}(X^i_*) \le \check{u}(x^\sigma) + \varepsilon,$$

where X^i_* is the projection image of X^i and $\sum_0^M \mu_i X^i_* = x^\sigma$. Replacing R^n, Ω, X^i, π in (18) by R^n_*, Ω_*, X^i_*, π_* and arguing as before we obtain $\sum_0^n \lambda_i u(x^i) = u(x^\sigma)$. □

LEMMA 18. *Let* $u \in C(\overline{\Omega})$ *where* Ω *is a bounded convex domain. The lower convex envelope* $\check{u}$ *belongs to* $C(\overline{\Omega})$.

REMARK. The following example demonstrates that $\check{u}$ is not continuous in $\overline{\Omega}$ when u is not continuous in $\overline{\Omega}$. Hence the requirement that u is continuous in $\overline{\Omega}$ is necessary.

EXAMPLE. Let $\Omega = (0, 1)$. Then $\overline{\Omega} = [0, 1]$. For the function u, $u(0) = 1$, $u(x) = 1$ for $0 < x \le 1$, $\check{u}(x) = u(x)$ in $[0, 1]$ is not continuous.

PROOF. It has been established in Lemma 16 that $\check{u} \in C(\Omega)$. So we need to show $\check{u}$ is continuous up to the boundary $\partial\Omega$. First we claim that when $\partial\Omega$ contains a flat subset ω this subset must be convex. Notice that ω is flat if for any two distinct x, z in ω there exists y in ω lying on the line segment connecting x and z. In case ω is not convex, one can find some point $w \notin \omega$ lying on the line segment passing $\overline{xyz}$. We are going to show that this is impossible.

By the convexity of $\overline{\Omega}$, $\overline{xz} \in \overline{\Omega}$ when $x, t \in \partial\Omega$. Hence $w \in \overline{\Omega}$. If $w \in \Omega$, we can draw a small ball B centered at w which is contained in Ω. Without loss of generality we may assume w lies between x and y. The cone V with vertex z and base B is contained in Ω. Obviously $y \in V$ and so $y \in \Omega$. This is in conflict with the assumption $y \in \omega$. Hence $w \in \partial\Omega$. As a result the line segment $\overline{xz} \subseteq \omega$ and ω is convex.

Now we show $\check{u} \in C(\overline{\Omega})$ by an induction on the dimension; we assume that $\check{u}$ is continuous in any closed convex subset of dimension less than n. Since $\overline{\omega}$ is such a subset, by the induction hypothesis the convex envelope $\check{u} \in C(\overline{\omega})$. For any $x_0 \in \overline{\omega}$ and $x \in \Omega$, $x \to x_0$, we have by Lemma 17 that

$$\check{u}(x) = \sum_0^n \lambda_i u(x^i), \qquad x = \sum_0^n \lambda_i x^i,$$

$$\lambda_i \geq 0 \quad (0 \leq i \leq n), \qquad \sum_0^n \lambda_i = 1.$$

As $x \to x_0$, $\{x^i\}$ and $\{\lambda_i\}$ are uniformly bounded. We can extract convergent subsequences. For simplicity assume both sequences converge. Taking the limit we get $x_0 = \sum_0^n \lambda_i^0 x_0^i$. Observe that $\lambda_i^0 = 0$ if $x_0^i \notin \overline{\varpi}$. Let us suppose that this holds for $N_0 < i \leq n$. In other words, we have

$$x_0 = \sum_0^{N_0} \lambda_i^0 x_0^i, \qquad \lambda_i^0 > 0 \quad (0 \leq i \leq N_0), \qquad \sum_0^{N_0} \lambda_i^0 = 1.$$

As u is bounded and uniformly continuous in $\overline{\Omega}$, we have

$$\lim_{x \to x_0} \check{u}(x) = \sum_0^{N_0} \lambda_i^0 u(x_0^i) = \check{u}(x_0),$$

which means, no matter along which path in Ω $x \to x_0$, we have

$$\lim_{x \to x_0} \check{u}(x) = \check{u}(x_0).$$

By induction we conclude $\check{u} \in C(\Omega \cup \varpi)$. So $\check{u} \in C(\overline{\Omega})$. □

LEMMA 19. *Let Ω be a bounded convex subset of $\mathbb{R}^n$. Suppose u is a Lipschitz continuous function in $\overline{\Omega}$ with Lipschitz constant K and it vanishes on $\partial\Omega$. Then $\check{u}$ is also Lipschitz continuous in $\overline{\Omega}$ with Lipschitz constant not more than $(n+1)K$.*

REMARK. The following example shows that the theorem ceases to be valid if the condition $u|_{\partial\Omega} = 0$ is removed.

EXAMPLE. Let $\overline{\Omega}$ $(\subseteq \mathbb{R}^2)$ be the triangle with vertices $(1, 0)$, $(-1, 0)$ and $(0, \delta)$ where δ is a small positive number. For $u(x, y) = 1 - x^2$, $(x, y) \in \overline{\Omega}$, its lower convex envelope $\check{u}(x, y) = y/\delta$, $(x, y) \in \overline{\Omega}$. The Lipschitz constant of u is 2 but the Lipschitz constant of $\check{u}$ is $1/\delta$.

PROOF. First we claim that $\forall x \in \Omega$, $|\check{u}(x)| \leq (n+1)K\, d(x, \partial\Omega)$.

By Lemma 17 there are $N \leq n$ and $x^i \in \overline{\Omega}$, $\lambda_i > 0$ $(0 \leq i \leq N)$, $\sum_0^N \lambda_i = 1$ such that

$$x = \sum_0^N \lambda_i x^i, \qquad \check{u}(x) = \sum_0^N \lambda_i u(x^i).$$

Denote the hyperplane spanned by $x^1, \ldots, x^N$ by π. The extended line segment of $\overline{x^0 x}$ intersects π at y and $\partial\Omega$ at z. Then y lies between x

and z. Therefore

$$\lambda_0 = d(x, \pi)/d(x^0, \pi) = \overline{xy}/\overline{x^0y} \le \overline{xz}/\overline{x^0z},$$

$$\lambda_0|u(x^0)|/d(x, \partial\Omega) \le K(\overline{xz}/\overline{x^0z})(d(x^0, \partial\Omega/d(x, \partial\Omega)).$$

Let $w \in \partial\Omega$ be the point with $d(x, w) = d(x, \partial\Omega)$. The extended line of $\overline{zw}$ cuts the line passing x^0 and parallel to $\overline{xw}$ at w_1. $w_1 \notin \Omega$. So $\overline{x^0w_1} \ge d(x^0, \partial\Omega)$ and

$$(\overline{xz}/\overline{x^0z})(d(x^0, \partial\Omega)/d(x, \partial\Omega)) \le (\overline{xz}/\overline{x^0z})(\overline{x^0w_1}/\overline{xw_1}) = 1,$$

which implies $\lambda_0|u(x^0)|/d(x, \partial\Omega) \le K$.

Similar estimates can be derived for $\lambda_i|u(x^i)|/d(x, \partial\Omega)$ $(1 \le i \le N)$. Therefore

$$\frac{|\check{u}(x)|}{d(x, \partial\Omega)} \le \sum_0^N \lambda_i \frac{|u(x^i)|}{d(x, \partial\Omega)} \le (n+1)K.$$

Next we claim $\check{u}$ satisfies a Lipschitz condition in $\overline{\Omega}$. Let $x^1, x^2 \in \Omega$. Suppose the extended line of $\overline{x^1x^2}$ intersects $\partial\Omega$ at x^3 and the extended line of $\overline{x^2x^1}$ intersects $\partial\Omega$ at x^0. By the convexity of $\check{u}$

$$\begin{aligned}\left|\frac{\check{u}(x^2) - \check{u}(x^1)}{x^2 - x^1}\right| &\le \max\left\{\frac{|\check{u}(x^3) - \check{u}(x^2)|}{|x^3 - x^2|}, \frac{|\check{u}(x^1) - \check{u}(x^0)|}{|x^1 - x^0|}\right\} \\ &= \max\left\{\frac{|\check{u}(x^2)|}{|x^3 - x^2|}, \frac{|\check{u}(x^1)|}{|x^1 - x^0|}\right\} \\ &\le (n+1)K \max\left\{\frac{d(x^2, \partial\Omega)}{|x^3 - x^2|}, \frac{d(x^1, \partial\Omega)}{|x^1 - x^0|}\right\} \\ &\le (n+1)K.\end{aligned}$$

Clearly this estimate remains valid when one of x^1 and x^2 belongs to $\partial\Omega$. The proof of this lemma is complete. □

Let Ω be a bounded convex subset of $\mathbb{R}^n$ and $Q = \Omega \times (0, T)$. For $u \in C(\overline{Q})$ let

$$\tilde{u}(x, t) = \inf_{0 \le \tau \le t} u(x, \tau), \quad \forall (x, t) \in \overline{Q},$$

and let $\check{u}$ be the lower convex envelope of $\check{u}$.

LEMMA 20. *For $u \in C(Q)$, $\check{u}$ is nonincreasing in t and $\tilde{u}$, $\check{u} \in C(\overline{Q})$. When u is Lipschitz continuous in $\overline{Q}$ with Lipschitz constant K, both $\tilde{u}$ and $\check{u}$ are Lipschitz continuous in $\overline{Q}$. Moreover, the Lipschitz constants of $\tilde{u}$ in x, $\tilde{u}$ and $\check{u}$ in t, do not exceed K. When $u|_{\partial\Omega\times[0,T]} = 0$, the Lipschitz constant of $\check{u}$ in x does not exceed $(n+1)K$.*

PROOF. It is easy to show that $\tilde{u} \in C(\overline{Q})$ if $u \in C(\overline{Q})$. It follows from Lemma 18 that $\breve{u}(\cdot, t) \in C(\overline{\Omega})$ for all $t \in [0, T]$. The continuity of $\breve{u}$ in t is not hard to establish. For $x \in \Omega$, $t_1 \le t_2$, by Lemma 17 there exist $x^i \in \overline{\Omega}$ and $\lambda_i \ge 0$ $(0 \le i \le n)$, $\sum_0^n \lambda_i = 1$ such that $x = \sum_0^n \lambda_i x^i$, $\breve{u}(x, t_1) = \sum_0^n \lambda_i \tilde{u}(x^i, t_1)$. Therefore

$$\breve{u}(x, t_1) = \sum_0^n \lambda_i \tilde{u}(x^i, t_1) \ge \sum_0^n \lambda_i \tilde{u}(x^i, t_2) \ge \breve{u}(x, t_2).$$

It is easy to see that $\tilde{u}$ satisfies a Lipschitz condition in x and t with a Lipschitz constant not exceeding the Lipschitz constant K of u. On the other hand, for $t_1 \le t_2$, by Lemma 17 we can find $x^i \in \overline{\Omega}$ and $\mu_i \ge 0$ $(0 \le i \le n)$, $\sum_0^n \mu_i = 1$ such that $x = \sum_0^n \mu_i x^i$, and $\breve{u}(x, t_2) = \sum_0^n \mu_i \tilde{u}(x^i, t_2)$. Therefore

$$0 \ge \breve{u}(x, t_2) - \breve{u}(x, t_1) \ge \sum_0^n \mu_i \tilde{u}(x^i, t_2) - \sum_0^n \mu_i \tilde{u}(x^i, t_1) \ge -K(t_2 - t_1).$$

When $u|_{\partial\Omega \times [0, T]} = 0$, we can use Lemma 19 to conclude that $\breve{u}$ satisfies a Lipschitz condition in x with a Lipschitz constant not exceeding $(n+1)K$. □

LEMMA 21. *Let Ω be convex and $Q = \Omega \times (0, T)$. Let u be Lipschitz continuous in $\overline{Q}$ with constant K. Then $\partial \breve{u}(x, t)/\partial t$ exists almost everywhere in $\overline{Q}$ and it satisfies $0 \ge \partial \breve{u}/\partial t \ge -K$.*

PROOF. By Lemma 20, for each $x \in \overline{\Omega}$, $\breve{u}$ is Lipschitz continuous in t with Lipschitz constant not more than K. Hence it is absolutely continuous with respect to t. So $\partial \breve{u}/\partial t$ exists almost everywhere and $0 \ge \partial \breve{u}/\partial t \ge -K$. □

§4. Several Aleksandrov maximum principles

THEOREM 22. *Let Ω be a bounded domain with piecewise C^1-boundary and let Ω_0 be a convex domain containing Ω. Suppose $u \in C^{2,1}(\overline{Q})$, $Q = \Omega_0 \times (0, T]$, $u|_{\partial^* Q} \ge 0$ and $u(x, t)$ attains its negative minimum at (x^0, T). Then*

$$(-u(x^0, T))^{n+1} \le \frac{(n+1)R^n}{\kappa_n} \int_{Q_{u, \Omega_0}} -u_t \det(u_{ij})\, dx\, dt,$$

where R is the diameter of Ω_0, κ_n is the volume of the n-dimensional unit ball and Q_{u,Ω_0} is the negative contact set of u in Q_0, i.e.,

$$Q_{u,\Omega_0} = \{(x, t)\colon Q \cap \{\breve{v} = v\}\}, \qquad v = \min\{0, u\}.$$

PROOF. v is not $C^{2,1}$ in $\overline{\{v < 0\}} \backslash \{v = 0\}$. Rewriting v as u, we have

$$\tilde{u}|_{\partial^* Q} = \breve{u}|_{\partial^* Q} = \tilde{u}|_{\partial^* Q_0} = \breve{u}|_{\partial^* Q_0} = 0.$$

First we extend $\check{u}$ to $\Omega_0 \times \mathbb{R}$ by setting $\check{u} = 0$ $(t < 0, x \in \overline{\Omega}_0)$ and $\check{u} = \check{u}(x, T)$ $(t > T, x \in \overline{\Omega}_0)$. Since $\check{u}$ is Lipschitz continuous in (x, t) and $\check{u}|_{\partial\Omega_0 \times \mathbb{R}} = 0$, the extended function is still convex in x and nonincreasing in t. Next we extend $\check{u}$ as a Lipschitz continuous function in $\mathbb{R}^{n+1}$ using Lemma 14. The resulting $\check{u}$ is independent of t outside $\Omega_0 \times \mathbb{R}$. Furthermore, its Lipschitz constant does not exceed cK where c is a constant only depending on $\partial\Omega_0$ and K is the Lipschitz constant of u in $\overline{Q}$. Now we consider smooth approximations of $\check{u}$ which are constructed in x and t separately, although they have the same radius of regularization η_k. As in Lemma 15 we obtain a smooth sequence $\{\check{u}_k\}$ of which each element is convex in x and nonincreasing in t and a sequence of smooth domains $\{\Omega_k\}$ (noticing that $\check{u}$ is independent of t outside $\Omega_0 \times \mathbb{R}$) such that $\check{u}_k$ is independent of k outside $\Omega_0 \times \mathbb{R}$, $\partial\Omega_k = \{x\colon \check{u}_k(x, t) \equiv \check{u}_k(x) = 0\}$, $\Omega_k \supseteq \Omega_{k+1}$ $(k = 1, 2, \dots)$, $\bigcap_{k=1}^{\infty} \Omega_k = \Omega_0$, $\check{u}_k(x, t)$ converges uniformly to $\check{u}(x, t)$ in every compact subset of $\mathbb{R}^{n+1}$, $\partial\check{u}_k/\partial t$ converges to $\partial\check{u}/\partial t$ in $\mathbb{R}^{n+1}$ almost everywhere and $|\partial\check{u}_k(x, t)/\partial t| \le cK$. Writing $Q^k = \Omega_k \times (0, T]$, we have by Theorem 1 that

$$
\begin{aligned}
[-\check{u}_k(x^0, T)]^{n+1} &\le \frac{(n+1)R_k^n}{\kappa_n} \int_{Q_k} \frac{-\partial\check{u}_k(x, t)}{\partial t} \det\left(\frac{\partial^2\check{u}_k}{\partial x_i \partial x_j}\right) dx\, dt \\
&= \frac{(n+1)R_k^n}{\kappa_n} \int_{Q^k} \omega(\check{u}_k, dx)\tilde{\omega}(\check{u}_k, dt), \qquad (19)
\end{aligned}
$$

where R_k is the diameter of Ω_k, $\omega(\check{u}_k, dx) = \det(\partial^2\check{u}_k/\partial x_i\partial x_j)\, dx$ and $\tilde{\omega}(\check{u}_k, dt) = -(\partial\check{u}_k/\partial t)\, dt$. For $k \ge 3$,

$$
\begin{aligned}
&\int_{Q_k} \omega(\check{u}_k, dx)\tilde{\omega}(\check{u}_k, dt) - \int_{Q_0} \omega(\check{u}, dx)\tilde{\omega}(\check{u}, dt) \\
&= \int_{Q^1} \rho(x)(\omega(\check{u}_k, dx) - \omega(\check{u}, dx))\tilde{\omega}(\check{u}_k, dt) \\
&\quad + \int_{Q^1} \rho(x)(\omega(\check{u}, dx) - \omega(\check{u}_m, dx))(\tilde{\omega}(\check{u}_k, dt) - \tilde{\omega}(\check{u}, dt)) \\
&\quad + \int_{Q^1} \rho(x) \det(\check{u}_{m_{ij}})\, dx(\tilde{\omega}(\check{u}_k, dt) - \tilde{\omega}(\check{u}, dt)) \\
&= \mathrm{I} + \mathrm{II} + \mathrm{III} \qquad (20)
\end{aligned}
$$

where $\rho \in C(\overline{\Omega}_1)$, $= 1$ for $x \in \overline{\Omega}_3$, $= 0$ for $x \in \overline{\Omega}_1 \backslash \Omega_2$ and $0 < \rho < 1$ for $x \in \Omega_2 \backslash \overline{\Omega}_3$.

$$
\begin{aligned}
|I| &\le cK \int_0^T \left| \int_{\Omega_1} \rho(x)(\omega(\check{u}_k, dx) - \omega(\check{u}, dx)) \right| dt \\
&\le cKT \max_t \left| \int_{\Omega_1} \rho(x)(\omega(\check{u}_k, dx) - \omega(\check{u}, dx)) \right| \\
&= cKT \left| \int_{\Omega_1} \rho(x)\omega(\check{u}_k(x, t^*), dx) - \int_{\Omega_1} \rho(x)\omega(\check{u}(x, t^*), dx) \right|
\end{aligned}
$$

(t^* is where $\max_t$ is attained). We claim $\lim_{k\to\infty} I = 0$. Suppose on the contrary there are $\delta > 0$, k_i and t_i^* such that $k_i \to \infty$ and

$$
\left| \int_{\Omega_1} \rho(x)[\omega(\check{u}_{k_i}(x, t_i^*), dx) - \omega(\check{u}(x, t_i^*), dx)] \right| \ge \delta,
$$

for $i = 1, 2, \ldots$. Since $0 \le t_i^* \le T$, $\{t_i^*\}$ has a subsequence converging to t_0. For simplicity let us assume $t_i^* \to t_0$. In view of Lemma 13

$$
\int_{\Omega_1} \rho(x)\omega(\check{u}(x, t_i^*), dx) \to \int_{\Omega_1} \rho(x)\omega(\check{u}(x, t_0), dx);
$$

and

$$
\int_{\Omega_1} \rho(x)\omega(\check{u}_{k_i}(x, t_i^*), dx) \to \int_{\Omega_1} \rho(x)\omega(\check{u}(x, t_0), dx).
$$

This is a contradiction. So one must have $\lim_{k\to\infty} I = 0$.

From

$$
|\,\mathrm{II}\,| \le 2cK \int_0^T \left| \int_{Q^1} \rho(x)(\omega(\check{u}, dx) - \omega(\check{u}_m, dx)) \right| dt
$$

one can see that $|\,\mathrm{II}\,|$ admits an upper bound uniformly in k. Arguing as before one concludes $\lim_{m\to\infty} \mathrm{II} = 0$.

We perform integration by parts in t in III. For fixed m, it is not hard to get $\mathrm{III} = O(\max_{\overline{Q}'} |\widetilde{u}_k - \check{u}|)$. Therefore, first letting $k \to \infty$ and then $m \to \infty$ we obtain $\mathrm{I} + \mathrm{II} + \mathrm{III} \to 0$, or,

$$
\begin{aligned}
\lim_{k\to\infty} \int_{Q_k} (\widetilde{u}_k, dx)\widetilde{\omega}(\widetilde{u}_k, dt) &= \int_{Q_0} \omega(\check{u}, dx)\widetilde{\omega}(\check{u}, dt) \\
&= \int_{Q_1+Q_2+Q_3} \omega(\check{u}, dx)\widetilde{\omega}(\check{u}, dt).
\end{aligned}
$$

Here

$$
\begin{aligned}
Q_1 &= \{(x, t) \in Q_0 \colon \check{u} < \widetilde{u}\}, \\
Q_2 &= \{(x, t) \in Q_0 \colon \check{u} = \widetilde{u} < u\}, \\
Q_3 &= \{(x, t) \in Q_0 \colon \check{u} = \widetilde{u} = u < 0\}.
\end{aligned}
$$

We have $\omega(\check{u}, dx) = 0$ in Q_1, $\widetilde{\omega}(\check{u}, dt) = 0$ in Q_2, and $\widetilde{\omega}(\check{u}, dt) = -(\partial u/\partial t)\, dt$, $\omega(\check{u}, dx) = \det(u_{ij})\, dx$ in Q_3. Notice that the above integration over Q_0 is equal to the one over $Q_1 + Q_2 + Q_3$ because $\widetilde{\omega}(\check{u}, dt)$

has no concentrated measure and hence concentrated measure, if it exists, is attributed to $\omega(\check{u}, dx)$. However, u is continuously differentiable on the portions of $\partial Q_1 \cap \partial Q_2$, $\partial Q_1 \cap \partial Q_3$, $\overline{\{u<0\}} \backslash \{u < 0\}$ in Q, and, if $\omega(\check{u}, dx)$ has concentrated measure in $\partial Q_2 \cap \partial Q_3$, then it has concentrated measure in t since $(\partial Q_2 \cap \partial Q_3) \cap Q$ forms a countable set in the t-direction. Therefore, both $\omega(\check{u}, dx)$ and $\tilde{\omega}(\check{u}, dt)$ have no concentrated measure in $\partial Q_2 \cap \partial Q_3$. Letting $k \to \infty$ in (19) we obtain

$$[-u(x^0, T)]^{n+1} \le \frac{(n+1)R^n}{\kappa_n} \int_{Q_3} -u_t \det(u_{ij})\, dx\, dt. \quad \square$$

THEOREM 23. *Let Ω be a bounded domain whose boundary is piecewise C^1 and let Ω_0 be any convex domain containing Ω. For $u \in C(\overline{Q}) \cap W^{2,1}_{n+1,\mathrm{loc}}(Q)$, $Q = \Omega \times (0, T]$ and $0 \le t^0 \le T$,*

$$\inf_{Q \cap \{t \le t^0\}} u \ge \inf_{\partial^* Q \cap \{t \le t^0\}} u - \left[\frac{(n+1)R^n}{\kappa_n} \int_q -u_t \det(u_{ij})\, dx\, dt\right]^{1/(n+1)},$$

where $q = Q_{u - \inf_{\partial^* Q} u, \Omega_0} \cap \{t \le t^0\}$, $Q_{u - \inf_{\partial^* Q} u, \Omega_0}$ is the negative contact set of $v = u - \inf_{\partial^* Q} u$ in $Q_0 = \Omega_0 \times (0, T]$ and R is the diameter of Q_0.

PROOF. First we extend u to $\overline{Q}_0$ so that $u \ge \inf_{\partial^* Q} u$ in $\overline{Q}_0 \backslash Q$. By setting

$$u(x, t) = \begin{cases} u(x, 0), & t < 0, \\ u(x, T), & t > T, \end{cases}$$

we obtain an extension of u in $\overline{\Omega}_0 \times \mathbb{R}$. Finally we further extend u to be a continuous function in $\mathbb{R}^{n+1}$. Let $\{u_m\}$ $(m = 1, 2, \dots)$ be a sequence of smooth regularizations of u which converges to u in $W^{2,1}_{n+1}(Q_\delta) \cap C(\overline{Q}_\delta)$ for each $\overline{Q}_\delta \subseteq Q$.

We construct Ω_l $(l = 1, 2, \dots)$ so that it satisfies: Each Ω_l is convex, $\partial\Omega_l \in C^\infty$, $\Omega_{l+1} \subseteq \Omega_l$, $\bigcup_{l=1}^\infty \Omega_l = \Omega_0$, and $d(\Omega_l, \partial\Omega) \ge 1/l$. The construction of Ω_l can be found in Lemmas 14 and 15. We have $u_m \to u$ in $W^{2,1}_{n+1}(Q^l) \cap C(\overline{Q}^l)$ where $Q^l = \Omega_l \times (1/l, T]$. Writing

$$v^l = u - \inf_{Q \backslash Q^l} u, \qquad v^l_m = v^l + \|u_m - u\|_{C(Q^l)} + u_m - u,$$

clearly $v^l_m|_{\partial^* Q^l} \ge 0$. By Theorem 22

$$\inf_{Q^l} v^l_m \ge -\left[\frac{(n+1)R^n}{\kappa_n} \int_{Q_{v^l_m}} -u_{mt} \det(u_{mij})\, dx\, dt\right]^{1/(n+1)}.$$

Using Lemma 18 it is easy to show that $\lim_{m\to\infty} Q_{v^l_m} = Q_{v^l}$. Therefore, for any $\varepsilon > 0$, $Q_{v^l_m} \subseteq \{(x, t) \in Q\colon d\{(x, t), Q_{v^l}\} \le \varepsilon\} \equiv Q^\varepsilon_{v^l}$ for large m.

Consequently

$$\inf_{Q^l} v_m^l \geq -\left[\frac{(n+1)R^n}{\kappa_n}\int_{Q^\varepsilon_{v^l}} -u_{mt}\det(u_{mij})\,dx\,dt\right]^{1/(n+1)}.$$

First letting $m \to \infty$ and then $\varepsilon \to 0$

$$\begin{aligned}\inf_{Q^l} v^l &\geq -\left[\frac{(n+1)R^n}{\kappa_n}\int_{Q_{v^l,\Omega_0}} -u_t\det(u_{ij})\,dx\,dt\right]^{1/(n+1)}\\ &\geq -\left[\frac{(n+1)R^n}{\kappa_n}\int_{Q_{v,\Omega_0}} -u_t\det(u_{ij})\,dx\,dt\right]^{1/(n+1)}.\end{aligned}$$

Notice that for the last inequality we have used $Q_{v^l,\Omega_0} \subseteq Q_{v,\Omega_0}$. Letting $l \to \infty$ and replacing Q by $Q \cap \{t \leq t^0\}$ we conclude the proof of this theorem. □

In the following we prove several versions of Aleksandrov maximum principle.

THEOREM 24. *Let Ω be a bounded domain with piecewise C^1-boundary and let Ω_0 be any bounded convex domain containing Ω. Consider the operator*

$$\mathscr{L}_0 u = a_{ij}(x, t)u_{ij} - u_t, \qquad (x, t) \in Q = \Omega_0 \times (0, T],$$

where (a_{ij}) is symmetric and positive semidefinite in Q, $a_{ij} \in L^\infty(Q)$, $1 \leq i, j \leq n$, and $\det(a_{ij}) \geq \mu_1 > 0$. For $u \in W^{2,1}_{n+1,\mathrm{loc}}(Q) \cap C(\overline{Q})$,

$$\inf_{Q\cap\{t\leq t^0\}} u(x, t) \geq \inf_{\partial^* Q\cap\{t\leq t^0\}} u - \left[\frac{R^n}{(n+1)\kappa_n}\int_q \frac{(\mathscr{L}_0 u)_+^{n+1}}{\det(a_{ij})}\,dx\,dt\right]^{n+1}$$

for all $0 \leq t^0 \leq T$. Here R is the diameter for Ω_0 and $q = Q_{u-\inf_{\partial^* Q} u,\Omega_0} \cap \{t \leq t^0\}$.

PROOF. From $a_{ij} \in L^\infty(Q)$, $\det(a_{ij}) \geq \mu_1$ and $u \in W^{2,1}_{n+1,\mathrm{loc}}(Q)$ one has

$$\frac{(\mathscr{L}_0 u)_+}{\det^{1/(n+1)}(a_{ij})} \in L^{n+1}_{\mathrm{loc}}(Q).$$

The theorem is trivial if $\|(\mathscr{L}_0 u)_+/\det^{1/(n+1)}(a_{ij})\|_{L^{n+1}(Q)} = \infty$. Hence we may always assume

$$\|(\mathscr{L}_0 u)_+/\det^{1/(n+1)}(a_{ij})\|_{L^{n+1}(Q)} < \infty.$$

From the elementary inequality asserting that the arithmetic mean is greater than or equal to the geometric mean we know

$$\mathscr{L}_0 u \geq (n+1)[\det(a_{ij})(-u_t)\det(u_{ij})]^{1/(n+1)} \tag{23}$$

on $Q_{u-\inf_{\partial^* Q} u, \Omega_0}$. Thus on the very subset

$$-u_t \det(u_{ij}) \le \frac{1}{(n+1)^{n+1}} \frac{(\mathscr{L}_0 u)_+^{n+1}}{\det(a_{ij})}.$$

The theorem now follows from Theorem 23. □

Next we deal with more general linear parabolic operators with measurable coefficients:

$$\mathscr{L} u = a_{ij}(x, t) u_{ij} + b_i(x, t) u_i + c(x, t) u - u_t.$$

THEOREM 25. *Suppose in addition to the assumptions of Theorem 24 that* $b \in L^{n+2+\delta}(Q)$ $(\delta > 0)$, $c \in L^{n+1}(Q)$, $c \le 0$ *a.e. in* Q, *and there exists a positive constant* N *such that* $|b| \le -NRc$ *a.e. in* Q ($|b| = (\sum b_i^2)^{1/2}$ *and* R *is the diameter of* Ω). *For* $u \in W^{2,1}_{n+1, \mathrm{loc}}(Q) \cap C(\overline{Q})$,

$$\inf_{Q \cap \{t \le t^0\}} u(x, t) \ge - \sup_{\partial^* Q \cap \{t \le t^0\}} u_- - \left[\frac{(N+1)R}{(n+1)\kappa_n^{1/n}}\right]^{n/(n+1)} \left[\int_q \frac{(\mathscr{L} u)_+^{n+1}}{\det(a_{ij})} \, dx \, dt\right]^{1/(n+1)}.$$

Here the origin is assumed to be in Ω,

$$u_- = -\min\{u, 0\}, \quad q = Q_{u + \sup_{\partial^* Q} u_-, B_{(N+1)R}} \cap \{t \le t^0\},$$

and $B_{(N+1)R}$ *is a ball with radius* $(N+1)R$ *centered at the origin.*

REMARK. Writing $v = u + \sup_{\partial^* Q} u_-$, $Q_{v, B_{(N+1)R}}$ is the negative contact set of v in $B_{(N+1)R} \times [0, T]$. The negative contact set $\Gamma_{v, B_{(N+1)R}}$ of this v ($v|_{\partial B_{(N+1)R} \times [0,T]} = 0$) is contained in the negative contact set Γ_v of the function v ($v|_{\partial \Omega_0 \times [0,T]} = 0$). However, the coefficient of the last term in the Aleksandrov maximum principle becomes larger. These two negative contact sets are always contained in Q.

PROOF. From an embedding theorem $u_i \in L^{(n+1)(n+2)-\delta}_{\mathrm{loc}}(Q)$ and so $(\mathscr{L} u)_+ \in L^{n+1}_{\mathrm{loc}}(Q)$. Without loss of generality we assume

$$\|(\mathscr{L} u)_+ / \det{}^{1/(n+1)}(a_{ij})\|_{L^{n+1}(Q)} < \infty.$$

Applying Theorem 23 to the function $v = u + \sup_{\partial^* Q} u_-$ yields

$$\inf_Q u \ge - \inf_{\partial^* Q} u_- - \left[\frac{(N+1)^n R^n (n+1)}{\kappa_n} \int_{Q_{v, B_{(N+1)R}}} -u_t \det(u_{ij}) \, dx \, dt\right]^{1/(n+1)}. \tag{24}$$

For $(y, t) \in Q_{v, B_{(N+1)R}}$,

$$v(y, t) + \nabla u(y, t)(x - y) \le v(x, t) \le 0, \quad \forall x \in B_{(N+1)R}, \; t \in [0, T].$$

Picking $x = y + NR\nabla u(y, t)/|\nabla u(y, t)| \in B_{(N+1)R}$, from this inequality we have

$$u(y, t) + NR|\nabla u(y, t)| \le 0, \quad \forall (y, t) \in Q_{v, B_{(N+1)R}}.$$

As a result $b_i u_i + cu \ge c(u + NR|\nabla u|) \ge 0$ a.e. in $Q_{v, B_{(N+1)R}}$. By virtue of (23)

$$(n+1)[\det(a_{ij})(-u_t)\det(u_{ij})]^{1/(n+1)} \le \mathscr{L}_0 u \le \mathscr{L} u$$

a.e. in $Q_{v, B_{(N+1)R}}$. Putting this estimate back to (24) the desired result follows. □

We have the following Aleksandrov maximum principle for quasilinear equations.

THEOREM 26. *Let Ω be a bounded domain in $\mathbb{R}^n$ with piecewise C^1-boundary and $Q = \Omega \times (0, T]$. Suppose $u \in W^{2,1}_{n+1}(Q)$ satisfies*

$$\mathscr{L} u = a_{ij}(x, t, u, Du)u_{ij} - u_t - B(x, t, u, Du) \le 0,$$

($Du = D_x u$) where the symmetric positive definite matrix (a_{ij}) satisfies $\det(a_{ij}) \ge \mu_1 > 0$ for some constant μ_1, $a_{ij}(x, t, u, Du) \in L^\infty(Q)$, and $B(x, t, u, Du) \in L^{n+1}(Q)$. Then

$$u(x, t^0) \ge - \sup_{\partial^* Q \cap \{t \le t^0\}} u_- - \left[\frac{R}{(n+1)\kappa_n^{1/n}}\right]^{n/(n+1)} \left[\int_{Q \cap \{t \le t^0\}} \frac{B_+^{n+1}}{\det(a_{ij})}\, dx\, dt\right]^{1/(n+1)}$$

PROOF. Let $\{u^k\}$ be a sequence of smooth approximations of u. We have

$$\begin{aligned} \mathscr{L} u^k &= a_{ij}(x, t, u, Du)u^k_{ij} - u^k_t \\ &\le B(x, t, u, Du) + a_{ij}(x, t, u, Du)(u^k - u)_{ij} - (u^k - u)_t. \end{aligned}$$

Using $N = 0$ in Theorem 25 and letting $k \to \infty$ the desired inequality follows. □

§5. Bony maximum principles

The Bony maximum principle can be deduced from the Aleksandrov maximum principle. For elliptic equations the Bony maximum principle was discussed in [37] and [38]. The Bony maximum principle presented below is the parabolic version of the original one.

LEMMA 27. *Let Ω be a bounded domain in $\mathbb{R}^n$ and $Q = \Omega \times (0, T]$. Suppose $u \in W^{2,1}_{n+1,\text{loc}}(Q)$ attains a strict local minimum at (x^0, T), $x^0 \in \Omega$. Then $\exists \delta_0 > 0$ and a positive function $\varepsilon(\tau)$, $\tau \in (0, \delta_0]$ satisfying $\varepsilon(\tau) \to 0$ (when $\tau \to 0$) such that the set*

$$\begin{aligned} A_\delta = \{(y, \eta) \in Q_\delta(x^0, T) \cap Q\colon\ & u_t(y, \eta) \le 0,\ (u_{ij})(y, \eta) \ge 0, \\ & D_x u \text{ exists at } (y, \eta) \text{ and } |D_x u(y, \eta)| \le \varepsilon(\delta)\} \end{aligned}$$

has positive measure for $\forall \delta \in (0, \delta_0]$.

PROOF. From an embedding theorem $u \in C(Q)$. As u attains a strict local minimum at (x^0, T), we can find $\delta_0 > 0$ such that $\delta_0 < \frac{1}{2}d(x^0, \partial\Omega)$ and

$$\frac{2}{d(x^0, \partial\Omega)}\left(\inf_{\partial^* Q_\delta(x^0, T)} u - u(x^0, T)\right) \equiv \varepsilon(\delta) > 0, \quad \forall \delta \in (0, \delta_0],$$

where

$$Q_\delta(x^0, t^0) = \{(x, t): |x - x^0| < \delta, t^0 - \delta^2 < t \le t^0\}.$$

Clearly $\varepsilon(\delta) \to 0$ as $\delta \to 0$. Write $v(x, t) = u(x, t) - \inf_{\partial^* Q_\delta(x^0, T)} u$, and

$$Q^\delta = \Omega_0 \times (T - \delta^2, T], \qquad \Omega_0 = B_{\delta + \frac{1}{2}d(x^0, \partial\Omega)}(x^0).$$

The negative contact set Q_{v, Ω_0} of $v(x, t)$ in Q^δ contains a small neighborhood of x^0. The following situations hold in Q_{v, Ω_0}: v satisfies a uniform Lipschitz condition, and hence $D_x v$ exists a.e.; v is monotonic in t, and hence v_t exists a.e., $v_t \le 0$. The directional derivatives of v in x always exist and are monotonic. Therefore $D_x^2 v$ exists a.e. and $v_{ij}\xi_i\xi_j \ge 0$, $\forall \xi \in \mathbb{R}^n$. In other words, $(v_{ij}) \ge 0$ a.e.. Finally, since $Q_{v, \Omega_0} \subseteq Q_\delta(x^0, T)$ and $|v(x, t)|$ has $\varepsilon(\delta)/d(x^0, \partial\Omega)$ as its upper bound, analogous to the proof of Lemma 3 we obtain

$$|D_x v(y, t)| = |D_x u(y, t)| \le \frac{\varepsilon(\delta) d(x^0, \partial\Omega)}{2d(y, \partial\Omega_0)} \le \varepsilon(\delta)$$

for a.e. $(y, t) \in Q_{v, \Omega_0}$. As a result $\operatorname{meas}(Q_{v, \Omega_0} \setminus A_\delta) = 0$. By Theorem 23 we have

$$\begin{aligned} 0 &< (\varepsilon(\delta)/d(x^0, \partial\Omega))^{n+1} = (-v(x^0, T))^{n+1} \\ &\le \frac{(n+1)(\delta + (1/2)d(x^0, \partial\Omega))^n}{\kappa_n} \int_{Q_{v, \Omega_0}} -u_t \det u_{ij}\, dx\, dt. \end{aligned}$$

Hence $\operatorname{meas} Q_{v, \Omega_0} > 0$,

$$\operatorname{meas}(Q_{v, \Omega_0} \cap A_\delta) = \operatorname{meas} Q_{v, \Omega_0} - \operatorname{meas}(Q_{v, \Omega_0} \setminus A_\delta) > 0,$$

and $\operatorname{meas} A_\delta > 0$. □

THEOREM 28 (BONY MAXIMUM PRINCIPLE FOR PARABOLIC EQUATIONS). *Let $u \in W^{2,1}_{n+1, \mathrm{loc}}(Q)$ and*

$$\mathscr{L}u = a_{ij}(x, t)u_{ij} + b_i(x, t)u_i + c(x, t)u - u_t,$$

where a_{ij}, b_i, c are bounded measurable, $(a_{ij}) \ge 0$ and $c \le 0$. Suppose u attains a nonpositive minimum at (x^0, t^0). Then

$$\lim_{(y, \eta) \to (x^0, t^0)} \operatorname{ess\,inf} \mathscr{L}u \ge 0.$$

PROOF. Let $v(x, t) = u(x, t) - (|x - x^0|^4 + |t - t^0|^2)$. Then v attains a strict local minimum at (x^0, t^0). By Lemma 27

$$\lim_{(y, \eta)\to(x^0, t^0)} \operatorname{ess\,inf} |\nabla u(y, \eta)| = 0, \qquad \lim_{(y, \eta)\to(x^0, t^0)} \operatorname{ess\,inf} \frac{\partial u}{\partial t}(y, \eta) \leq 0,$$

and

$$\lim_{(y, \eta)\to(x^0, t^0)} \operatorname{ess\,inf} \lambda(y, \eta) \geq 0.$$

Here $\lambda(y, \eta)$ is the smallest eigenvalue of the matrix $(u_{ij}(y, \eta))$. Now the theorem follows from these inequalities and our assumptions on the coefficients of the operator. □

CHAPTER VIII

The Density Theorem and Its Applications

The Harnack inequality and the Hölder estimates play basic roles in the study for quasilinear parabolic and elliptic equations. For equations whose principal parts are in divergence form effective methods deriving them include Ladyzhenskja's estimate (including de Giorgi's) and Moser's estimate. Both estimates have been covered in [49] and the first part of this book.

For quasilinear or fully nonlinear equations (parabolic and elliptic) with nondivergence principal parts an effective method is the density theorem for some differential inequality with measurable coefficients. From it one can swiftly derive Harnack's inequality and the relevant Hölder estimates.

Literature on the density theorem for parabolic and elliptic equations includes [39], [40], and [41].

§1. The statement of the density theorem

Let Ω be a bounded domain in $\mathbb{R}^n$, $0 < T < T^* < \infty$, $Q = \Omega \times (0, T]$, and $Q^* = \Omega \times (0, T^*]$. Consider the linear parabolic operator with bounded measurable coefficients

$$\mathscr{L}u = a_{ij}(x, t)u_{ij} + b_i(x, t)u_i + c(x, t)u - u_t.$$

It is assumed that the coefficients of $\mathscr{L}$ satisfy

$$\lambda|\xi|^2 \le a_{ij}\xi_i\xi_j \le \Lambda|\xi|^2, \quad \forall\xi \in \mathbb{R}^n, \; 0 < \lambda \le \Lambda < \infty,$$
$$\left(\sum b_i^2\right)^{1/2}, \qquad |c| \le \Lambda$$

in Q^*. Our aim is to establish the following property of $\mathscr{L}$.

THEOREM 1. *Let $u \in W^{2,1}_{n+1}(Q^*)$, $u \ge 0$, $\mathscr{L}u \le 0$ almost everywhere in Q^*. Suppose for some $\beta \in (0, 1)$*

$$|\{(x, t) \in Q : u(x, t) \ge 1\}| \ge \beta|Q|, \tag{1}$$

where $|A|$ is the measure of the set A. For $\delta > 0$, denote $\{x \in \Omega : d(x, \partial\Omega) > \delta\}$ by Ω_δ. Then

$$\min_{\substack{x\in\overline{\Omega}_\delta \\ (1-\theta\beta)T\le t\le T^*}} u \ge \gamma\ (> 0) \qquad (0 < \theta < 1), \tag{2}$$

where γ *only depends on* β, n, λ, Λ, θ, δ, *an upper bound of* $\operatorname{diam}\Omega$, $T/(\operatorname{diam}\Omega)^2$, *and* $T^*/(\operatorname{diam}\Omega)^2$.

REMARK 1. By the Sobolev embedding theorem $u \in C(Q^*)$.

REMARK 2. As

$$\frac{1}{|Q|}|\{(x, t) \in Q: u(x, t) \geq 1\}|$$

is the density of the set $u \geq 1$ in Q, Theorem 1 can be called a density theorem.

The proof of this theorem depends on several lemmas. Before stating them we add a few comments on the conditions and the conclusions of the theorem. First, the conclusion of this theorem can be deduced from the special case

$$\min_{x\in\overline{\Omega}_\delta} u(x, T) \geq \gamma. \tag{3}$$

This is because, under (3), for all $(1-\theta\beta)T \leq \widetilde{T} \leq T^*$ we have

$$|\{(x, t) \in \Omega \times (0, \widetilde{T}]: u(x, t) \geq 1\}| \geq \min\left\{\theta, T/T^*\right\} \beta|\Omega|\widetilde{T}.$$

Replacing $\min\{\theta, T/T^*\}\beta$ by β we get (2) from (3). Hence in proving (3) we may replace Q^* in the assumption by Q.

On the other hand, the density condition in the theorem suffices to be valid in a rectangular cylinder instead of the whole Q. Write

$$k_R(x^0) = \{x: |x_i - x_i^0| < R, 1 \leq i \leq n\}.$$

(1) is equivalent to: There exists $k_R(x^0) \subseteq \Omega$ such that

$$|\{(x, t) \in K_{R,T}: u(x, t) \geq 1\}| \geq \beta|K_{R,T}|, \tag{4}$$

where $K_{R,T} = k_R(x^0) \times (0, T)$. This is because, when (4) holds, (1) also holds with β replaced by $\beta|k_R|/|\Omega|$. Conversely, when (1) is valid, consider the family of disjoint rectangles k_δ^i $(1 \leq i \leq N)$ with length $\delta/(2\sqrt{n})$ satisfying $\Omega_{\delta/2} \subseteq \bigcup_{i=1}^N \overline{k}_\delta^i \subseteq \Omega$. For δ so small that $|\Omega \setminus \Omega_{\delta/2}| \leq (\beta/2)|\Omega|$, there is at least one $k_\delta^{i_0}$ satisfying

$$|\{(x, t) \in K_{\delta,T}^{i_0}: u \geq 1\}| \geq \frac{\beta}{2}|K_{\delta,T}^{i_0}|.$$

So (4) (replacing β by $\beta/2$) holds.

The proof of the density theorem can be outlined as follows. The density of $u \geq 1$ in $K_{R,T}$ satisfies $\geq \beta$. So decreasing the lower bound of u increases the density, i.e., the density of $u \geq \alpha$ $(0 < \alpha < 1)$ is higher than that of β. When α is small enough, it can be shown that the density is close to 1. A main step in proving this fact is provided by Lemma 5. As soon as the density is close to 1, one can derive

$$\min_{|x-x^0|\leq R/2} u(x, (1-\beta/2)T) \geq \gamma_1 \ (>0)$$

by Lemma 4. From $u(x, (1-\beta/2)T) \geq \gamma_1$, $|x-x^0| \leq R/2$ one derives $\min_{x\in\overline{\Omega}_\delta} u(x, T) \geq \gamma$ (>0). This is the content of Lemma 2.

Before proving the lemmas, we point out that one may take $x^0 = 0$ and write $k_R(0)$ simply as k_R. Moreover we rescale in the t-direction by replacing t by $(R^2/T)t - R^2$, and change the coefficients a_{ij}, b_i, c correspondingly. Let $K_R = k_R \times (-R^2, 0)$ be the standard rectangle. Under these simplifications the assumptions of the density theorem now read as

$$\begin{gathered} u \in W^{2,1}_{n+1}(\Omega \times (-R^2, 0)), \\ u \geq 0, \quad \mathscr{L}u \leq 0 \quad \text{a.e. in } \Omega \times (-R^2, 0), \\ |\{(x, t) \in K_R : u \geq 1\}| \geq \beta |K_R|. \end{gathered} \tag{5}$$

The fact that in the density theorem γ depends on the diameter of Ω only through its upper bound is very important. This is because in using this result to derive the Hölder estimates one needs γ to be uniform for arbitrarily small R. To prove this assume that the density theorem holds for R_0. Making the change of variables $\tilde{x} = (R/R_0)x$ and $\tilde{t} = (R^2/R_0^2)t$, the operator $\mathscr{L}u$ becomes

$$\widetilde{\mathscr{L}}u = a_{ij}u_{ij} + \frac{R}{R_0}b_iu_i + \frac{R^2}{R_0^2}cu - u_t.$$

From $(\sum b_i^2)^{1/2} \leq \Lambda$ and $-\Lambda \leq c \leq 0$ we have

$$\frac{R}{R_0}\left(\sum b_i^2\right)^{1/2} \leq \Lambda, \qquad -\Lambda \leq \frac{R^2}{R_0^2}c \leq 0, \quad \forall R \leq R_0.$$

Hence the assumptions of the density theorem continue to hold for all $R \leq R_0$, and the density theorem holds for all $R \leq R_0$.

In the next section we shall prove the density theorem starting with (5).

§2. Several lemmas and the proof of the density theorem

LEMMA 2. *Let $u \in W^{2,1}_{n+1}$ satisfy $u \geq 0$, $\mathscr{L}u \leq 0$ a.e. in $\{|x| \leq X_2\} \times [T_1, T_2]$. Also $\inf_{|x|\leq X_1, t=T_1} u \geq 1$ for X_1, T_1 satisfying $0 < C^{(1)} \leq (T_2 - T_1)/(X_2^2 - X_1^2) \leq C^{(2)} < \infty$. Then*

$$\inf_{\substack{|x|\leq X_2/2 \\ t=T_2}} u \geq C_1 \left(\frac{X_1}{X_2}\right)^{C_2},$$

where the positive constants C_1, C_2 only depend on n, λ, Λ, $C^{(1)}$, $C^{(2)}$, and D (an upper bound of X_2).

PROOF. Since $\mathscr{L}u - \Lambda u \leq \mathscr{L}u \leq 0$, without loss of generality one can assume $-\Lambda \leq c \leq 0$. Let $\tau = t - T_1 + LX_1^2$, where $L = (T_2 - T_2)/(X_2^2 - X_1^2)$.

Then $t = T_1$ and $t = T_2$ correspond to $\tau = LX_1^2$ and $\tau = LX_2^2$ respectively. Consider the function

$$v(x, \tau) = \left(1 - \frac{L|x|^2}{\tau}\right)^m \tau^{-M},$$

where m, M are positive constants, $m \geq 1$. We have

$$\begin{aligned} &\tau^{M+1}\left(1 - \frac{L|x|^2}{\tau}\right)^{2-m} \mathscr{L}v \\ &= -2Lm(a_{ii} + b_i x_i)\left(1 - \frac{L|x|^2}{\tau}\right) + 4m(m-1)\frac{L^2 a_{ij} x_i x_j}{\tau} \\ &\quad + (M + C\tau)\left(1 - \frac{L|x|^2}{\tau}\right)^2 - Lm\frac{|x|^2}{\tau}\left(1 - L\frac{|x|^2}{\tau}\right). \end{aligned}$$

Let us denote $\sigma = L|x|^2/\tau$ and $C = 2n\Lambda(1 + D)$. For $M \geq 4C^{(2)}D^2\Lambda$, the right-hand side of this inequality satisfies

$$\begin{aligned} &\geq -LmC(1-\sigma) + 4Lm(m-1)\lambda\sigma \\ &\quad + (M - C^{(2)}D^2\Lambda)(1-\sigma)^2 - m\sigma(1-\sigma) \\ &\geq 2Lm(m-1)\lambda\sigma - LmC(1-\sigma) + \frac{M}{4}(1-\sigma)^2 \\ &\quad + \frac{1}{2}[4Lm(m-1)\lambda\sigma^2 + M(1-\sigma)^2 - 2m\sigma(1-\sigma)] \\ &\geq \begin{cases} Lm[(m-1)\lambda - C], & \frac{1}{2} \leq \sigma \leq 1, \\ \frac{M}{16} - LmC, & 0 \leq \sigma \leq \frac{1}{2}, \end{cases} \end{aligned}$$

provided $LM \geq (1 - 1/m)/4\lambda$. Choosing m and M so that $m \geq 1 + C/\lambda$ and $M \geq \max\{16C^{(2)}mC, (1-1/m)/4\lambda C^{(1)}, 4C^{(2)}D^2\Lambda\}$, we get $\mathscr{L}v \geq 0$.

Now we apply the Aleksandrov maximum principle (notice that the condition $(\sum b_i^2)^{1/2} \leq -N_1 c$ in the Aleksandrov maximum principle is not fulfilled. We need to add εu to both sides of the equation and let $\varepsilon \to 0$ after applying the maximum principle) to the function

$$w = \begin{cases} u - (LX_1^2)^M v, & L|x|^2/\tau < 1, \\ u, & L|x|^2/\tau \geq 1, \end{cases}$$

in the domain

$$(x, \tau) \in \{|x| < X_2\} \times (LX_1^2, LX_2^2)$$

to conclude $w \geq 0$ here. Therefore,

$$u|_{\tau = LX_2^2, |x| \leq X_2/2} \geq \left(1 - \frac{1}{4}\right)^m \left(\frac{X_1}{X_2}\right)^{2M} \geq C_1\left(\frac{X_1}{X_2}\right)^{C_2}. \quad \square$$

LEMMA 2′. *Let* $u \in W^{2,1}_{n+1}$ *satisfy* $u \geq 0$, $\mathscr{L}u \leq 0$ *a.e. in* $\{x\colon d(x, \overline{x^1x^2}) < X_2\} \times [T_1, T_2]$. *Also* $\inf_{|x-x^1|\leq X_1, t=T_1} u \geq 1$ *for* $|x^2 - x^1| \leq k(X_2 - X_1)$, $0 < C^{(1)} \leq (T_2 - T_1)/(X_2^2 - X_1^2) \leq C^{(2)} < \infty$. *Then*

$$\inf_{\substack{|x-x^2|\leq X_2/2 \\ t=T_2}} u \geq C_1 \left(\frac{X_1}{X_2}\right)^{C_2},$$

where C_1 *and* C_2 *depend only on* n, λ, Λ, $C^{(1)}$, $C^{(2)}$, D, *and* k.

The proof of this lemma is similar to that of Lemma 2. Only we should change v to

$$v(x, \tau) = \left[1 - \frac{L}{\tau}\left|x - x^2 + \frac{x^2 - x^1}{T_2 - T_1}(LX_2^2 - \tau)\right|^2\right]^m \tau^{-M},$$

and choose m, M so that

$$m \geq 1 + (C + k/C^{(1)})/\lambda,$$
$$M \geq \max\{16C^{(2)}m(C + k/C^{(1)}), (1 - 1/m)/4\lambda C^{(1)}, 4C^{(2)}D^2\Lambda\}.$$

In the following we write $K_R(x^0, t^0) = k_R(x^0) \times [t^0 - R^2, t^0]$ and $K_R = K_R(0, 0)$.

LEMMA 3. *Let* $u \in W^{2,1}_{n+1}(K_{1+\theta})$ *satisfy* $u \geq 0$, $\mathscr{L}u \leq 0$ *a.e. in* $K_{1+\theta}$ $(0 < \theta < 1)$ *where* $0 \leq c \leq \Lambda$. *Then there exists a constant* μ, $0 < \mu < 1$, *such that for any* $K_R(x^1, t^1) \subseteq K_1$ *satisfying* $|K_R(x^1, t^1) \cap \{u \geq 1\}| \geq \mu|K_R(x^1, t^1)|$,

$$u \geq C \quad \text{in } \{k_{3R}(x^1) \times [t^1, t^1 + 7R^2]\} \cap K_1,$$

where C *only depends on* n, λ, Λ, μ, *and* θ.

PROOF. Introduce the change of variables $x - x^1 = RX$ and $t - t^1 = R^2T$ and denote the new operator by $\widetilde{\mathscr{L}}$. Let

$$v = \begin{cases} u - [1 - e^{L(|X|^2 - 2T)}]^2, & |X|^2 < 2T, \\ u, & |X|^2 \geq 2T, \end{cases}$$

where the constant $L \geq 1$ will be chosen later. On the parabolic boundary $\Gamma = \{|X| = 1, 0 \leq T \leq \frac{1}{2}\} \cup \{T = 0, |X| \leq 1\}$, $v > 0$. Also

$$\begin{aligned}(\widetilde{\mathscr{L}} - cR^2)v &\leq (8L^2a_{ij}X_iX_j + 4La_{ii} + b_iRX_iL + 4L)e^{L(|X|^2-2T)} \\ &\quad - [16L^2a_{ij}X_iX_j + 8L(a_{ii} + b_iRX_i) + 4L]e^{2L(|X|^2-2T)} \\ &\leq C_1L^2e^{L|X|^2-2T)} \leq C_1L^2[2e^{L(|X|^2-2T)} - e^{2L(|X|^2-2T)}] \\ &= C_1L^2(v + 1 - u),\end{aligned}$$

provided $L \geq 1$ and $|X|^2 \leq 2T$. Therefore

$$\widetilde{\mathscr{L}}v - (cR^2 + C_1L^2)v \leq \begin{cases} C_1L^2(1-u)_+, & |X|^2 < 2T, \\ 0, & |X|^2 \geq 2T. \end{cases}$$

By the Aleksandrov maximum principle (let $B_r = \{x\colon |x| < r\}$)

$$\begin{aligned} v &\geq v_-|_\Gamma - C_2\left\{\frac{1}{2|B_1|}\int_{B_1\times[0,1/2]}[\widetilde{\mathscr{L}}v - (R^2c + C_1L^2)v]_+^{n+1}\,dX\,dT\right\}^{1/(n+1)} \\ &\geq -C_2C_1L^2\left[\frac{1}{2|B_1|}\int_{B_1\times[0,1/2]}(1-u)_+^{n+1}\,dX\,dT\right]^{1/(n+1)} \\ &\geq -C_3L^2\left|B_1\times\left[0,\frac{1}{2}\right]\cup\{u\leq 1\}\right| \\ &\geq -C_3L^2|\{k_1\times(0,1)\}\cap\{u\leq 1\}| \geq -C_4L^2(1-\mu). \end{aligned}$$

As a result

$$\begin{aligned} u|_{B_{1/2}\times\{T=1/2\}} &\geq (1-e^{-3L/4})^2 - C_4L^2(1-\mu) \geq 1 - \frac{1}{9L^2/64} - C_4L^2(1-\mu) \\ &= 1 - \frac{16}{3}[C_4(1-\mu)]^{1/2} \end{aligned}$$

if we choose L so that $(9C_4/64)L^4(1-\mu) = 1$.

Taking $\mu \geq 1 - 1/1024C_4$, in the (X, T)-coordinate

$$u|_{B_{1/2}\times\{T=1/2\}} \geq \frac{1}{2},$$

and in the (x, t)-coordinate it becomes

$$u|_{B_{R/2}(x^1)\times\{t=t^2-R^2/2\}} \geq \frac{1}{2}.$$

Now if we take an arbitrary point x^2 in $k_{3R}(x^1)\cap k_1$, $X_1 = \theta R/2$, $X_2 = \theta R$, $T_1 = t^1 - R^2/2$, and $t^1 \leq T_2 \leq \min\{0, t^1 + 7R^2\}$, this lemma follows from Lemma 2′. □

Similarly one can prove

LEMMA 4. *Let $u \in W^{2,1}_{n+1}(K_1)$ satisfy $u \geq 0$, $\mathscr{L}u \leq 0$ a.e. in K_1 where $0 \leq c \leq \Lambda$. There exists μ, $0 < \mu < 1$, such that*

$$u|_{B_{1/2}\times\{t=0\}} \geq 1 - C(1-\mu)^{1/2},$$

provided $|K_1\cap\{u\geq 1\}| \geq \mu|K_1|$. Here C depends only on n, λ, and Λ.

LEMMA 5. *Let Γ be a measurable subset of K_1 with positive measure and let $\widetilde{\Gamma}$ satisfy*

$$\begin{aligned} \widetilde{\Gamma} \supseteq \Gamma \cup \{k_{3R}(x^0)\times[t^0, t^0+7R^2)\cap K_1 \colon K_R(x^0, t^0) \subseteq K_1, \\ |\Gamma\cap K_R(x^0, t^0)| \geq \mu|K_R(x^0, t^0)|\}. \end{aligned} \tag{6}$$

Then either

$$|\widetilde{\Gamma}| \geq \left(1 + \frac{1-\mu}{12}\right)|\Gamma|,$$

or there exists $K_R(x^0, 0) \subseteq K_1$ *such that*

$$R \geq \frac{1}{4}|\Gamma|^{1/2}, \qquad |\Gamma \cap K_R(x^0, 0)| \geq \mu|K_R(x^0, 0)|. \tag{7}$$

PROOF. If $|\Gamma \cap K_1| \geq \mu|K_1|$, (7) holds. Otherwise $|\Gamma \cap K_1| < \mu|K_1|$. We divide K_1 equally into 2^{n+2} many rectangles $K_{1/2}(y_i, \tau_i)$ where $y_i = \pm 1/2$ $(1 \leq i \leq n)$ and $\tau_i = -3/4, -1/2, -1/4, 0$. For those $K_{1/2}$'s satisfying $|\Gamma \cap K_{1/2}| < \mu|K_{1/2}|$ we apply the same division to get $K_{1/4}(\cdots)$ and so on.

First we look at one of the largest rectangles $K^0_{2^{-k}}$ such that $|\Gamma \cap K^0_{2^{-k}}| \geq \mu|K^0_{2^{-k}}|$ (that is to say, k is the smallest integer for which this inequality holds). Let $\widetilde{K}^0_{2^{-k+1}}$ be the precedent rectangle of $K^0_{2^{-k}}$ and let $\widetilde{K}^l_{2^{-k+1}}$ $(l = 0, 1, \dots, L)$ be all the rectangles, touching one by one in the t-direction, with the same property. In other words, $|\Gamma \cap \widetilde{K}^l_{2^{-k+1}}| < \mu|\widetilde{K}^l_{2^{-k+1}}|$ and there is at least one $K^{0,l}_{2^{-k}} \subseteq \widetilde{K}^l_{2^{-k+1}}$ such that

$$|\Gamma \cap K^{0,l}_{2^{-k}}| \geq \mu|K^{0,l}_{2^{-k}}| \qquad (l = 0, 1, \dots, L).$$

Moreover, for each j, $1 \leq j \leq 2^{n+2}$, $K^j_{2^{-k}} \subseteq \widetilde{K}^{-1}_{2^{-k+1}} \cup \widetilde{K}^{L+1}_{2^{-k+1}}$, one has $|\Gamma \cap K^j_{2^{-k}}| < \mu|K^j_{2^{-k}}|$. Denote the small rectangles in the upper layer of $\widetilde{K}^{-1}_{2^{-k+1}}$ by $K^{-1,j}_{2^{-k}}$ $(1 \leq j \leq 2^n)$ and the small rectangles in the lower layer of $\widetilde{K}^{L+1}_{2^{-k+1}}$ by $K^{L+1,j}_{2^{-k}}$ $(1 \leq j \leq 2^n)$. From $|\Gamma \cap K^{0,l}_{2^{-k}}| \geq \mu|K^{0,l}_{2^{-k}}|$ $(0 \leq l \leq L)$ one has $\widetilde{K}^l_{2^{-k+1}} \subseteq \widetilde{\Gamma}$ $(1 \leq l \leq L)$ and $K^{L+1,j}_{2^{-k}} \subseteq \widetilde{\Gamma}$ $(1 \leq j \leq 2^n)$. Therefore

$$\begin{aligned}
&\left|\widetilde{\Gamma} \cap \left(\bigcup_{l=0}^{L} \widetilde{K}^l_{2^{-k+1}} \cup \bigcup_{i=1}^{2^n} K^{L+1,j}_{2^{-k}}\right)\right| - \frac{1-\mu}{4}\left|\bigcup_{j=1}^{2^n} K^{-1,j}_{2^{-k}}\right| \\
&\quad \geq \left|\Gamma \cap \left(\bigcup_{l=0}^{L} \widetilde{K}^l_{2^{-k+1}} \cup \bigcup_{j=1}^{2^n} K^{L+1,j}_{2^{-k}}\right)\right| \\
&\qquad + (1-\mu)\left|\bigcup_{l=1}^{L} \widetilde{K}^l_{2^{-k+1}} \cup \bigcup_{j=1}^{2^n} K^{L+1,j}_{2^{-k}}\right| \\
&\qquad - \frac{1-\mu}{4}\left|\bigcup_{j=1}^{2^n} K^{-1,j}_{2^{-k}}\right| \\
&\quad \geq \left[1 + \frac{3/16 + L}{5/4 + L}(1-\mu)\right]\left|\Gamma \cap \left(\bigcup_{l=0}^{L} \widetilde{K}^l_{2^{-k+1}} \cup \bigcup_{j=1}^{2^n} K^{L+1,j}_{2^{-k}}\right)\right|.
\end{aligned} \tag{8}$$

Notice that when $K^{-1,j}_{2^{-k}}$ lies outside of K_1 (i.e., it is under $t = -1$), the relevant term in (8) should be deleted. On the other hand, if $K^{L+1}_{2^{-k}}$ lies outside K_1 (i.e., it is above $t=0$), (8) should be replaced by

$$(1-\mu)\left|\bigcup_{j=1}^{2^n} K_{2^{-k}}^{L+1,j}\right| + \left|\widetilde{\Gamma}\cap\bigcup_{l=0}^{L}\widetilde{K}_{2^{-k+1}}^{l}\right| - \frac{1-\mu}{4}\left|\bigcup_{j=1}^{2^n} K_{2^{-k}}^{-1,j}\right| \geq \left[1+\frac{3/16+L}{1+L}(1-\mu)\right]\left|\Gamma\cap\left(\bigcup_{l=0}^{L}\widetilde{K}_{2^{-k+1}}^{l}\right)\right|. \tag{9}$$

In the remaining part of K_1 we consider those rectangles satisfying $|\Gamma\cap K_{2^{-k}}|\geq\mu|K_{2^{-k}}|$ from small k to large k and obtain analogues of (8). In case $K_{2^{-k}}^{-1,j}$ lies outside K_1 or has been handled in a smaller k, its corresponding term should not be contained in (8). In case $K_{2^{-k}}^{L+1,j}$ lies outside K_1 or has been handled in a smaller k, we replace (8) by (9).

Summing all rectangles involved in (8) and (9), since

$$\frac{1-\mu}{4}\left|\bigcup_{j=1}^{2^n} K_{2^{-k}}^{-1,j}\right| \geq (1-\mu)\left|\bigcup_{j=1}^{2^n} K_{2^{-k}}^{L+1,j}\right|$$

holds for $k_1>k$, it follows that

$$(1-\mu)|\overline{\Gamma}|+|\widetilde{\Gamma}|\geq\left[1+\frac{3}{20}(1-\mu)\right]\left|\Gamma\cap\left(\bigcup_k\widetilde{K}_{2^{-k+1}}\right)\right|,$$

where $\overline{\Gamma}$ is the sum of all rectangles just above $t=0$.

By the measurability of Γ, $|\Gamma\cap K_R(y,\tau)|/|K_R(y,\tau)|\to 1>\mu$ as $R\to 0$ for a.e. $(y,\tau)\in\Gamma$. Therefore $|\Gamma\setminus\bigcup_k\widetilde{K}_{2^{-k+1}}|=0$, that is to say, $|\Gamma\cap(\bigcup_k\widetilde{K}_{2^{-k+1}})|=|\Gamma|$. We conclude

$$(1-\mu)|\overline{\Gamma}|+|\widetilde{\Gamma}|\geq\left[1+\frac{3}{20}(1-\mu\right]|\Gamma|.$$

When the maximal height of $\overline{\Gamma}$ exceeds $|\Gamma|/16$, there is at least one rectangle in K_1 whose top face lies on $t=0$, density $\geq\mu$, and width $\geq|\Gamma|^{1/2}/4$. So (7) holds. Otherwise

$$|\widetilde{\Gamma}|\geq\left[1+\frac{3}{20}(1-\mu)-\frac{1}{16}(1-\mu)\right]|\Gamma|>\left(1+\frac{1-\mu}{12}\right)|\Gamma|. \qquad\square$$

Now we are ready for the proof of the density theorem (Theorem 1).

PROOF. Without loss of generality we may assume $R=1+\theta$ in condition (5) where $\theta=\theta(n,\beta)$ is determined by $|K_{1+\theta}|=(1+\beta/2)|K_1|$. So $u\geq 0$, $\mathscr{L}u\leq 0$ almost everywhere in $K_{1+\theta}$ and

$$|\{(x,t)\in K_1(0,-2\theta-\theta^2): u(x,t)\geq 1\}|\geq\beta|K_{1+\theta}|-\frac{\beta}{2}|K_1|\geq\frac{\beta}{2}|K_1|.$$

Further we may assume $0\leq c\leq\Lambda$, for otherwise we consider v, $u=e^{-\Lambda t}v$, instead of u.

Take $\Gamma = K_1(0, -2\theta-\theta^2)\cap\{u \geq 1\}$ and $\widetilde{\Gamma} = K_1(0, -2\theta-\theta^2)\cap\{u \geq C\}$, where C is the constant in Lemma 3. By Lemma 3, Γ and $\widetilde{\Gamma}$ satisfy (6) for Lemma 5. Hence by this lemma either

$$|\widetilde{\Gamma}| \geq \left(1+\frac{1-\mu}{12}\right)|\Gamma| \geq \frac{\beta}{2}\left(1+\frac{1-\mu}{12}\right)|K_1|,$$

or there exists $K_R(x^0, 0)$ satisfying (7). In the first case we take

$$\Gamma = K_1(0, -2\theta-\theta^2)\cap\{u \geq C\},$$
$$\widetilde{\Gamma} = K_1(0, -2\theta-\theta^2)\cap\{u \geq C^2\}, \ldots.$$

In general, as long as the second case has not appeared, we have

$$\Gamma = K_1(0, -2\theta-\theta^2)\cap\{u \geq C^{k-1}\},$$
$$\widetilde{\Gamma} = K_1(0, -2\theta-\theta^2)\cap\{u \geq C^k\}, \qquad k = 1, 2, \ldots,$$

which implies

$$|\widetilde{\Gamma}| \geq \frac{\beta}{2}\left(1+\frac{1-\mu}{12}\right)^k |K_1|.$$

Therefore the second case must happen when

$$k \leq \log\frac{2\mu}{\beta}\Big/\log\left(1+\frac{1-\mu}{12}\right).$$

In other words, there is k_0 satisfying

$$1 \leq k_0 \leq \log\frac{2\mu}{\beta}\Big/\log\left(1+\frac{1-\mu}{12}\right)$$

and $K_R(x^0, 0) \subseteq K_1$ such that

$$R \geq \frac{1}{4}|K_1(0, -2\theta-\theta^2)\cap\{u \geq C^{k_0-1}\}|^{1/2} \geq \frac{1}{4}\left(\frac{\beta}{2}|K_1|\right)^{1/2},$$

$$\mu|K_R(x^0, 0)| \leq |K_R(x^0, 0)\cap\{u \geq C^{k_0-1}\}| \leq |K_R(x^0, 0)\cap\{u \geq C_1\}|$$

for $C_1 = (\beta/2\mu)^{\log(1/C)/\log(1+(1-\mu)/12)}$.

In view of Lemma 4, $u \geq C_2$ on $k_{R/2}(x^0)\times\{t = -\theta-\theta^2/2\}$. By taking $T_1 = -\theta-\theta^2/2$, $T_2 = 0$, $X_2 = \min\left\{\frac{1}{16}\beta^{1/2}, \frac{\delta}{2}\right\}$, $X_1 = X_2/2$, $x^1 = x^0$, and x^2 any point in Ω_δ in Lemma 2′ we obtain $\inf_{x\in\Omega_\delta, t=0} u \geq \gamma$.* The proof of Theorem 1 is complete. □

*The assumptions in Lemma 2′ may not be fulfilled when Ω is nonconvex. The proof must be slightly modified as follows. One can always draw a curve from $(x^0, -2\theta-\theta^2)$ to $(x^2, 0)$ in $\Omega_\delta\times(-2\theta-\theta^2, 0)$ which is increasing in t. We can change it to be a polygonal line in $-\frac{1}{2}(k+1)\delta \leq t \leq -\frac{1}{2}k\delta$ $(0 \leq k \leq \frac{2}{\delta}(2\theta+\theta^2))$ and apply Lemma 2′ to each line segment in this polygonal line.

§3. The Harnack inequality for parabolic equations with measurable coefficients

The density theorem has several applications, one of which is the following.

THEOREM 6. *Let $Q = \Omega \times (0, T]$, where Ω is a bounded domain in $\mathbb{R}^n$. Consider the operator with measurable coefficients in Q,*

$$\mathscr{L} = a_{ij}\frac{\partial^2}{\partial x_i \partial x_j} + b_i \frac{\partial}{\partial x_i} + c - \frac{\partial}{\partial t},$$

satisfying

$$\lambda|\xi|^2 \le a_{ij}\xi_i\xi_j \le \Lambda|\xi|^2, \quad \forall \xi \in \mathbb{R}^n, \ 0 < \lambda \le \Lambda < \infty,$$
$$\left(\sum b_i^2\right), \qquad |c| \le \Lambda, \quad \forall (x, t) \in Q.$$

Let $u \in W^{2,1}_{n+1}(Q)$ be a nonnegative function satisfying $\mathscr{L}u = 0$ almost everywhere in Q. For (x^1, t^1), $(x^2, t^2) \in Q_\delta = \Omega_\delta \times [\delta, T]$, $\Omega_\delta = \{x : d(x, \partial\Omega) > \delta\}$, and $T \ge t^2 \ge t^1 + \eta$ $(\eta > 0)$, there exists a positive constant C such that $u(x^2, t^2) \ge Cu(x^1, t^1)$, where C depends only on n, λ, Λ, δ, η, and $\operatorname{diam} Q$.

PROOF. It suffices to look at the domain $\Omega \times [t^1 - \delta, T]$. Without loss of generality we take $x^1 = 0$, $t^1 = 0$, $\delta = 2$. So $K_1 \subseteq Q$.

The theorem clearly holds if $u(0, 0) = 0$. So we may assume $u(0, 0) > 0$. Letting m be a constant to be determined later we consider the largest root, R_0, of the function

$$g(R) = u(0, 0)(1 - R)^{-m} - 2 \max_{\overline{K_R(0,0)}} u = 0$$

in $0 \le R \le 1$. Since $g(0) \le 0$ and $g(1 - 0) = \infty$, R_0, $0 < R_0 < 1$, is uniquely defined. Let $\max_{\overline{K_{R_0}(0,0)}} u = u(x^0, t^0)$. We have

$$\sup_{K_{(1-R_0)/2}(x^0, t^0)} u \le \frac{1}{2}u(0, 0)\left(1 - \frac{1 + R_0}{2}\right)^{-m} = 2^{m-1}u(x^0, t^0),$$

because $K_{(1-R_0)/2}(x^0, t^0) \subseteq K_{(1-R_0)/2}(0, 0)$. In case

$$\left| K_{(1-R_0)/2}(x^0, t^0) \cap \left\{ \frac{2^{m-1}u(x^0, t^0) - u}{(2^{m-1} - 1/2)u(x^0, t^0)} \ge 1 \right\} \right| \ge \mu |K_{(1-R_0)/2}|,$$

by Lemma 4 we have

$$\left. \frac{2^{m-1}u(x^0, t^0) - u}{(2^{m-1} - 1/2)u(x^0, t^0)} \right|_{B(1-R_0)/4 \times \{t = t^0\}} \ge 1 - C(1 - \mu)^{1/2}.$$

In particular at (x^0, t^0) we have

$$\frac{2^{m-1}-1}{2^{m-1}-1/2} \geq 1 - C(1-\mu)^{1/2},$$

which is impossible when μ is sufficiently close to 1. Therefore, we can find $\mu_0 \in (0, 1)$ such that

$$\left| K_{(1-R_0)/2}(x^0, t^0) \cap \left\{ \frac{2^{m-1}u(x^0, t^0) - u}{(2^{m-1}-1/2)u(x^0, t^0)} \geq 1 \right\} \right| < \mu_0 |K_{(1-R_0)/2}|.$$

In other words,

$$\left| K_{(1-R_0)/2}(x^0, t^0) \cap \left\{ \frac{u(x, t)}{u(x^0, t^0)} \leq \frac{1}{2} \right\} \right| < \mu_0 |K_{(1-R_0)/2}|$$

and

$$\left| K_{(1-R_0)/2}(x^0, t^0) \cap \left\{ \frac{2u(x, t)}{u(x^0, t^0)} > 1 \right\} \right| \geq (1-\mu_0)|K_{(1-R_0)/2}|$$

hold. Using the density theorem (Theorem 1) we obtain

$$\left. \frac{2u(x, t)}{u(x^0, t^0)} \right|_{B_{(1-R_0)/4}(x^0) \times \{t=t^0\}} \geq \gamma > 0.$$

When Ω is convex, $\{x : d(x, \overline{0x^2}) \leq 1\} \subseteq \Omega$. Applying Lemma 2$'$ to $x^1 = x^0$, $T_1 = t^0$, $x_1 = (1-R_0)/4$, $x^2 = x^2$, $T_2 = t^2$, $X_2 = 1$, $k = (4/3)\operatorname{diam}\Omega$ yields

$$\begin{aligned} u(x^2, t^2) &\geq C\left(\frac{1-R_0}{4}\right)^{C_2} \inf_{B_{(1-R_0)/4}(x^0) \times \{t=t^0\}} u \\ &\geq \frac{C\gamma}{2} u(x^0, t^0) \left(\frac{1-R_0}{4}\right)^{C_2} = \frac{C\gamma}{4^{C_2+1}} u(0, 0)(1-R_0)^{C_2-m}, \end{aligned}$$

where C and C_2 depend on η. Thus the Harnack inequality

$$u(x^2, t^2) \geq \frac{C\gamma}{4^{C_2+1}} u(0, 0)$$

holds by choosing $m = C_2$. When Ω is not convex, a slight modification is needed. See the remark at the end of the proof of Theorem 1. □

§4. A Hölder estimate for the solution of a quasilinear parabolic equation

It is possible to derive a Hölder estimate for the solution of a linear parabolic equation with measurable coefficients from the density theorem. Actually, such an estimate is also available for the solution of a quasilinear parabolic equation satisfying a natural structure condition.

THEOREM 7. *Let* $Q = \Omega \times (0, T]$, *where* Ω *is a bounded domain in* $\mathbb{R}^n$. *Consider the equation with measurable coefficients*

$$a_{ij}(x, t, u, Du)u_{ij} + b(x, t, u, Du) - u_t = 0, \qquad Du = (u_1, \dots, u_n),$$

satisfying

$$|b(x, t, z, p)| \le b_0|p|^2 + b_1(x, t)|p| + b_2(x, t)|z| + g(x, t),$$
$$\lambda|\xi|^2 \le a_{ij}\xi_i\xi_j \le \Lambda|\xi|^2, \quad \forall \xi \in \mathbb{R}^n, \ 0 < \lambda \le \Lambda < \infty,$$

where b_0 *is a constant,* $b_1 \in L^{2(n+1)}(Q)$, *and* b_2, $g \in L^{n+1}(Q)$.

Let u *be a function in* $W^{2,1}_{n+1}(Q)$ *satisfying the above equation almost everywhere in* Q. *Suppose further that it is bounded,* $|u| \le M$. *Then for* (x^1, t^1), (x^2, t^2) *in a parabolic-closed interior subdomain of* Q

$$|u(x^2, t^2) - u(x^1, t^1)| \le C(|x^2 - x^1|^2 + |t^2 - t^1|)^{\alpha/2}.$$

Here C *and* α *only depend on* n, λ, Λ, $\operatorname{diam} Q$, M, b_0, $\|b_1\|_{L^{2(n+1)}(Q)}$, $\|b_2\|_{L^{n+1}(Q)}$, $\|g\|_{L^{n+1}(Q)}$, *and the distance between this subdomain and* $\partial^* Q$ ($\partial^* Q$ *is the parabolic boundary*).

Notice that in the linear case M can be estimated in terms of other given quantities. So the Hölder exponent and constant do not depend on M.

PROOF. Fix (x^0, t^0) in this parabolic-closed interior subdomain and let $K_R(x^0, t^0) \subseteq Q$. Denote

$$\max_{K_{R/2}(x^0, t^0)} u = u_1 = u(x^1, t^1), \qquad \min_{K_{R/2}(x^0, t^0)} u = u_2 = u(x^2, t^2),$$

and $u_0 = (u_1 + u_2)/2$. Without loss of generality we can assume $|\{(x, t) \in K_R(x^0, t^0) : u \le u_0\}| \ge |K_R|/2$, for otherwise we can replace u by $-u$. Letting $\widetilde{K}_R = K_R(x^0, t^0) \times \{t < t^1\}$, we have

$$\begin{aligned} |\{(x, t) \in \widetilde{K}_R : u \le u_0\}| &\ge \frac{1}{2}|K_R| - \left|k_R(x^0) \times \left(t^0 - \frac{R^2}{4}, t^0\right)\right| \\ &= \frac{1}{4}|K_R| \ge \frac{1}{4}|\widetilde{K}_R|. \end{aligned}$$

Putting

$$v = \frac{\lambda}{2b_0}\left\{\exp\left[\frac{2b_0}{\lambda}(u - u_0)\right] - 1\right\}$$

into the equation we see that v satisfies the equation

$$\begin{aligned} a_{ij}v_{ij} &- \frac{2b_0}{\lambda}\exp\left[\frac{-2b_0}{\lambda}(u - u_0)\right] a_{ij}v_i v_j + \exp\left[\frac{2b_0}{\lambda}(u - u_0)\right] \\ &\cdot b\left(x, t, u, \exp\left[\frac{-2b_0}{\lambda}(u - u_0)\right] Dv\right) - v_t = 0. \end{aligned}$$

Consequently, for $\mathscr{L}v = a_{ij}v_{ij} - v_t$,

$$\mathscr{L}v \geq -\exp\left[\frac{-2b_0}{\lambda}(u-u_0)\right]\left(\frac{b_1^2}{4b_0} + b_2 M + g\right) \geq -K(b_1^2 + b_2 + g),$$

where K depends only on b_0, λ, and M.

Now we claim that for a further restricted R

$$\max_{\widetilde{K}_R} v > \left(1+\gamma\left(\frac{1}{4}\right)\right)\frac{\lambda}{2b_0}\left\{\exp\left[\frac{2b_0}{\lambda}(u_1-u_0)\right]-1\right\}, \tag{10}$$

provided $\operatorname{osc}_{K_{R/2}} u \geq R^{n/2(n+1)}$. Here $\gamma(\frac{1}{4})$ is the constant $\gamma(\beta)$ $(\beta = \frac{1}{4})$ for the density theorem in $\widetilde{K}_R$. For if, on the contrary, (10) does not hold, there exists $w \geq 0$ in $\widetilde{K}_R$ satisfying $|\{(x, t) \in \widetilde{K}_R : w \geq 1\}| \geq \frac{1}{4}|\widetilde{K}_R|$ and

$$\mathscr{L}w \leq \frac{2b_0 K(b_1^2 + b_2 + g)}{[1+\gamma(1/4)]\lambda\{\exp[2b_0(u_1-u_0)/\lambda]-1\}}.$$

All assumptions in the density theorem are fulfilled except the condition $\mathscr{L}w \leq 0$. Let's show that Lemma 2, Lemma 2′, and Lemma 3 remain valid for sufficiently small R. Hence the density theorem is applicable in the present situation. When $C_1 \leq X_1/R \leq X_2/R \leq C_2$, modify the proof of Lemma 2 as follows:

$$u|_{\tau=LX_2^2,\,|x|\leq X_2/2} \geq \left(1-\frac{1}{4}\right)^m\left(\frac{X_1}{X_2}\right)^{2M} - \frac{KR^{n/(n+1)}\|b_1^2+b_2+g\|_{L^{n+1}(Q)}}{R^{n/2(n+1)}}.$$

So Lemma 2 holds for small R. Similarly the same modification works for Lemmas 2′ and 3. By the density theorem

$$w|_{x\in k_{R/2}(x^0),\,t=t'} \geq \gamma\left(\frac{1}{4}\right)$$

for $R \leq R_0$. In particular,

$$w(x^1, t^1) = 1 - \frac{1}{1+\gamma(1/4)} = \frac{\gamma(1/4)}{1+\gamma(1/4)} \geq \gamma\left(\frac{1}{4}\right).$$

Contradiction holds. So (10) must be valid.

Denote the value of u in which the left-hand side of (10) attains its maximum by u_3 and set

$$\frac{u_3-u_0}{u_1-u_0} = 1+\sigma, \qquad \sigma \geq 0.$$

A lower bound for σ can be obtained in the following way. Letting

$$1+\kappa = \exp\left[\frac{2b_0}{\lambda}(u_1-u_0)\right],$$

from

$$\exp\left[(1+\sigma)\frac{2b_0}{\lambda}(u_1-u_0)\right]-1 \geq \left(1+\gamma\left(\frac{1}{4}\right)\right)\left\{\exp\left[\frac{2b_0}{\lambda}(u_1-u_0)\right]-1\right\}$$

we have

$$1+\gamma\left(\frac{1}{4}\right) \le \frac{(1+\kappa)^{1+\sigma}-1}{\kappa} \le (1+\sigma)(1+\kappa)^{\sigma} \le (1+\sigma)\exp\left[\frac{4b_0 M}{\lambda}\sigma\right],$$

and so

$$\log\left(1+\gamma\left(\frac{1}{4}\right)\right) \le \log(1+\sigma) + \frac{4b_0 M}{\lambda}\sigma \le \left(1+\frac{4b_0 M}{\lambda}\right)\sigma.$$

That is to say,

$$\sigma \ge \frac{\log(1+\gamma(1/4))}{1+4b_0 M/\lambda} > 0.$$

Therefore, when $R \le R_0$ and $\operatorname{osc}_{K_{R/2}} u \ge R^{n/2(n+1)}$,

$$\operatorname*{osc}_{K_R} u \ge \operatorname*{osc}_{\widetilde{K}_R} u \ge u_3 - u_2 = \left(1+\frac{\sigma}{2}\right)(u_1 - u_2) = \left(1+\frac{\sigma}{2}\right)\operatorname*{osc}_{K_{R/2}} u.$$

In general, for $R \le R_0$,

$$\operatorname*{osc}_{K_{R/2}} u \le \frac{1}{1+\sigma/2}\operatorname*{osc}_{K_R} u + R^{n/2(n+1)}.$$

Now, a Hölder estimate of u in this subdomain can be obtained by a standard argument. □

A Hölder estimate for u in the whole region Q will be discussed later.

§5. A Hölder estimate for the solution for a quasilinear parabolic system

Denote by $d(P_1, P_2) = (|x^2 - x^1|^2 + |t^2 - t^1|)^{1/2}$ the parabolic distance between two points $P_1(x^1, t^1)$ and $P_2(x^2, t^2)$ in $Q = \Omega \times (0, T]$. We consider the following system of linear parabolic operators in Q

$$\mathscr{L}_l = a_{ij}(x, t)\frac{\partial^2}{\partial x_i \partial x_j} + b_i^l(x, t)\frac{\partial}{\partial x_i} + c(x, t) - \frac{\partial}{\partial t} \qquad (1 \le l \le N),$$

which satisfies $\lambda|\xi|^2 \le a_{ij}^l \xi_i \xi_j \le \Lambda|\xi|^2$, $\forall \xi \in \mathbb{R}^n$, $0 < \lambda \le \Lambda < \infty$, and $|b_i^l|$, $|c^l| \le K_1$, $\forall (x, t) \in Q$, for $i, j = 1, \dots, n$ and $l = 1, \dots, N$. Let $a_+ = \max\{a, 0\}$ and $a_- = -\min\{a, 0\}$.

THEOREM 8. *Let $u_l \in W_{n+1}^{2,1}(Q)$, $\sup_Q |u_l| \le M_0$, satisfy $\mathscr{L}_l u_l \ge -K_2$ almost everywhere in Q. Suppose there exists $f_l \in C^1(\mathbb{R})$, $f_l' > 0$ in $\mathbb{R}$ such that*

$$\delta \sum_l [f_l(u_l(P_2)) - f_l(u_l(P_1))]_+ - \sum_l [f_l(u_l(P_2)) - f_l(u_l(P_1))]_- \le K_2 d_{P_1 P_2}^{-\gamma} d^{\gamma}(P_1, P_2),$$

for some $\gamma \in (0, 1]$ and $\delta > 0$ whenever P_1, P_2 in Q satisfy

$$4d(P_1, P_2) \le d_{P_1 P_2} = \min\{d(P_1, \partial^* Q), d(P_2, \partial^* Q)\}.$$

Then for every P_1, P_2 *in* Q

$$\sum_l |u_l(P_2) - u_l(P_1)| \le M\, d_{P_1P_2}^{-\beta}\, d^\beta(P_1, P_2)(K_2 + M_0),$$

where $\beta = \min\{\alpha_0, \gamma\}$, $\alpha_0 = \alpha_0(N, n, \lambda, \Lambda, \delta) \in (0, 1)$. *When* $f_l(w) = w$, M *depends only on* N, n, λ, Λ, δ, γK_1. *In general it further depends on* M_0 *and positive upper and lower bounds for* f_l' *in* $[-M_0, M_0]$.

PROOF. The general case can be reduced to the special case $f_l(t) = t$ $(1 \le l \le N)$ by applying the mean-value theorem to f_l. (Change δ and K_2 if necessary.) So we only prove the theorem for this special case. (Moreover, from $\mathscr{L}_l u_l \ge -K_2$ we deduce $(\mathscr{L}_l - 2K_1)u_l \ge -K_2 - 2K_1M_0$ and hence $(\mathscr{L}_l - 2K_1)(u_l - 2M_0 - K_2/K_1) \ge 0$. So we can rewrite $C^l - 2K_1$, $u_l - 2M_0 - K_2/K_1$, $3M_0 + K_2/K_1$, and $3K_1$ as C^l, u_l, M_0, and K_1, and reduce the general case to the special case $\mathscr{L}_l u_l \ge 0$. So it suffices to prove the theorem for this special case.

For a fixed $(x^0, t^0) \in Q$, let $Q_{R_0}(x^0, t^0) \subseteq Q$ with $d(Q_{R_0}(x^0, t^0), \partial^* Q) \ge 4R_0$ and $R \le R_0$. To simplify notation we shift (x^0, t^0) to the origin. Let

$$\sup_{Q_R} u_l(x, t) = M_l(R), \qquad \inf_{Q} u_l(x, t) = m_l(R),$$

$$M_l(R) - m_l(R) = \omega_l(R),$$

$$\widetilde{Q}_R = B_R \times \left\{-R^2 < t < -\frac{R^2}{4}\right\},$$

$$\Gamma_l = \left\{(x, t) \in \widetilde{Q}_R \colon u_l(x, t) \le (1 - \varepsilon)M_l\left(\frac{R}{2}\right) + \varepsilon m_l\left(\frac{R}{2}\right)\right\},$$

where $\varepsilon \in (0, 1)$ will be chosen later.

$$A = \left\{l \colon |\Gamma_l| \ge \frac{1}{N+1}|\widetilde{Q}_R|\right\} \quad \text{and} \quad B = \{1, 2, \dots, N\}\backslash A.$$

For each $l \in A$, $M_l(R) - u_l \ge 0$, $\mathscr{L}_l(M_l(R) - u_l) \le 0$, and

$$\left|\left\{(x, t) \in \widetilde{Q}_R \colon \frac{M_l(R) - u_l}{M_l(R) - [(1-\varepsilon)M_l(R/2) + \varepsilon m_l(R/2)]} \ge 1\right\}\right|$$
$$= |\Gamma_l| \ge \frac{1}{N+1}|\widetilde{Q}_R|.$$

By the density theorem (Theorem 1) there exists $\eta \in (0, 1)$ such that

$$M_l(R) - u_l \ge \eta\{M_l(R) - [(1-\varepsilon)M_l(R/2) + \varepsilon m_l(R/2)]\}$$

for all $(x, t) \in Q_{R/2}$. Therefore

$$M_l(R) - M_l(R/2) \ge \eta\{M_l(R) - [(1-\varepsilon)M_l(R/2) + \varepsilon m_l(R/2)]\},$$

i.e.,

$$(1-\eta)M_l(R) \ge (1-\eta)M_l(R/2) + \varepsilon\eta\omega_l(R/2).$$

Adding $-(1-\eta)m_l(R)$ to both sides of the inequality we obtain, after noticing $m_l(R) \le m_l(R/2)$,

$$(1-\eta)\omega_l(R) \ge (1-\eta+\varepsilon\eta)\omega_l(R/2),$$

or

$$\omega_l\left(\frac{R}{2}\right) \le \frac{1-\eta}{1-\eta+\varepsilon\eta}\omega_l(R), \quad \forall l \in A. \tag{11}$$

As $|\widetilde{Q}_R \setminus \bigcup_{l\in B}\Gamma_l| \ge |\widetilde{Q}_R|/N+1 > 0$, there exists $P_0 \in \widetilde{Q}_R \setminus \bigcup_{l\in B}\Gamma_l$ such that

$$u_l(P_0) \ge (1-\varepsilon)M_l(R/2) + \varepsilon m_l(R/2), \quad \forall l \in B. \tag{12}$$

For a fixed $\tilde{l} \in B$, let $\widetilde{P} \in \overline{Q}_{R/2}$ with $u_{\tilde{l}}(\widetilde{P}) = m_{\tilde{l}}(R/2)$. From the assumptions of the theorem (now $f_l(t) = t$) we have

$$K_2R_0^{-\gamma}d^\gamma(P_0, \widetilde{P}) \ge \delta(u_{\tilde{l}}(P_0) - u_{\tilde{l}}(\widetilde{P}))_+ - \sum_{l=1}^N (u_l(P_0) - u_l(\widetilde{P}))_- . \tag{13}$$

We estimate the two terms on the right-hand side of (13) as follows. By (12) and the choice of $\widetilde{P}$

$$u_{\tilde{l}}(P_0) - u_{\tilde{l}}(\widetilde{P}) \ge (1-\varepsilon)M_{\tilde{l}}(R/2) + \varepsilon m_{\tilde{l}}(R/2) - m_{\tilde{l}}(R/2) = (1-\varepsilon)\omega_{\tilde{l}}(R/2).$$

Next we see that for $l \in B$

$$u_l(P_0) - u_l(\widetilde{P}) \ge (1-\varepsilon)M_l(R/2) + \varepsilon m_l(R/2) - M_l(R/2) \ge -\varepsilon\omega_l(R/2),$$

and for $l \in A$, clearly,

$$u_l(P_0) - u_l(\widetilde{P}) \ge m_l(R) - M_l(R) = -\omega_l(R).$$

Putting these into (13) we get

$$K_2R_0^{-\gamma}R^\gamma \ge \delta(1-\varepsilon)\omega_{\tilde{l}}(R/2) - \varepsilon\sum_{l\in B}\omega_l(R/2) - \sum_{l\in A}\omega_l(R).$$

Summing this inequality over $\tilde{l} \in B$ results in

$$NK_2R_0^{-\gamma}R^\gamma \ge [\delta(1-\varepsilon) - N\varepsilon]\sum_{l\in B}\omega_l(R/2) - N\sum_{l\in A}\omega_l(R).$$

Now we choose ε by letting $\delta(1-\varepsilon) - N\varepsilon = \delta/2$. Then

$$\sum_{l\in B}\omega_l\left(\frac{R}{2}\right) \le \frac{2N}{\delta}\sum_{l\in A}\omega_l(R) + \frac{2N}{\delta}K - 2R_0^{-\gamma}R^\gamma .$$

Together with (11) this implies

$$\sum_{l=1}^N \omega_l\left(\frac{R}{2}\right) \le \left(\frac{1-\eta}{1-\eta+\varepsilon\gamma} + \frac{2N}{\delta}\beta\right)\sum_{l\in A}\omega_l(R)$$
$$+ \frac{2N}{\delta}\beta K_2R_0^{-\gamma}R^\gamma + (1-\beta)\sum_{l\in B}\omega_l(R),$$

where β is an arbitrary number in $(0, 1)$. We choose it so that

$$\frac{1-\eta}{1-\eta+\varepsilon\gamma}+\frac{2N}{\delta}\beta=1-\beta,$$

i.e.,

$$\beta=\left(\frac{2N}{\delta}+1\right)^{-1}\frac{\varepsilon\eta}{1-\eta+\varepsilon\gamma}.$$

Then

$$\sum_{l=1}^{N}\omega_l(R/2)\le\tilde{\eta}\sum_{l=1}^{N}\omega_l(R)+K_2R_0^{-\gamma}R^{\gamma},$$

where

$$\tilde{\eta}=\frac{1-\eta+\frac{2N}{2N+\delta}\varepsilon\gamma}{1-\eta+\varepsilon\gamma}.$$

We take $\gamma_1=\gamma$ when $(1+\tilde{\eta})/2\le 1/2^{\gamma}$ and γ_1 so that $2^{-\gamma_1}=(1+\tilde{\eta})/2$ when $(1+\tilde{\eta})/2>1/2^{\gamma}$. Setting

$$\psi(R)=\sum_{l=1}^{N}\omega_l(R)+\frac{2K_2}{1-\tilde{\eta}}R_0^{-\gamma_1}R^{\gamma_1},$$

we have

$$\psi(R/2)\le\tilde{\eta}\psi(R)=2^{-\beta}\psi(R),\qquad \beta=-\log\tilde{\eta}/\log 2.$$

It follows that

$$\psi(R)\le\frac{1}{\tilde{\eta}}\left(\frac{R}{R_0}\right)^{\beta}\psi(R_0),\quad \forall R\le R_0.$$

Therefore

$$\sum_{l=1}^{N}\omega_l(R)\le\frac{1}{\tilde{\eta}}R_0^{-\beta}R^{\beta}\left(\frac{2K_2}{1-\tilde{\eta}}+\sum_{l=1}^{N}\omega_l(R_0)\right).$$

Recall that (x^0, t^0) is an arbitrary point in Q. The proof of the theorem is complete. □

THEOREM 9. *Let* $u_l\in W_{n+1}^{2,1}(Q)$ $(1\le l\le N)$, $Q=\Omega\times(0, T]$, *and* g_i $(1\le i\le n)$ *be functions defined in* Q *satisfying*

$$\left|\mathscr{L}u_l-\sum_{i=1}^{n}g_iu_{lx_i}\right|\le K_2\left(1+\sum_{l,i}u_{lx_i}^2\right)^{\alpha},\qquad \alpha\in(0, 1),$$

and

$$\sum_{i=1}^{n}g_i^2\le K_2\left(1+\sum_{l,i}u_{lx_i}^2\right),$$

where

$$\mathscr{L} = a_{ij}(x, t)\frac{\partial^2}{\partial x_i \partial x_j} + b_i(x, t)\frac{\partial}{\partial x_i} + c(x, t) - \frac{\partial}{\partial t},$$

$$\lambda|\xi|^2 \le a_{ij}\xi_i\xi_j \le \Lambda|\xi|^2, \quad \forall \xi \in \mathbb{R}^n,$$

$$0 < \lambda \le \Lambda < \infty, \quad \text{and} \quad |b_i|, |c| \le K_1.$$

Then there exist constants $\beta \in (0, 1)$ *and* $M > 0$ *such that*

$$\sum_{l=1}^{N} |u_l(P_2) - u_l(P_1)| \le M\, d_{P_1P_2}^{-\beta}\, d^\beta(P_1, P_2), \quad \forall P_1, P_2 \in Q.$$

In fact β *depends only on* n, λ, Λ, N; M *depends only on* n, λ, Λ, N, α, K_1, K_2, *and* $\sup_Q |u_l| = [u_l]_0$ $(1 \le l \le N)$.

REMARK. The requirement $\alpha < 1$ in this theorem is essential. The theorem ceases to hold when $\alpha = 1$. Consider the example due to E. Heinz. The system

$$u_{1t} - u_{1xx} = u_1(u_{1x}^2 + u_{2x}^2), \qquad u_{2t} - u_{2xx} = u_2(u_{1x}^2 + u_{2x}^2)$$

admits a pair of solutions $u_1 = \cos mx$, $u_2 = \sin mx$, $\forall x \in \mathbb{R}$. Since m can be arbitrarily large, one does not expect a uniform Hölder estimation on u_1 and u_2 in any compact subset of Ω.

PROOF. Let $v_l = \exp(\eta u_l + \sigma \sum_{m=1}^N u_m^2)$, where η, σ will be determined later. We have

$$\begin{aligned}
\frac{1}{v_l}\mathscr{L}v_l &= c + \eta\mathscr{L}u - cu_l + 2\sigma \sum u_m(\mathscr{L}u_m - cu_m) \\
&\quad + 2\sigma \sum a_{ij}u_{mx_i}u_{mx_j} + \sum a_{ij}\left(\eta u_{lx_i} + 2\sigma \sum u_m u_{mx_i}\right) \\
&\quad \cdot \left(\eta u_{lx_j} + 2\sigma \sum u_m u_{mx_j}\right) \\
&\ge c\left(1 - \eta u_l - 2\sigma \sum u_m^2\right) + \sum g_i\left(\eta u_{lx_i} + 2\sigma \sum u_m u_{mx_i}\right) \\
&\quad - K_2\left(\eta + 2\sigma \sum |u_m|\right)\left(1 + \sum_{l,i} u_{lx_i}^2\right)^\alpha + 2\sigma\lambda \sum u_{mx_i}^2 \\
&\quad + \lambda \sum \left(\eta u_{lx_i} + 2\sigma \sum u_m u_{mx_i}\right)^2 \\
&\ge 2\sigma\lambda \sum u_{mx_i}^2 - \frac{1}{2\lambda}\sum g_i^2 - K_2\left(\eta + 2\sigma \sum |u_m|\right) \\
&\quad \cdot \left(1 + \sum_{l,i} u_{lx_i}^2\right)^\alpha + c\left(1 - \eta u_l - 2\sigma \sum u_m^2\right).
\end{aligned}$$

Using the inequality satisfied by g_i and letting $\sigma = K_2/2\lambda^2$ we have

$$\frac{1}{v_l}\mathscr{L}v_l \geq \frac{K_2}{\lambda}\sum u_{mx_i}^2 - \frac{K_2^2}{2\lambda} - K_2\left(\eta + \frac{K_2}{2\lambda^2}\sum|u_m|\right)\left(1+\sum_{l,i}u_{lx_i}^2\right)^\alpha$$
$$+c\left(1-\eta u_l - \frac{K_2}{\lambda^2}\sum u_m^2\right) \geq -C,$$

where C depends only on n, N, λ, K_1, K_2, η, α, and $\max_{1\leq l\leq N}[u_l]_0$.

Without loss of generality we take

$$\sum_{l=1}^N u_l(x, t) = 0, \quad \forall(x, t)\in Q. \tag{14}$$

(Otherwise we let $u_{N+k} = -u_k$ $(1\leq k\leq N)$ and consider $\{u_l\}_1^{2N}$ instead of $\{u_l\}_1^N$.)

By taking $f_l(w) = \log w$ $(1\leq l\leq N)$ in Theorem 8 we obtain

$$\begin{aligned}
I &\equiv \delta\sum(\log v_l(P_2) - \log v_l(P_1))_+ - \sum(\log v_l(P_2) - \log v_l(P_1))_- \\
&\leq \delta\left[\eta\sum(u_l(P_2)-u_l(P_1))_+ + N\sigma\sum|u_m^2(P_2) - u_m^2(P_1)|\right] \\
&\quad + \left[\eta\sum(u_l(P_2)-u_l(P_1))_- + N\sigma\sum|u_m^2(P_2) - u_m^2(P_1)|\right] \\
&\leq \delta\eta\sum(u_l(P_2)-u_l(P_1))_+ - \eta\sum(u_l(P_2)-u_l(P_1))_- \\
&\quad + 2(\delta+1)N\sigma\max_{1\leq l\leq N}[u_l]_0\sum|u_l(P_2)-u_l(P_1)| \\
&= [\delta\eta + 2(\delta+1)N\sigma\max[u_l]_0]\sum(u_l(P_2)-u_l(P_1))_+ \\
&\quad - [\eta - 2(\delta+1)N\sigma\max[u_l]_0]\sum(u_l(P_2)-u_l(P_1))_-.
\end{aligned}$$

We choose δ so that the brackets in the two terms above are equal, that is,

$$\delta = \frac{\eta - 2NK_2\max[u_l]_0/\lambda^2}{\eta + 2NK_2\max[u_l]_0/\lambda^2}.$$

For sufficiently large η, δ becomes positive. By (14) $I\leq 0$. We may now appeal to Theorem 8 to conclude that there exist positive constants M and $\beta\in(0, 1)$ such that

$$|v_l(P_2) - v_l(P_1)| \leq M\, d_{P_1P_2}^{-\beta}\, d^\beta(P_1, P_2), \qquad 1\leq l\leq N.$$

Consequently

$$\begin{aligned}
&\eta|u_l(P_2)-u_l(P_1)| - 2\sigma\max[u_m]_0\sum|u_m(P_2)-u_m(P_1)| \\
&\quad \leq M_1\, d_{P_1P_2}^{-\beta}\, d^\beta(P_1, P_2).
\end{aligned}$$

Summing over l we have

$$\left(\eta - \frac{K_2 N}{\lambda^2}\max[u_m]_0\right)\sum |u_l(P_2) - u_l(P_1)| \le M_1\, d_{P_1P_2}^{-\beta}\, d^{\beta}(P_1\,,\, P_2)\,.$$

The theorem follows after taking $\eta = 1 + 2K_2 N \max[u_m]_0/\lambda^2$. □

A question worth further investigation is whether it is possible to derive appropriate density theorems for degenerate parabolic equations.

CHAPTER IX

Fully Nonlinear Parabolic Equations

Let Ω be a bounded domain in $\mathbb{R}^n$ and $Q=\Omega\times(0,T]$. Let $F(x,t,z,p,r)$ be a function defined in $S = Q \times \mathbb{R} \times \mathbb{R}^n \times \mathscr{S}^n$ where $\mathscr{S}^n$ is the family of all $n \times n$ real symmetric matrices. In this chapter we present a study of the fully nonlinear parabolic equation

$$u_t - F(x, t, u, Du, D^2u) = 0, \tag{1}$$

which includes *a priori* estimation for its solution ([42], [43]) and unique solvability of the first boundary value problem, that is, with

$$u|_{\partial^* Q} = \varphi \tag{2}$$

being prescribed along the parabolic boundary $\partial^* Q = \partial\Omega\times[0, T]\cup\Omega\times\{t = 0\}$ under certain conditions on F ([44], [45]). Precise assumptions on F will be given while we proceed. Roughly speaking, they are

(i) F is twice continuously differentiable with respect to x, z, p, r, and continuously differentiable with respect to t in S.

(ii) There are λ, Λ with $0 < \lambda \le \Lambda < \infty$ such that

$$\lambda|\xi|^2 \le \frac{\partial F}{\partial r_{ij}}\xi_i\xi_j \le \Lambda|\xi|^2, \quad \forall\xi \in \mathbb{R}^n,$$

in S.

(iii) F is convex in r, i.e.,

$$\frac{\partial^2 F}{\partial r_{ij}\partial r_{kl}}\eta_{ij}\eta_{kl} \le 0, \quad \forall\eta_{ij} \in \mathscr{S}^n.$$

(iv) F satisfies a natural structure condition.

Principal examples of fully nonlinear equations are the Bellman equation $F(x, t, u, Du, D^2u) = \sup_{\nu\in N}[a_{ij}(\nu)u_{ij} + b_i(\nu)u_i + c(\nu)u + f(\nu, x)] = 0$ and the Monge-Ampère equation $F = \det(u_{ij}) + f(x, t, u, Du) = 0$. However, the Bellman equation is not smooth and the Monge-Ampère equation degenerates on the boundary. To treat them we need to approximate them by fully nonlinear equations discussed above. We shall not consider the approximation here.

§1. A uniform bound for a solution u and an interior estimate for $D_x u$

To begin with we derive a bound on a solution of (1) and (2).

Theorem 1. *Suppose that* $(\partial F/\partial r_{ij})\xi_i\xi_j \geq 0$, $\forall \xi \in \mathbb{R}^n$ *in* S, *and*

$$uF(x, t, u, 0, 0) \leq \mu_1 u^2 + \mu_2, \quad \forall (x, t, u) \in Q \times \mathbb{R},$$

where μ_1, μ_2 *are positive constants. Let* u *be a solution of* (1) *and* (2) *in* $C^{2,1}(Q) \cap C(\overline{Q})$. *Then*

$$\max_{\overline{Q}} |u| \leq \inf_{\lambda > \mu_1} \left\{ e^{\lambda T} \max \left[\left(\frac{\mu_2}{\lambda - \mu_1} \right)^{1/2}, \sup_{\partial^* Q} |\varphi| \right] \right\} \equiv M_0.$$

Proof. We can write (1) in the form

$$u_t - a_{ij} u_{ij} = F(x, t, u, Du, 0),$$

where

$$a_{ij} = \int_0^1 \frac{\partial F}{\partial r_{ij}}(x, t, u, Du, \tau D^2 u)\, d\tau.$$

The desired estimate follows by considering the maximum of v^2, $v = ue^{-\lambda t}$ $(\lambda > \mu_1)$, in $\overline{Q}$. □

Denote

$$[u]_{k,j} = \sup_{P \in Q} d_P^{k+j} |D_x^j u(P)|,$$

and

$$[u]_{k, j+\beta} = \sup_{P_1, P_2 \in Q} d_{P_1 P_2}^{k+j+\beta} \frac{|D_x^j u(P_2) - D_x^j u(P_1)|}{d^\beta(P_1, P_2)},$$

where $d_P = d(P, \partial^* Q)$, $d_{P_1 P_2} = \min\{d_{P_1}, d_{P_2}\}$, and $d(P_1, P_2) = (|x^2 - x^1|^2 + |t^2 - t^1|)^{1/2}$ for $P_i = (x^i, t^i)$, $i = 1, 2$. When $k = 0$ we write $[u]_{0,j}$ and $[u]_{0, j+\beta}$ simply as $[u]_j$ and $[u]_{j+\beta}$ respectively.

Theorem 2. *Let* u *be a solution of* (1) *in* Q *with* $\sup_Q |u| \leq M_0$. *Suppose that* F *satisfies*

$$\lambda |\xi|^2 \leq \frac{\partial F}{\partial r_{ij}} \xi_i \xi_j \leq \Lambda |\xi|^2, \quad \forall \xi \in \mathbb{R}^n, \quad \forall (x, t, z, p, r) \in S;$$

$$|F(x, t, z, p, 0)| \leq \mu_1 (1 + |p|^2), \quad \forall (x, t, z, p) \in Q \times [-M_0, M_0] \times \mathbb{R}^n.$$

Then there exist $\beta \in (0, 1)$ *and* $M_\beta > 0$ *such that* $[u]_\beta \leq M_\beta$, *where* β, M_β *depend only on* $n, \lambda, \Lambda, \mu_1, M_0$, *and* diam Q.

Proof. As in the proof of Theorem 1, (1) can be written as

$$a_{ij} u_{ij} + F(x, t, u, Du, 0) - u_t = 0,$$

where

$$a_{ij} = \int_0^1 \frac{\partial F}{\partial r_{ij}}(x, t, u, Du, \tau D^2 u)\, d\tau.$$

Letting

$$g_i = -\frac{F(x, t, u, Du, 0)}{1+|Du|^2} u_i \qquad (1 \le i \le n),$$

we have

$$a_{ij}u_{ij} - u_t - g_i u_i = -\frac{F(x, t, u, Du, 0)}{1+|Du|^2}.$$

According to Theorem 9 of the previous chapter there are $\beta \in (0, 1)$ and $C > 0$ such that

$$|u(P_2) - u(P_1)| \le C d_{P_1P_2}^{-\beta} d^{\beta}(P_1, P_2). \quad \square$$

Next we derive an interior estimate for $D_x u$.

THEOREM 3. *Let $D_x u \in C^{2,1}(Q)$ where u solves* (1) *in Q and $[u]_0 \le M_0$, $[u]_\beta \le M_\beta$. Suppose that F satisfies*

$$\lambda|\xi|^2 \le \frac{\partial F}{\partial r_{ij}} \xi_i \xi_j \le \Lambda |\xi|^2, \quad \forall \xi \in \mathbb{R}^n,$$

$$|F| + |F_z| + |F_p|(1+|p|) + |F_x|(1+|p|)^{-1} \le \mu_2(1 + |p|^2 + |r|).$$

Then $[u]_1 \le C$ for some C depending only on $n, \lambda, \Lambda, \mu_2, M_0, \beta, M_\beta$, and $\operatorname{diam} Q$.

PROOF. Without loss of generality we may assume $M = [u]_1 < \infty$ for otherwise we can replace Q by the parabolic-closed domain $Q_\varepsilon \subseteq Q$ and then let $\varepsilon \to 0$. The theorem is true when $M < 2\sqrt{2}$. So we can assume $M \ge 2\sqrt{2}$.

There exists $\overline{P}(\overline{x}, \overline{t}) \in Q$ such that $M/2 \le d_{\overline{P}} |D_x u(\overline{P})| \le M$, or

$$\frac{M}{2d_{\overline{P}}} \le |D_x u(\overline{P})| \le \frac{M}{d_{\overline{P}}}.$$

Consider the cylinder $Q_{d_{\overline{P}}/M}(\overline{P})$. For all $P \in Q_{d_{\overline{P}}/M}(\overline{P})$ we have

$$d_{\overline{P}} - \sqrt{2}\frac{d_{\overline{P}}}{M} \le d_P \le d_{\overline{P}} + \sqrt{2}\frac{d_{\overline{P}}}{M}.$$

As $M \ge 2\sqrt{2}$, it follows that

$$\frac{1}{2} d_{\overline{P}} \le d_P \le \frac{3}{2} d_{\overline{P}}.$$

Incidentally we have

$$Q_{d_{\overline{P}}/M}(\overline{P}) \subseteq Q, \qquad |D_x u(P)| \le \frac{M}{d_P} \le 2\frac{M}{d_{\overline{P}}}.$$

We consider the function

$$v(x, t) = \eta^2(x, t)|D_x u|^2 + \alpha \frac{M^2}{d_{\overline{P}}^2}(u(x, t) - u(\overline{P}))^2, \qquad (x, t) \in \overline{Q_{d_{\overline{P}}/M}(\overline{P})},$$

where α is a positive constant to be chosen and η is a cut-off function in $Q_{d_{\overline{P}}/M}(\overline{P})$ satisfying

$$0 \le \eta(x, t) \le 1, \qquad \eta(\overline{P}) = 1, \qquad \eta|_{\partial^* Q_{d_{\overline{P}}/M}(\overline{P})} = 0,$$

$$|D_x \eta| \le C\frac{M}{d_{\overline{P}}}, \qquad |\eta_t| \le C\frac{M^2}{d_{\overline{P}}^2}, \quad \text{and} \quad |D_x^2 \eta| \le C\frac{M^2}{d_{\overline{P}}^2}.$$

Suppose that v attains its maximum over $\overline{Q_{d_{\overline{P}}/M}(\overline{P})}$ at $P_0(x^0, t^0)$. Then

$$\begin{aligned}\frac{M^2}{4d_{\overline{P}}^2} &\le |D_x u(\overline{P})|^2 = v(\overline{P}) \le v(P_0) \\ &= \eta^2(P_0)|D_x u(P_0)|^2 + \alpha\frac{M^2}{d_{\overline{P}}^2}(u(P_0) - u(\overline{P}))^2 \\ &\le \eta^2(P_0)|D_x u(P_0)|^2 + \alpha\frac{M^2}{d_{\overline{P}}^2}M_\beta^2\left(\frac{d_{\overline{P}}}{M}\right)^{2\beta}.\end{aligned}$$

We consider the following two cases separately.

CASE I. $\alpha M_\beta^2(d_{\overline{P}}/M)^{2\beta} \ge 1/8$. In this case $M \le (8\alpha M_\beta^2)^{1/2\beta} d_{\overline{P}}$.

CASE II. $\alpha M_\beta^2(d_{\overline{P}}/M)^{2\beta} < 1/8$. We have

$$\frac{M^2}{8d_{\overline{P}}^2} \le \eta^2(P_0)|D_x u(P_0)|^2 \le \eta^2(P_0)\cdot 4\frac{M^2}{d_{\overline{P}}^2}.$$

So $\eta^2(P_0) \ge 1/32$. Notice this also implies that P_0 does not belong to $\partial^* Q_{d_{\overline{P}}/M}$.

Consider the parabolic operator

$$\mathscr{L}v = \frac{\partial F}{\partial r_{ij}}v_{ij} + \frac{\partial F}{\partial p_i}v_i - v_t.$$

At P_0

$$\begin{aligned}0 \ge \mathscr{L}v = |D_x u|^2\mathscr{L}(\eta^2) + \eta^2\mathscr{L}(|D_x u|^2) + 2\frac{\partial F}{\partial r_{ij}}(\eta^2)_i(|D_x u|^2)_j \\ + 2\alpha\frac{M^2}{d_{\overline{P}}^2}(u - u(\overline{P}))\mathscr{L}u + 2\alpha\frac{M^2}{d_{\overline{P}}^2}\frac{\partial F}{\partial r_{ij}}u_i u_j,\end{aligned}$$

$$\begin{aligned}
|\mathscr{L}(\eta^2)| &= \left|\frac{\partial F}{\partial r_{ij}}(\eta^2)_{ij} + \frac{\partial F}{\partial P_i}(\eta^2)_i - (\eta^2)_t\right| \\
&\le C_1(1+|p|^2+|r|), \\
\mathscr{L}(|D_x u|^2) &= 2u_k \mathscr{L} u_k + 2\frac{\partial F}{\partial r_{ij}} u_{ki} u_{kj}, \\
|\mathscr{L} u_k| &= \left|\frac{\partial F}{\partial r_{ij}}(u_{ij})_k + \frac{\partial F}{\partial P_i}(u_i)_k - \frac{dF}{dx_k}\right| = \left|\frac{\partial F}{\partial u}u_k + \frac{\partial F}{\partial x_k}\right| \\
&\le C_2(1+|p|)(1+|p|^2+|r|). \\
|\mathscr{L} u| &= \left|\frac{\partial F}{\partial r_{ij}}u_{ij} + \frac{\partial F}{\partial P_i}u_i - F\right| \le C_3(1+|p|^2+|r|).
\end{aligned}$$

Combining with the fact $\eta^2(P_0) \ge 1/32$ we obtain

$$\begin{aligned}
\frac{\alpha\lambda}{2}|p|^4 + \frac{\lambda}{16}|r|^2 &\le 2\alpha \frac{M^2}{d_{\overline{P}}^2}\frac{\partial F}{\partial r_{ij}}u_i u_j + 2\eta^2 \frac{\partial F}{\partial r_{ij}} u_{ki}u_{kj} \\
&\le |D_x u|^2 |\mathscr{L}(\eta^2)| + 2\eta^2 |u_k \mathscr{L} u_k| + 2\left|\frac{\partial F}{\partial r_{ij}}(\eta^2)_i(|D_x u|^2)_j\right| \\
&\quad + 2\alpha\frac{M^2}{d_{\overline{P}}^2}|(u-u(\overline{P}))\mathscr{L} u| \\
&\le C_4(1+|p|^2)(1+|p|^2+|r|) + C_5 \alpha M_\beta |p|^{2-\beta}(1+|p|^2+|r|) \\
&\le C_6|p|^4 + \frac{\lambda}{32}|r^2| + C_7\alpha M_\beta |p|^{-\beta}(|p|^4+|r|^2),
\end{aligned}$$

provided $|p| \ge 1$. When $|p| < 1$, $M^2 \le 8d_{\overline{P}}^2$ and the theorem clearly holds.

Consequently

$$\left(\frac{\alpha\lambda}{2} - C_6\right)|p|^4 + \frac{\lambda}{32}|r|^2 \le C_7 \alpha M_\beta |p|^{-\beta}(|p|^4+|r|^2).$$

Taking $\alpha = \frac{2C_6}{\lambda} + \frac{1}{16}$ we conclude

$$\frac{M^2}{8d_{\overline{P}}^2} \le |p|^2 \le \left(\frac{32C_7\alpha M_\beta}{\lambda}\right)^{2/\beta}.$$

Therefore in both cases

$$M \le \max\left\{(8\alpha M_\beta^2)^{1/2\beta},\ \sqrt{8},\ \sqrt{8}\left(\frac{32C_7\alpha M_\beta}{\lambda}\right)^{1/\beta}\right\} d_{\overline{P}}. \quad \square$$

§2. An interior estimate for the second derivatives

The estimate $[u]_1 \le M_1$ obtained in the previous section is not in the right form for this section. Since we are concerned with interior estimation, we can replace Q by a smaller domain $Q_\varepsilon = \Omega_\varepsilon \times (\varepsilon^2, T]$ in which $|D_x u| \le M_1/\varepsilon$ holds. For simplicity we denote Q_ε, M_1/ε by Q, M_1 respectively. As a preliminary step to the estimation for the second derivatives we derive a C^α-estimate for $D_x u$.

THEOREM 4. *Let u be a solution of (1) in Q with $D_x u \in C^{2,1}(Q)$, $[u]_0 \le M_0$, and $D_x u \le M_1$. Suppose that F satisfies*

$$\lambda|\xi|^2 \le \frac{\partial F}{\partial r_{ij}}\xi_i\xi_j \le \Lambda|\xi|^2, \quad \forall \xi \in \mathbb{R}^n$$

and

$$|F| + |F_z| + |F_p| + |F_x| \le \mu_3(1 + |r|)$$

in $Q \times [-M_0, M_0] \times \{|p| \le M_1\} \times \mathcal{S}^n$. Then there are $\beta \in (0, 1)$ and $C > 0$ such that $[u]_{1+\beta} \le C$. Here β and C depend only on n, λ, Λ, u_3, and $\operatorname{diam}\Omega$.

PROOF. Let $v^k = u_k$ $(1 \le k \le n)$ and

$$\mathcal{L} = \frac{\partial F}{\partial r_{ij}} \frac{\partial^2}{\partial x_i \partial x_j} - \frac{\partial}{\partial t}.$$

Differentiating (1) with respect to x_k yields

$$\mathcal{L} v^k + \frac{\partial F}{\partial p_i} v_i^k = -\frac{\partial F}{\partial z} v^k - \frac{\partial F}{\partial x_k}.$$

Letting $g_i = -\partial F/\partial p_i$ we have

$$|\mathcal{L} v^k - g_i v_i^k| \le C\left(1 + \sum_{i=1}^n |Dv^i|\right), \qquad 1 \le k \le n,$$

and

$$\sum_{i=1}^n g_i^2 \le C\left(1 + \sum_{k=1}^n |Dv^k|^2\right).$$

By Theorem 9 of the previous chapter a Hölder estimate for $D_x u$ follows. □

Next we derive an upper estimate for $D_x^2 u$. We shall write the directional derivatives as

$$u_\xi = \sum_{i=1}^n u_i \xi_i,$$

and

$$u_{\xi\xi} = \sum_{i=1}^n u_{\xi i}\xi_i = \sum_{i,j=1}^n u_{ij}\,\xi_i\xi_j.$$

THEOREM 5. *Let u be a solution of (1) in Q with $D_x^2 u \in C^{2,1}(Q)$, $[u]_0 \le M_0$, $[u]_1 \le M_1$, $[u]_{1+\beta} \le M_{1+\beta}$. Suppose in addition to the conditions in Theorem 4 F further satisfies*

$$F_{\xi\xi} \le \mu_4 \left\{ \left(1 + \sum_{i,j} |u_{ij}|^2\right)^{3/2} (|\tilde{x}|^2 + |\tilde{u}|^2) + \left(1 + \sum_{i,j} |u_{ij}|\right) \sum_i |\tilde{u}_i|^2 \right.$$
$$\left. + \left[\left(1 + \sum_{i,j} |u_{ij}|\right) \cdot (|\tilde{x}| + |\tilde{u}|) + \sum_i |\tilde{u}_i|\right] \sum_{i,j} |\tilde{u}_{ij}| \right\}$$

for $\forall (x, t, u, p, r) \in Q \times [-M_0, M_0] \times \{|p| \le M_1\} \times \mathcal{S}^n$, and $\forall \xi = (\tilde{x}, \tilde{u}, \tilde{u}_i, \tilde{u}_{ij}) \in \mathbb{R}^{2n+1} \times \mathcal{S}^n$. Then

$$\sup_{P \in Q} \left\{ d_P^2 \max_{|\xi|=1} [u_{\xi\xi}(P)]_+ \right\} \le C,$$

where C depends only on n, λ, Λ, μ_3, μ_4, M_0, M_1, $M_{1+\beta}$, β, and $\operatorname{diam} Q$.

REMARK 1. Taking $\tilde{x} = \tilde{u} = \tilde{u}_1 = \cdots = \tilde{u}_n = 0$, we see that $F_{\xi\xi} \le 0$, i.e., F is concave in r.

REMARK 2. $\max_{|\xi|=1} u_{\xi\xi}$ is the largest eigenvalue of the matrix (u_{ij}).

PROOF. Let

$$M = \sup_{P \in Q} \left\{ d_P^2 \max_{|\xi|=1} (u_{\xi\xi}(P))_+ \right\}.$$

We may assume $2\sqrt{2} \le M < \infty$. So

$$\max_{|\xi|=1} (u_{\xi\xi}(P))_+ = \max_{|\xi|=1} u_{\xi\xi}(P).$$

There is $\overline{P}(\overline{x}, \overline{t}) \in Q$ such that

$$\frac{M}{2d_{\overline{P}}^2} \le \max_{|\xi|=1} u_{\xi\xi}(\overline{P}) \le \frac{M}{d_{\overline{P}}^2}.$$

Consider the function

$$\eta^2 \left(\max_{|\xi|=1} u_{\xi\xi}\right)^2 + \alpha \frac{M^2}{d_{\overline{P}}^4} |D_x u - D_x u(\overline{P})|^2$$

in the cylinder $\overline{Q_{d_{\overline{P}}/M}(\overline{P})}$ where α is a constant to be specified and η is a cut-off function as in Theorem 3. Let $P_0(x^0, t^0)$ be a point in $\overline{Q_{d_{\overline{P}}/M}(\overline{P})}$ where the above function takes a maximum. We consider two separate cases.

CASE I. $\alpha |D_x u - D_x u(\overline{P})|^2 \ge \frac{1}{8}$ AT P_0. We have

$$M \le (8\alpha M_{1+\beta}^2)^{1/2\beta} d_{\overline{P}}^2.$$

CASE II. $\alpha|D_x u - D_x u(\overline{P})|^2 < \frac{1}{8}$ AT P_0. Then

$$\frac{M^2}{8d_{\overline{P}}^4} \le \eta^2 \left(\max_{|\xi|=1} u_{\xi\xi}\right)^2 \le \eta^2 \frac{16M^2}{d_{\overline{P}}^4}.$$

Consequently

$$\eta^2 \ge \frac{1}{128}, \qquad P_0 \in Q_{d_{\overline{P}}/M}(\overline{P}).$$

The maximization of the auxiliary function involves the maximum of another function. It is more convenient to regard it as the maximization of a single function. To this end, we observe that the maximal value is still the maximal value for

$$\eta^2 u_{\xi\xi}^2 + \alpha \frac{M^2}{d_{\overline{P}}^4} |D_x u - D_x u(\overline{P})|^2$$

in $Q_{d_{\overline{P}}/M}(\overline{P}) \times \{|\xi| = 1\}$, $u_{\xi\xi} \ge 0$. It is also the maximal value for the function

$$-(R - \eta u_{\xi\xi})^2 + \eta^2 u_{\xi\xi}^2 + \alpha \frac{M^2}{d_{\overline{P}}^4} |D_x u - D_x u(\overline{P})|^2$$

in $Q_{d_{\overline{P}}/M}(\overline{P}) \times \{|\xi| = 1\} \times \{0 \le R < \infty\}$. Finally, the function

$$v(x, t, \xi) = 2\eta u_{\xi\xi} - |\xi|^4 + \alpha \frac{M^2}{d_{\overline{P}}^4} |D_x u - D_x u(\overline{P})|^2$$

has the same maximal value in $Q_{d_{\overline{P}}/M}(\overline{P}) \times \mathbb{R}^n$, which it attains at (x^0, t^0, ξ^0) for some ξ^0.

At the maximum $\partial v/\partial x_i = 0$, $\partial v/\partial \xi_i = 0$ $(1 \le i \le n)$, $\partial v/\partial t \ge 0$ and a seminegative definite condition holds.

From $\partial v/\partial \xi_i = 0$ we get $|\xi^0|^2 = \eta(P_0) \max_{|\xi|=1} u_{\xi\xi}(P_0)$. So $|\xi^0| \le 4M/d_{\overline{P}}^2$. The aforementioned negative semidefinite condition is

$$(v_{ij}\hat{x}_i\hat{x}_j + 2v_{i\xi_j}\tilde{x}_i\tilde{\xi}_j + v_{\xi_i\xi_j}\tilde{\xi}_i\tilde{\xi}_j) \le 0, \quad \forall \tilde{x}, \tilde{\xi} \in \mathbb{R}^n.$$

To get rid of $\tilde{\xi}$ in this expression we proceed in the following way. Observe that for $\varepsilon > 0$,

$$[v_{ij}\tilde{x}_i\tilde{x}_j + 2v_{i\xi_j}\tilde{x}_i\tilde{\xi}_j + (v_{\xi_i\xi_j} - \varepsilon\delta_{ij})\tilde{\xi}_i\tilde{\xi}_j] < 0$$

for $\tilde{x}$, $\tilde{\xi} \in \mathbb{R}^n$, $|\tilde{\xi}| \ne 0$. At the point where this expression attains a minimum,

$$v_{i\xi_j}\tilde{x}_i = (\varepsilon\delta_{ij} - v_{\xi_i\xi_j})\tilde{\xi}_i, \qquad 1 \le i \le n.$$

Putting this into the previous expression we have

$$(v_{ij} + v_{i\xi_k}(\varepsilon\delta_{kl} - v_{\xi_k\xi_l})^{-1} v_{j\xi_l})\tilde{x}_i\tilde{x}_j < 0$$

for $\tilde{x} \in \mathbb{R}^n$, $|\tilde{x}| \neq 0$. Using these inequalities we see that at (x^0, t^0, ξ^0)

$$\mathscr{L}v + \frac{\partial F}{\partial r_{ij}} v_{i\xi_k} (\varepsilon\delta_{kl} - v_{\xi_k\xi_l})^{-1} v_{j\xi_l} \leq 0,$$

where

$$\mathscr{L}v = \frac{\partial F}{\partial r_{ij}} v_{ij} + \frac{\partial F}{\partial p_i} v_i - v_t.$$

Now we estimate each term on the left of this inequality as follows.

$$\mathscr{L}v = 2\eta \mathscr{L} u_{\xi\xi} + 2u_{\xi\xi}\mathscr{L}\eta + 4\frac{\partial F}{\partial r_{ij}}\eta_i u_{\xi\xi j} + 2\alpha\frac{M^2}{d_{\overline{P}}^4}(u_k - u_k(\overline{P}))\mathscr{L}u_k$$

$$+ 2\alpha\frac{M^2}{d_{\overline{P}}^4}\frac{\partial F}{\partial r_{ij}} u_{ki}u_{kj},$$

$$\mathscr{L}u_{\xi\xi} = \frac{\partial F}{\partial r_{ij}}u_{\xi\xi ij} + \frac{\partial F}{\partial p_i}u_{\xi\xi i} - \frac{d^2F}{d\xi^2},$$

$$\frac{dF}{d\xi} = \frac{\partial F}{\partial r_{ij}}u_{ij\xi} + \frac{\partial F}{\partial p_i}u_{i\xi} + \frac{\partial F}{\partial z}u_\xi + F_\xi = F_{\tilde{\xi}},$$

$$\frac{d^2F}{d\xi^2} = \frac{\partial F}{\partial r_{ij}}u_{ij\xi\xi} + \frac{\partial F}{\partial p_i}u_{i\xi\xi} + \frac{\partial F}{\partial z}u_{\xi\xi} + F_{\tilde{\xi}\tilde{\xi}},$$

where $\tilde{\xi} = (\xi, u_\xi, u_{i\xi}, u_{ij\xi})$ is a $(n+1)^2$-vector. We have

$$\begin{aligned}\mathscr{L}u_{\xi\xi} &= -\frac{\partial F}{\partial z}u_{\xi\xi} - F_{\tilde{\xi}\tilde{\xi}} \geq -C(1+|r|^3)\\ &\quad - \mu_4\{(1+|r|^3)(|\xi|^2 + |u_\xi|^2) + (1+|r|)\sum |u_{i\xi}|^2\\ &\qquad + [(1+|r|)(|\xi| + |u_\xi|) + \sum |u_{i\xi}|]\sum |u_{ij\xi}|\}\\ &\geq -C[1+|r|^4 + (1+|r|^{3/2})|D_x^2 u_\xi|], \qquad |r| = (\sum u_{ij}^2)^{1/2}.\end{aligned}$$

Also

$$|\mathscr{L}\eta| \leq C\left[\frac{M^2}{d_{\overline{P}}^2} + (1+|r|)\frac{M}{d_{\overline{P}}}\right] \leq C(1+|r|^2),$$

$$|\mathscr{L}u_k| = \left|\frac{\partial F}{\partial z}u_k + \frac{\partial F}{\partial x_k}\right| \leq C(1+|r|),$$

and

$$\begin{aligned}|\eta_i u_{\xi\xi_j}| &\leq C\frac{M}{d_{\overline{P}}}(1+|r|^{1/2})|D_x^2 u_\xi|\\ &\leq C(1+|r|^{3/2})|D_x^2 u_\xi|.\end{aligned}$$

By diagonalizing the matrix $(\varepsilon\delta_{kl} - v_{\xi_k\xi_l})$

$$\zeta^T(\varepsilon\delta_{kl} - v_{\xi_k\xi_l})^{-1}\zeta \geq \frac{|\zeta|^2}{\operatorname{tr}(\varepsilon\delta_{kl} - v_{\xi_k\xi_l})}$$

for any $\zeta \in \mathbb{R}^n$. Therefore

$$\frac{\partial F}{\partial r_{ij}} v_{i\xi_k}(\varepsilon\delta_{kl} - v_{\xi_k\xi_l})^{-1} v_{j\xi_l} \geq \lambda v_{i\xi_k}(\varepsilon\delta_{kl} - v_{\xi_k\xi_l})^{-1} v_{i\xi_l}$$
$$\geq \lambda \frac{\sum |v_{i\xi_k}|^2}{\operatorname{tr}(\varepsilon\delta_{kl} - v_{\xi_k\xi_l})}.$$

Using

$$v_{i\xi_k} = 4\left(\eta \frac{\partial^2 u_\xi}{\partial x_i \partial x_k} + \eta_i \frac{\partial u_\xi}{\partial x_k}\right)$$

and

$$v_{\xi_k\xi_l} = 4\left(\eta \frac{\partial^2 u}{\partial x_k \partial x_l} - |\xi|^2 \delta_{kl} - 2\xi_k \xi_l\right),$$

we have

$$\sum |v_{i\xi_k}|^2 \geq 8|D_x^2 u_\xi|^2 - C(1 + |r|^5),$$

$$0 < \operatorname{tr}(\varepsilon\delta_{kl} - v_{\xi_k\xi_l}) = n\varepsilon + 4[(n+2)|\xi|^2 - \eta\Delta u]$$
$$\leq n\varepsilon + 4(n+2)|\xi|^2 \leq C|r|,$$

if we take $\varepsilon = |r|$. Combining all these inequalities we obtain

$$\alpha|r|^4 + C_1 \frac{|D_x^2 u_\xi|^2 - C_2(1 + |r|^5)}{|r|}$$
$$\leq C_3(1 + |r|^{3/2})|D_x^2 u_\xi| + C_4(1 + |r|^4) + C_5\alpha(1 + |r|^3).$$

So we conclude that $|r| \leq C_6$ by choosing $\alpha = C_1C_2 + C_3/4C_1^2 + C_4 + 1$. As a consequence $M \leq C_7 d_P^{\frac{2}{P}}$.

So we have bounded M in either case. The proof of this theorem is complete. □

Next we consider the interior estimation for u_t.

THEOREM 6. *Let u be a solution of* (1) *in Q, with $u_t \in C^{2,1}(Q)$, $[u]_0 \leq M_0$, and $|D_x u| \leq M_1$. Suppose in addition to the hypotheses in Theorem 4 that F further satisfies $|F_t| \leq \mu_5(1 + |r|^2)$ in $Q \times [-M_0, M_0] \times \{|p| \leq M_1\} \times \mathcal{S}^n$. Then*

$$[u_t]_{3,0} \leq C$$

for some constant C depending solely on n, λ, Λ, μ_3, μ_5, M_0, M_1, and diam Q.

PROOF. Let $M = [u_t]_{3,0}$. Without loss of generality we assume $M < \infty$. Fix $\overline{P}$ such that

$$\frac{M}{2d_{\overline{P}}^3} \le |u_t(\overline{P})| \le \frac{M}{d_{\overline{P}}^3}.$$

Consider the function

$$v(x, t) = \eta^2 u_t^2 + \alpha \frac{M}{d_{\overline{P}}^4} |D_x u|^2$$

in the cylinder $\overline{Q_{d_{\overline{P}}/2}(\overline{P})}$. Suppose that it attains its maximum at $P_0(x^0, t^0) \in \overline{Q_{d_{\overline{P}}/2}(\overline{P})}$. Then

$$\tfrac{1}{2} d_{\overline{P}} \le d_P \le \tfrac{3}{2} d_{\overline{P}},$$

and

$$u_t^2(P_0) \le \frac{M^2}{d_P^6} \le \frac{16M^2}{d_{\overline{P}}^6},$$

$$\begin{aligned}\frac{M^2}{4d_{\overline{P}}^4} &\le u_t^2(\overline{P}) \le \eta^2(P_0) u_t^2(P_0) + \frac{\alpha M}{d_{\overline{P}}^4} |D_x u(P_0)|^2 \\ &\le \eta^2(P_0) \frac{64M^2}{d_{\overline{P}}^6} + \frac{\alpha M_1^2 M}{d_{\overline{P}}^4}\end{aligned}$$

at P_0.

CASE I. $\alpha M_1^2 M / d_{\overline{P}}^6 \ge M^2 / 8d_{\overline{P}}^4$. Then $M \le 8M_1^2 d_{\overline{P}}^2 \alpha$.

CASE II. $\alpha M_1^2 M / d_{\overline{P}}^6 < M^2 / 8d_{\overline{P}}^4$. Then $\eta^2(P_0) \ge 1/512$, $P_0 \in Q_{d_{\overline{P}}/2}(\overline{P})$ and $u_t^2(P_0) \ge M^2 / 8d_{\overline{P}}^6$.

Let

$$\mathscr{L} v = \frac{\partial F}{\partial r_{ij}} v_{ij} + \frac{\partial F}{\partial p_i} v_i - v_t.$$

At P_0 we have

$$\begin{aligned}&2\eta^2 \frac{\partial F}{\partial r_{ij}} u_{ti} u_{tj} + 2\alpha \frac{M}{d_{\overline{P}}^4} \frac{\partial F}{\partial r_{ij}} u_{ki} u_{kj} \\ &\quad \le \left| u_t^2 \mathscr{L}(\eta^2) + 2\eta^2 u_t \mathscr{L}(u_t) + 2\frac{\partial F}{\partial r_{ij}} (\eta^2)_i (u_t^2)_j + 2\alpha \frac{M}{d_{\overline{P}}^2} u_k \mathscr{L}(u_k) \right|,\end{aligned}$$

as well as

$$u_t^2 [|\mathscr{L}(\eta^2)| + |D_x \eta|^2] \le C \frac{M}{d_{\overline{P}}^3} \left(\frac{1 + |r|^2}{d_{\overline{P}}} + \frac{1 + |r|}{d_{\overline{P}}^2} \right),$$

$$\begin{aligned}|u_t \mathscr{L}(u_t)| &= \left| u_t \left(\frac{\partial F}{\partial r_{ij}} u_{ijt} + \frac{\partial F}{\partial p_i} u_{it} - \frac{dF}{dt} \right) \right| = |u_t (F_u u_t + F_t)| \\ &\le \frac{4M}{d_{\overline{P}}^3} |F_u F + F_t| \le C \frac{M}{d_{\overline{P}}^3} (1 + |r|^2),\end{aligned}$$

and

$$|\mathscr{L}(u_k)| = \left|\frac{\partial F}{\partial u} u_k + \frac{\partial F}{\partial x_k}\right| \le c(1+|r|),$$

when $|r| < 1$, $|u_t| \le C$, and the theorem holds. When $|r| \ge 1$, the same conclusion holds by combining the inequalities above and choosing a suitable α. □

Finally we give an interior estimate for $D_x^2 u$.

THEOREM 7. *Under the hypotheses in Theorems* 5 *and* 6

$$[u]_{1,2} \le C,$$

where C *depends only on* $n, \lambda, \Lambda, \mu_3, \mu_4, \mu_5, M_0, M_1, M_{1+\beta}$, β *and* $\operatorname{diam} Q$.

PROOF. We write equation (1) in the form

$$u_t - F(x, t, u, u_i, 0) = a_{ij} u_{ij}, \tag{3}$$

where

$$a_{ij} = \int_0^1 \frac{\partial F}{\partial r_{ij}}(x, t, u, u_i, \tau u_{ij})\, d\tau.$$

For each (x, t), denote the eigenvalues and normalized eigenfunctions of the matrix (u_{ij}) by μ^k and $\xi^k = (\xi_1^k, \dots, \xi_n^k)$ respectively. We have $\mu^k = u_{\xi^k \xi^k}$ and $\mu^k \le C/d_P^2$ $(1 \le k \le n)$.

Let μ^1 be the smallest eigenvalue. Since the theorem is valid for $\mu^1 \ge 0$, we may assume $\mu^1 < 0$. Substituting $u_{ij} = \sum_{k=1}^n \mu^k \xi_i^k \xi_j^k$ into (3) yields

$$u_t - F(x, t, u, u_i, 0) \le \mu^1 a_{ij} \xi_i^1 \xi_j^1 + \frac{C}{d_P^2} \le \lambda \mu^1 + \frac{C}{d_P^2}.$$

Observe that

$$|u_t| \le \frac{C}{d_P^3}, \qquad |F(x, t, u, u_i, 0)| \le C|D_x u|^2 + C \le C.$$

Putting them into the inequality above we get

$$\lambda \mu^1 \ge -\frac{C}{d_P^3}.$$

So after all we have

$$|\mu^k| \le \frac{C}{d_P^3}, \qquad 1 \le k \le n;$$

$$|u_{ij}| = \left|\sum_{k=1}^n \mu^k \xi_i^k \xi_j^k\right| \le \frac{C}{d_P^3}.$$

The proof of the theorem is complete. □

§3. An interior Hölder estimate for the second derivatives

In this section we derive a $C^{2+\alpha,1+\alpha/2}$-interior estimate for u. We replace Q by $Q_\varepsilon = \Omega_\varepsilon \times (\varepsilon, T]$. However, for simplicity, we shall write Q_ε as Q (the bottom is still denoted by $t = 0$).

THEOREM 8. *Let u be a solution of* (1) *in Q with $u_t \in C^{2,1}(Q)$ and $\|u\|_{C^{2,1}(Q)} \le M_2$. Suppose that F satisfies*

$$\lambda|\xi|^2 \le \frac{\partial F}{\partial r_{ij}}\xi_i\xi_j \le \Lambda|\xi|^2, \qquad 0 < \lambda \le \Lambda < \infty,$$

for $\xi \in \mathbb{R}^n$, $(x, t, u, p, r) \in S$ and $|F|$, $|DF|$, $|D^2F| \le \mu_6$ in $Q \times \{|p| \le M_1\} \times \{\|r\| \le M_2\}$. Then there are $\beta \in (0, 1)$ and $C > 0$ such that $[u_t]_\beta \le C$.

Here β and C depend only on n, λ, Λ, μ_6, M_2, and $\operatorname{diam} Q$.

PROOF. By differentiating equation (1) with respect to t, we see that $v = u_t$ satisfies

$$\frac{\partial F}{\partial r_{ij}}v_{ij} + \frac{\partial F}{\partial P_i}v_i + \frac{\partial F}{\partial z}v + \frac{\partial F}{\partial t} - v_t = 0.$$

In other words, v is a solution for a parabolic equation with bounded measurable coefficients. The conclusion of the theorem now follows from Theorem 9 of the last chapter. □

In order to show that $D_x^2 u$ satisfies a Hölder condition we need a lemma on the decomposition of positive symmetric matrices.

Regard the collection of all $n \times n$ symmetric matrices $\mathcal{S}^n$ as a Euclidean space $\mathscr{A}$ under a Euclidean norm (that is, we don't consider the matrix multiplication). For $0 < \lambda < \infty$, set

$$\mathscr{A}[\lambda] = \{A \in \mathscr{A} : \lambda I < A < \lambda^{-1} I\}, \qquad (I \text{ is the identity matrix}).$$

LEMMA 9. *There exist a positive integer N, $\rho_0 > 0$, and unit vectors $\gamma_1, \dots, \gamma_N \in \mathbb{R}^n$ such that each A in $\mathscr{A}[\lambda]$ admits a representation of the form*

$$A = \sum_{i=1}^{N} \beta_i(A)\gamma_i \otimes \gamma_i, \qquad \rho_0 \le \beta_i(A) \le \rho_0^{-1}.$$

Here N, ρ_0, $\gamma_1, \dots, \gamma_N$ depend only on n and λ. Moreover $\gamma_i \otimes \gamma_i$ $(1 \le i \le N)$ span $\mathscr{A}$.

PROOF. It is easy to show that $\mathscr{A}[\lambda]$ is a bounded open convex subset of $\mathscr{A}$ and $\mathscr{A}[\lambda] \Subset \mathscr{A}[\lambda/2]$ (i.e., the closure of $\mathscr{A}[\lambda]$ is contained in $\mathscr{A}[\lambda/2]$). As a consequence there exists an open convex polytope π satisfying $\mathscr{A}[\lambda] \Subset \pi \subseteq \mathscr{A}[\lambda/2]$. π can be constructed in the following manner. Let $2d_1 = d(\mathscr{A}[3\lambda/4], \partial\mathscr{A}[\lambda/2]) > 0$ and consider the decomposition of $\mathscr{A}$ into cubes of length d_1 whose vertices are lattice points of $\mathscr{A}$. Let π be the convex polytope whose vertices consist of the vertices of the cubes with nonempty intersections with $\mathscr{A}[3\lambda/4]$. Then $\mathscr{A}[\lambda] \Subset \mathscr{A}[3\lambda/4] \subseteq \pi$. On the other hand, by the convexity of $\mathscr{A}[\lambda/2]$, $\pi \subseteq \mathscr{A}[\lambda/2]$.

Denote the vertices of π by $A_1, \dots, A_{N_1}$. Let $\lambda_i(k)$, $l_i(k)$ $(1 \le i \le n)$ be the eigenvalue and unit eigenvector of A_k respectively. We have $\lambda/2 < \lambda_i(k) < 2/\lambda$. Since $\mathscr{A}[\lambda] \Subset \pi$,

$$\mathscr{A}[\lambda] - \rho_0 \sum_{k=1}^{N_1} \sum_{i=1}^{n} l_i(k) \otimes l_i(k) \subseteq \pi$$

for small ρ_0. For a further restricted ρ_0, $\rho_0 + 2/\lambda \le 1/\rho_0$ holds.

As π is a convex polytope, for each $A \in \mathscr{A}[\lambda]$ there exist p_k $(1 \le k \le N_1)$, $p_k \ge 0$, $\sum p_k = 1$, such that

$$A - \rho_0 \sum_{k=1}^{N_1} \sum_{i=1}^{n} l_i(k) \otimes l_i(k) = \sum_{k=1}^{N_1} p_k A_k.$$

Therefore

$$A = \sum_{k=1}^{N_1} \sum_{i=1}^{n} (\rho_0 + p_k \lambda_i(k)) l_i(k) \otimes l_i(k).$$

Let $N = N_1 n$, $\beta_{ik} = \rho_0 + p_k \lambda_i(k)$, and re-order $\{l_i(k)\}_{1\le i\le n,\, 1\le k\le N_1}$ as $\{\gamma_i\}$ $(1 \le i \le N)$.

In view of $I \in \mathscr{A}[\lambda]$, $I + (1/2)(1/\lambda - 1)e_i \otimes e_i \in \mathscr{A}[\lambda]$ and $I + (1/3) \cdot (1/\lambda - 1)(e_i \pm e_j)/\sqrt{2} \otimes (e_i \pm e_j)/\sqrt{2} \in \mathscr{A}[\lambda]$ (e_i is a unit vector), we see that $\mathscr{A}[\lambda]$ spans $\mathscr{A}$. So $\mathscr{A}$ is spanned by $\gamma_i \otimes \gamma_i$ $(1 \le i \le N)$ too. □

THEOREM 10. *Let u be a solution of (1) in Q with $D_x^2 u \in C^{2,1}(Q)$, $\|u\|_{C^{2,1}(Q)} \le M_2$, $[u_t]_\beta \le M_{t\beta}$, $\beta \in (0, 1)$. Suppose that F fulfills all requirements in Theorem 8 as well as*

$$F_{\tilde{\xi}\tilde{\xi}} \le \mu_7[|\tilde{x}|^2 + |\tilde{u}|^2 + \sum |\tilde{u}_i|^2 + (|\tilde{x}| + |\tilde{u}| + \sum |\tilde{u}_i|) \sum |\tilde{u}_{ij}|]$$

for all $\tilde{\xi} = (\tilde{x}, \tilde{u}, \tilde{u}_i, \tilde{u}_{ij}) \in \mathscr{S}^n$. Then there are $\alpha \in (0, 1)$ and $C > 0$ such that

$$[D_x^2 u]_\alpha \le C$$

for some α, C depending solely on n, λ, Λ, μ_7, M_2, $M_{t\beta}$, β, and diam Q.

PROOF. Without loss of generality we take $\Lambda = 1/\lambda$. Consider

$$v^k = u_{\gamma_k \gamma_k} + \varepsilon \sum_{l=1}^{N} (M_2 + u_{\gamma_l \gamma_l})^2,$$

where N, ρ_0, $\gamma_1, \dots, \gamma_N$ are determined in Lemma 9 and ε is a positive number to be specified. As derived already in Theorem 5,

$$\mathscr{L} u_{\xi\xi} = -\frac{\partial F}{\partial z} u_{\xi\xi} - F_{\tilde{\xi}\tilde{\xi}}.$$

Therefore

$$\begin{aligned}\mathscr{L}v^k &= \mathscr{L}u_{\gamma_k\gamma_k} + 2\varepsilon\sum_l (M_2 + u_{\gamma_l\gamma_l})\mathscr{L}u_{\gamma_l\gamma_l} + 2\varepsilon\sum\frac{\partial F}{\partial r_{ij}}u_{\gamma_l\gamma_l i}u_{\gamma_l\gamma_l j}\\ &\geq -C(1+|D_x^3u|) + 2\varepsilon\lambda\sum_l |D_x u_{\gamma_l\gamma_l}|^2\\ &\geq -C_\varepsilon,\ (1\leq k\leq N).\end{aligned}$$

Notice that Lemma 9 has been used in the last inequality. As $\mathscr{A}[\lambda]$ spans $\mathscr{A}$, u_{ij} $(1\leq i, j\leq n)$ can be spanned by $u_{\gamma_l\gamma_l}$.

Next, from

$$(F(x, t, u, D_xu, D_x^2u) - u_t)|_{P_2} - (F(x, t, u, D_xu, D_x^2u) - u_t)|_{P_1} = 0,$$

and $[u_t]_\beta \leq M_{t\beta}$, $\|u\|_{C^{2,1}(Q)} \leq M_2$ we deduce

$$|a_{ij}u_{ij}(P_2) - a_{ij}u_{ij}(P_1)| \leq Kd_{P_1P_2}^{-\beta}d^\beta(P_1, P_2),$$

where the a_{ij}'s satisfy

$$\lambda|\xi|^2 \leq a_{ij}\xi_i\xi_j \leq \Lambda|\xi|^2, \quad \forall\xi\in\mathbb{R}^n.$$

By Lemma 9

$$\begin{aligned}Kd_{P_1P_2}^{-\beta}d^\beta(P_1, P_2) \geq a_{ij}(u_{ij}(P_2) - u_{ij}(P_1)) &= \sum_{k=1}^N \beta_k(U_{k^2} - U_{k^1})\\ &\geq \rho_0\sum_{k=1}^N (U_{k^2} - U_{k^1})_+ - \frac{1}{\rho_0}\sum_{k=1}^N (U_{k^2} - U_{k^1})_-\end{aligned}$$

where we have written U_{k^i} $(i = 1, 2; k = 1, 2, \ldots, N)$ for $u_{\gamma_k\gamma_k}(P_i)$. Also

$$v^k(P_2) - v^k(P_1) = U_{k^2} - U_{k^1} + \varepsilon\sum(U_{l^2} - U_{l^1})(2M_2 + U_{l^2} + U_{l^1}).$$

Therefore, for $0 < \delta \leq 1$ we have

$$\begin{aligned}&\delta\sum_{k=1}^N (v^k(P_2) - v^k(P_1))_+ - \sum_{k=1}^N (v^k(P_2) - v^k(P_1))_-\\ &\quad\leq \delta\sum(U_{k^2} - U_{k^1})_+ - \sum(U_{k^2} - U_{k^1})_- + C\varepsilon\sum|U_{k^2} - U_{k^1}|\\ &\quad= (\delta + C\varepsilon)\sum(U_{k^2} - U_{k^1})_+ - (1 - C\varepsilon)\sum(U_{k^2} - U_{k^1})_-\\ &\quad\leq \frac{2\rho_0}{2+\rho_0^2}\left[\rho_0\sum(U_{k^2} - U_{k^1})_+ - \frac{1}{\rho_0}\sum(U_{k^2} - U_{k^1})_-\right]\\ &\quad\leq Kd_{P_1P_2}^{-\beta}d^\beta(P_1, P_2),\end{aligned}$$

provided $C\varepsilon \leq \delta = \rho_0^2/(2+\rho_0^2)$. Appealing to Theorem 8 of the last chapter we conclude that there are $\alpha\in(0, \beta)$ and $C > 0$ such that

$$\sum|v^k(P_2) - v^k(P_1)| \leq Cd_{P_1P_2}^{-\alpha}d^\alpha(P_1, P_2).$$

From this inequality we further get

$$(1 - NC\varepsilon)\sum |U_{k^2} - U_{k^1}| \le C d_{P_1P_2}^{-\alpha} d^\alpha(P_1, P_2).$$

Taking ε so that $C\varepsilon = \min\{\delta, 1/2N\}$ we finally get

$$\sum |U_{k^2} - U_{k^1}| \le 2C d_{P_1P_2}^{-\alpha} d^\alpha(P_1, P_2).$$

Now the conclusion of the theorem follows since u_{ij} $(1 \le i, j \le n)$ can be written as a linear combination of $u_{\gamma_k\gamma_k}$ $(1 \le k \le N)$. □

§4. A near boundary Hölder estimate for u

In this section we investigate the behavior of the solution of (1) and (2) near the boundary. We shall divide our discussion into three regions: first near the bottom $\Omega \times \{t = 0\}$, second near the lateral boundary $\partial\Omega \times (0, T]$, and third near the intersection of the bottom and the lateral boundary $\partial\Omega \times \{t = 0\}$. Basically the last case can be treated at the same time with the second case. Let Ω be a bounded domain in $\mathbb{R}^n$ and $Q = \Omega \times (0, T]$. Suppose that the operator

$$L = a_{ij}\frac{\partial^2}{\partial x_i \partial x_j} - \frac{\partial}{\partial t} \tag{4}$$

satisfies

$$\lambda|\xi|^2 \le a_{ij}\xi_i\xi_j \le \Lambda|\xi|^2, \quad \forall \xi \in \mathbb{R}^n, \quad 0 < \lambda \le \Lambda < \infty$$

in Q.

LEMMA 11. *Let $v \in W^{2,1}_{n+1}(Q)$ satisfy $[v]_0 \le M_0$ and $\mathscr{L}v \ge -K_2(1 + |D_x v|^2)$ in Q where $\mathscr{L}$ is given in* (4). *Also suppose that $v(x, 0) \in C^\beta(\Omega)$ $(0 < \beta \le 1)$. Then there are $\beta_1 \in (0, \beta/2)$ and $C > 0$ such that for all $P(x, t) \in Q$,*

$$(v(x, t) - v(x, 0))_+ \le C([v]_{\beta, t=0} + 1) d_x^{-\beta} t^{\beta_1},$$

where

$$[v]_{\beta, t=0} = \sup_{x^1, x^2 \in \Omega} \frac{|v(x^2, 0) - v(x^1, 0)|}{|x^2 - x^1|^\beta}.$$

Here C depends only on n, λ, Λ, K_1, K_2, and M_0.

PROOF. Let

$$\eta(x) = \left(1 - \frac{|x - x^0|}{R_2}\right)^2 \quad \text{and} \quad w = \eta(e^{\sigma(v - v(x^0, 0))} - 1),$$

where $R < d(x^0) = d(x^0, \partial\Omega)$ and σ is to be specified. We have

$$\mathscr{L}w = \eta\sigma e^{\sigma(v - v(x^0, 0))}\left(\mathscr{L}v + \sigma a_{ij}v_iv_j + 2a_{ij}\frac{1}{\eta}\eta_i\eta_j\right)$$
$$+ [e^{\sigma(v - v(x^0, 0))} - 1]\mathscr{L}\eta.$$

Taking $\sigma = 2K_2/\lambda$, by the assumption on $\mathscr{L}v$ we get

$$\mathscr{L}w \geq -C_1 - C_2\frac{M_0}{R^2}.$$

By an application of the Aleksandrov maximum principle

$$\begin{aligned}
&\sup_{\substack{|x-x^0|\leq R\\ 0\leq t\leq \tau}} w(x, t)_+ \\
&\quad \leq \sup_{|x-x^0|\leq R} w(x, 0) + C_3 R^{n/(n+1)}\left(C_1 + C_2\frac{M_0}{R^2}\right)\left(\int_{\substack{|x-x^0|\leq R\\ 0\leq t\leq \tau}} dx\,dt\right)^{1/(n+1)} \\
&\quad \leq C_4(1 + [v]_{\beta, t=0})R^\beta + C_5\left(\frac{\tau}{R^2}\right)^{1/(n+1)}.
\end{aligned}$$

For $\tau < d_{x^0}^{2+(n+1)\beta}$, we choose $R = \tau^{1/[2+(n+1)\beta]}$. It follows that

$$[v(x^0, \tau) - v(x^0, 0)]_+ \leq C_6(1 + [v]_{\beta, t=0})d_{x^0}^{-\beta}\tau^{\beta/[2+(n+1)\beta]}.$$

When $\tau \geq d_{x^0}^{2+(n+1)\beta}$, this inequality obviously holds. The lemma is proved for $\beta_1 = \beta/[2 + (n + 1)\beta]$. □

When $\partial\Omega \in C^2$, by a change of variables a portion of $\partial\Omega$ is mapped into $x_n = 0$ and $|x| < R_0$. Also locally $\Omega \subseteq \{x_n > 0\}$.

LEMMA 12. *Let $v \in W_{n+1}^{2,1}(Q)$ satisfy $[v]_0 \leq M_0$ and $\mathscr{L}v \geq -K_2(1 + |D_x v|^2)$ in Q where $\mathscr{L}$ is given in* (4). *Suppose that*

$$v|_{x_n=0} \in C^{\beta, \beta/2}(\{|x| < R_0\} \times \{0 \leq t \leq T\})\ (0 < \beta < 1).$$

Then for $x_n > 0$, $|x| < R/2$, $t^0 + R^2/2 \leq t \leq t^0 + R^2$ $(t^0 \geq 0, R \leq R_0)$, there exists $\beta_1 \in (0, \beta)$ such that

$$(v(x, t) - v(x, t)|_{x_n=0})_+ \leq C_7(1 + [v]_{\beta, \beta/2, x_n=0})\left(\frac{x_n}{R}\right)^{\beta_1}; \tag{5}$$

when $x_n > 0$ and $|x| \leq R/2$,

$$(v(x, t) - v(x, t)|_{x_n=t=0})_+ \leq C_8(1 + [v]_{\beta, \beta/2, x_n=0})\left(\frac{x_n^2 + t}{R^2}\right)^{\beta_1/2}. \tag{6}$$

PROOF. Without loss of generality we prove the lemma for $x_1 = \cdots = x_{n-1} = 0$. For $t^0 + R^2/2 \leq t^1 \leq t^0 + R^2$, let

$$\eta(x, t) = \left(1 - \frac{|x|^2}{R^2}\right)^m \left(\frac{t - t^0}{R^2}\right)^m,$$

$$s = \frac{x_n}{R} + \varepsilon\left[\frac{(t - t^1)^2}{R^4} + \frac{\sum_{i=1}^{n-1} x_i^2}{R^2}\right] + \varepsilon_1,$$

and

$$w(x,t)=\eta(x,t)s^{-\alpha}(v-v^*)_+^m.$$

Here $v^*=v(0,\dots,0,t^1)$, $m(>2)$ is an even number, ε is a small positive number, the constant α satisfies $0<\alpha\le\min\{1/2(n+1),\, m\beta/4\}$ and ε_1 is a small positive number.

For $|x|<R$, $t^0<t\le t^0+R^2$, we have

$$w|_{x_n=0}\le C_1\left(\frac{R^2}{\varepsilon}\right)^{\alpha}\left[\frac{(v|_{x_n=0}-v^*)_+}{(\sum_{i=1}^{n-1}x_i^2+|t-t^1|^{\alpha/m})}\right]^m\le C_2[v]^m_{\beta,\beta/2,x_n=0}R^{\beta m},$$

$$\mathscr{L}w=s^{-\alpha}(v-v^*)_+^m\mathscr{L}\eta+\eta s^{-\alpha}\mathscr{L}(v-v^*)_+^m+\eta(v-v^*)_+^m\mathscr{L}s^{-\alpha}$$
$$+2a_{ij}\left[s^{-\alpha}\frac{\partial(v-v^*)_+^m}{\partial x_i}\eta_j+\eta\frac{\partial s^{-\alpha}}{\partial x_i}\frac{\partial(v-v^*)_+^m}{\partial x_j}+(v-v^*)_+^m\frac{\partial s^{-\alpha}}{\partial x_i}\eta_j\right],$$

$$\mathscr{L}\eta\ge -C_3\frac{1}{R^2}\eta^{1-(1/m)},$$

$$\begin{aligned}\mathscr{L}(v-v^*)_+^m&=m(m-1)(v-v^*)_+^{m-2}a_{ij}v_iv_j+m(v-v^*)_+^{m-1}\mathscr{L}v\\&\ge m(v-v^*)_+^{m-2}[(m-1)\lambda|D_xv|^2-2K_2M_0(1+|D_xv|^2)]\\&\ge C_3(v-v^*)_+^{m-2}|D_xv|^2-C_4\quad\left(\text{if } m\ge 1+\frac{4K_2M_0}{\lambda}\right),\end{aligned}$$

$$\begin{aligned}\mathscr{L}s^{-\alpha}&\ge\alpha(\alpha+1)s^{-\alpha-2}\lambda\frac{1+4\varepsilon^2\sum_{i=1}^{n-1}x_i^2}{R^2}-\alpha s^{-\alpha-1}\Lambda\frac{2n\varepsilon}{R^2}\\&\ge\alpha\frac{s^{-\alpha-2}}{R^2}(\lambda-4n\Lambda\varepsilon)\\&\ge\frac{\alpha\lambda}{2}\frac{s^{-\alpha-2}}{R^2}\quad\left(\text{if }\varepsilon\le\frac{\lambda}{8n\Lambda}\text{ and }\varepsilon_1\text{ is sufficiently small}\right),\end{aligned}$$

$$\begin{aligned}\left|2a_{ij}\frac{\partial(v-v^*)_+^m}{\partial x_i}\eta_j\right|&\le 2\Lambda m(v-v^*)_+^{m-1}|D_xv|\frac{2m\eta^{1-(1/m)}}{R}\\&\le\frac{C_3}{2}(v-v^*)_+^{m-2}\eta|D_xv|^2+C_5\frac{\eta^{1-(2/m)}}{R^2},\end{aligned}$$

$$\begin{aligned}\left|2a_{ij}\frac{\partial s^{-\alpha}}{\partial x_i}\frac{\partial(v-v^*)_+^m}{\partial x_j}\right|&\le C_6\frac{\alpha s^{-\alpha-1}}{R}|D_xv|(v-v^*)_+^{m-1}\\&\le\frac{C_3}{2}(v-v^*)_+^{m-2}s^{-\alpha}|D_xv|^2+\frac{C_7\alpha^2}{R^2}s^{-\alpha-2}(v-v^*)_+^m,\end{aligned}$$

and

$$\begin{aligned}\left|2a_{ij}\frac{\partial s^{-\alpha}}{\partial x_i}\eta_j\right|&\le\frac{C_8\alpha}{R^2}s^{-\alpha-1}\eta^{1-(1/m)}\\&\le\frac{1}{R^2}\left(\frac{a\lambda}{4}\eta s^{-\alpha-2}+C_9\alpha\eta^{1-(2/m)}s^{-\alpha}\right).\end{aligned}$$

With m, ε chosen as above and taking $\alpha \le \lambda/4C_7$, we get

$$\mathscr{L}w \ge -\frac{C_{10}}{R^2}\eta^{1-(2/m)}s^{-\alpha} \ge -\frac{C_{11}}{R^2}\left(\frac{R^2}{\sum_{i=1}^{n-1} x_i^2}\right)^{\alpha}.$$

By the Aleksandrov maximum principle

$$\begin{aligned}\sup_{\substack{|x|\le R,\, x_n\ge 0\\ t^0\le t\le t^0+R^2}} w(x,t) &\le C_2[v]^m_{\beta,\beta/2,x_n=0}R^{\beta m} + C_{12}R^{n/(n+1)+2\alpha-2}\\ &\quad\times\left[\int_{\substack{|x|\le R,\, x_n\ge 0\\ t^0\le t\le t^0+R^2}}\left(\sum x_i^2\right)^{-(n+1)\alpha}dx\,dt\right]^{1/(n+1)}\\ &\le C_{13}(1+[v]_{\beta,\beta/2,x_n=0})^m.\end{aligned}$$

Therefore,

$$\begin{aligned}&(v(0,\dots,0,x_n,t^1)-v(0,\dots,0,t^1)]_+\\ &\quad\le C_{14}(1+[v]_{\beta,\beta/2,x_n=0})\left(\frac{x_n}{R}+\varepsilon_1\right)^{\alpha/m}.\end{aligned}$$

Letting $\varepsilon_1\to 0$, (5) holds for $x_1=\dots=x_{n-1}=0$ with $\beta_1=\alpha/m$. To prove (6) for $x_1=\dots=x_{n-1}=0$ we first obtain the estimate for $[v(x,t)-v|_{x_n=0}]_+$ as in the proof of (5) (delete the factor $(t-t^0/R^2)^m$ in the definition of η). Next we use the assumption that v satisfies a Hölder condition on the boundary to get an estimate for $(v|_{x_n=0}-v|_{x_n=t=0})_+$. □

THEOREM 13. *Let $\partial\Omega\in C^2$ and let $u\in C^{2,1}(Q)\cap C(\overline{Q})$ be a solution of* (1), (2). *Suppose that $\varphi\in C^{\beta,\beta/2}(\partial^*Q)$, $[u]_0\le M_0$, $\|\varphi\|_{C^\beta(\partial^*Q)}\le M_\beta$, and F satisfies the conditions specified in Theorem* 2. *Then*

$$[u]_\alpha \le C(1+M_\beta),$$

where α, C depend only on n, λ, Λ, μ_1, M_0, β, and diam Q.

PROOF. For any $P_1(x^1,t^1)$, $P_2(x^2,t^2)\in Q$, from the interior estimate (Theorem 2) we know that there is $\beta_2\in(0,1)$ such that

$$|u(P_2)-u(P_1)|\le Cd^{-\beta_2}_{P_1P_2}d^{\beta_2}(P_1,P_2).$$

when $d(P_1,P_2)<d^2_{P_1P_2}$. It follows that

$$|u(P_2)-u(P_1)|\le Cd^{\beta_2/2}(P_1,P_2).$$

when $d(P_1,P_2)\ge d^2_{P_1P_2}$. We distinguish two cases.

CASE I. $d_{P_1P_2}=\min\{t^1,t^2\}^{1/2}$. Let $t^1=\min\{t^1,t^2\}$, say. We have

$$t^2=t^1+(t^2-t^1)\le d(P_1,P_2)+d^2(P_1,P_2)\le 2d(P_1,P_2).$$

We can write (1) as

$$\mathscr{L}u = a_{ij}u_{ij} - u_t = -F(x, t, u, D_x u, 0),$$

where

$$a_{ij} = \int_0^1 \frac{\partial F}{\partial r_{ij}}(x, t, u, D_x u, \tau D_x^2 u)\, d\tau.$$

As $\lambda|\xi|^2 \le a_{ij}\xi_i\xi_j \le \Lambda|\xi|^2$, and

$$|F(x, t, u, D_x u, 0)| \le K_2|D_x u|^2 + K,$$

we can apply Lemma 11 to u and $-u$ to conclude, when $d(P_1, P_2) < \min\{d_{x^1}, d_{x^2}\}^{2\beta/\beta_1}$,

$$|u(x^2, t^2) - u(x^2, 0)| \le Cd_{x^2}^{-\beta_2} t_2^{\beta_1} \le C\frac{(2d(P_1, P_2))^{\beta_1}}{(d(P_1, P_2))^{\beta_1/2}} = C_1 d^{\beta_1/2}(P_1, P_2),$$

and

$$|u(x^1, t^1) - u(x^1, 0)| \le C_1 d^{\beta_1/2}(P_1, P_2).$$

Therefore

$$\begin{aligned} |u(x^2, t^2) - u(x^1, t^1)| &\le |u(x^2, t^2) - u(x^2, 0)| + |u(x^1, t^1) - u(x^1, 0)| \\ &\quad + |u(x^2, 0) - u(x^1, 0)| \\ &\le Cd^\gamma(P_1, P_2), \qquad \gamma = \min\{\beta, \beta_{1/2}\}. \end{aligned}$$

When $d(P_1, P_2) > \min\{d_{x^1}, d_{x^2}\}^{2\beta/\beta_1}$. Let us assume $d_{x^1} = \min\{d_{x^1}, d_{x^2}\}$. It is easy to see that $d_{x^2} \le |x^2 - x^1| + d_{x^1}$. Thus $d_{x^2} \le 2d^{\beta_1/2\beta}(P_1, P_2)$ and

$$\max\{d_{x^1}, d_{x^2}, (t^1)^{1/2}, (t^1)^{1/2}\} \le 2d^{\beta_1/2\beta}(P_1, P_2).$$

It will be taken care of in the second case.

CASE II. $d_{P_1P_2} = \min\{d_{x^1}, d_{x^2}\}$. Let us assume $d_{x^1} = \min\{d_{x^1}, d_{x^2}\}$. Then

$$d_{x^2} \le |x^2 - x^1| + d_{x^1} \le d(P_1, P_2) + d^{1/2}(P_1, P_2) \le 2d^{1/2}(P_1, P_2).$$

When $\min\{t^1, t^2\} > d^{2/3}(P_1, P_2)$, we take $R = d^{1/3}(P_1, P_2)$ and $t^0 = t^i - d^{2/3}(P_1, P_2)$ $(i = 1, 2)$. Applying (5) to u and $-u$ yields

$$|u(x^i, t^i) - u(y^i, t^i)| \le C\left(\frac{d_{x^i}}{d^{1/3}(P_1, P_2)}\right)^{\beta_1} \le Cd^{\beta_1/6}(P_1, P_2)$$

$(d_{x^i} = |x^i - y^i|,\ y^i \in \partial\Omega,\ i = 1, 2)$. As

$$|y^2 - y^1| \le |x^2 - y^2| + |x^1 - y^1| + |x^2 - x^1| \le 4d^{1/2}(P_1, P_2),$$

$$\begin{aligned}|u(P_2)-u(P_1)| &\le |u(x^2, t^2)-u(y^2, t^2)|+|u(x^1, t^1)-u(y^1, t^1)|\\ &\quad + |u(y^2, t^2)-u(y^1, t^1)|\\ &\le 2Cd^{\beta_1/6}(P_1, P_2)+Cd^{\beta}(P_1, P_2)\\ &\le 3Cd^{\gamma}(P_1, P_2), \qquad \gamma=\min\{\beta_1/6, \beta\}.\end{aligned}$$

When $\min\{t^1, t^2\}\le d^{2/3}(P_1, P_2)$,

$$\max\{d_{x^1}, d_{x^2}, (t^1)^{1/2}, (t^2)^{1/2}\}\le 2d^{1/3}(P_1, P_2).$$

Let R be a constant. When $d(P_1, P_2)\le R^{\max\{6, 2\beta/\beta_1\}}$, by applying (6) to u and $-u$ we obtain a Hölder estimate for u. Such an estimate clearly is also available when $u\ge R$. So the theorem is valid in both cases. □

§5. Uniform and Hölder estimates for $D_x u$ near the boundary

Now we derive boundary, global uniform, and global $C^{\alpha, \alpha/2}$-estimates for $D_x u$.

THEOREM 14. *Let $\partial\Omega\in C^2$ and let $u\in C^{2,1}(Q)\cap C(\overline{Q})$ be a solution of* (1), (2). *Suppose that $\varphi\in C^{2,1}(\partial^* Q)$, $[u]_0\le M_0$, and F fulfills the conditions specified in Theorem* 4. *Then*

$$\left|\frac{\partial u}{\partial N}\right|_{\partial\Omega\times[0, T]}\le C(1+\|\varphi\|_{C^{2,1}(\partial\Omega\times[0, T])}+\|\varphi\|_{C^1(\Omega\times\{t=0\})}),$$

where N is the unit inner normal of $\partial\Omega$ and C depends only on n, λ, Λ, M_0, T, and the C^2-modulus of $\partial\Omega$.

PROOF. By a change of variables, a portion of $\partial\Omega$ is mapped to $\{x: x_n = 0, |x|<R_0\}$ and $\Omega\subseteq\{x: x_n>0\}$ locally. Let $Q_{R,T}=\{(x, t): |x_i|<R$ $(1\le i\le n-1)$, $0<x_n<R$, $0<t<T\}\subseteq Q$ for $0<R<R_0$. Consider the barrier function

$$v=\log\left(1+\frac{x_n}{R}+\frac{\varepsilon}{R^2}\sum_{i=1}^{n-1}x_i^2\right), \qquad \varepsilon>0 \text{ small},$$

and the function

$$w=Kv\pm(u(x, t)-\varphi(x_1, \dots, x_{n-1}, 0, t)),$$

where K will be chosen later. For sufficiently large $K=K_\varepsilon$, $w\ge 0$ on $\{(x, t)\in\partial Q_{R,T}: 0<t\le T\}$. Along where $t=0$ on $\partial Q_{R,T}$ one has

$$\frac{\partial w}{\partial x_n}\ge\frac{K}{1+x_n/R+(\varepsilon/R^2)\sum_{i=1}^{n-1}x_i^2}\pm\frac{\partial\varphi}{\partial x_n}\ge\frac{K}{2+(n-1)\varepsilon}-\left|\frac{\partial\varphi}{\partial x_n}\right|\ge 0.$$

Consequently

$$w|_{\partial Q_{R,T}\cap\{t=0\}}\ge 0.$$

Letting

$$\mathscr{L} = a_{ij}\frac{\partial^2}{\partial x_i \partial x_j} - \frac{\partial}{\partial t}, \qquad a_{ij} = \int_0^1 \frac{\partial F}{\partial r_{ij}}(x, t, u, D_x u, \tau D_x^2 u)\, d\tau,$$

we have

$$\pm\mathscr{L}u \le |F(x, t, u, D_x u, 0)| \le C(1 + |D_x u|^2).$$

If w attains a minimum in $Q_{R,T}$, at this minimum $w_x = 0$, and $\mathscr{L}w \ge 0$. However,

$$\begin{aligned} \mathscr{L}w|_{w_x=0} \le \frac{K}{R^3}\Bigg[& -\lambda\frac{1 + 4\varepsilon^2\sum_{i=1}^{n-1} x_i^2/R^2}{\left(1 + \frac{x_n}{R} + \frac{\varepsilon}{R^2}\sum_{i=1}^{n-1} x_i^2\right)^2} + \frac{\Lambda n\varepsilon}{\left(1 + \frac{x^n}{R} + \frac{\varepsilon}{R^2}\sum_{i=1}^{n-1} x_i^2\right)^2} \\ & + C\sum_{i=1}^{n}\left(\varphi_i \pm K\frac{v_i}{R}\right)^2\Bigg] \\ \le -\frac{K}{R^3} & \frac{\lambda - \Lambda n(n+1)\varepsilon - C_1 R}{(1 + x_n/R + (\varepsilon/R^2)\sum_{i=1}^{n-1} x_i^2)^2} + C_2 < 0 \end{aligned}$$

for sufficiently small ε and R. Hence w cannot attain its minimum in $Q_{R,T}$ and one must have $w \ge 0$ in $\overline{Q}_{R,T}$. Therefore, from $\partial w/\partial x_n|_{x=0} \ge 0$ we conclude $|\partial u/\partial x_n|_{x=0} \le K/R^2$. By the same argument we obtain $|\partial u/\partial x_n| \le K/R^2$ for $|x_i| \le R/2$ $(1 \le i \le n-1)$ and $x_n = 0$.

Going back to the original coordinate and using a finite covering argument the conclusion of the theorem follows. □

The derivation of a uniform bound for $D_x u$ in $\overline{Q}$ differs from the proof of Theorem 3 only at two places. First we should get rid of the factor $d_{\overline{P}}$, d_P, that is, replace them by 1. Second, now

$$v(x, t) = \eta^2(x, t)|D_x u|^2 + \alpha M^2(u(x, t) - u(\overline{P}))^2$$

may take a maximum on $\partial\Omega \times [0, T]$. When $\alpha M_\beta^2 M^{-2\beta} < 1/8$ and v attains a maximum on $\partial\Omega \times [0, T]$,

$$\frac{M^2}{8} \le \eta^2(P_0)|D_x u(P_0)|^2 \le M_1^2.$$

Here $D_x u|_{\partial^* Q} \le M_1$ is a known constant. Therefore M is bounded in this case.

Before we can derive a $C^{\alpha,\alpha/2}$-estimate for $D_x u$, we need to give a Hölder estimate for $\partial u/\partial N|_{\partial\Omega\times[0,T]}$ on $\partial\Omega \times [0, T]$. For this purpose several lemmas are needed.

Let $K_{R,T} = \{(x, t)\colon |x_i| < R\ (1 \le i \le n-1),\ 0 < x_n < R,\ 0 < t \le T\}$. Consider the following degenerate linear parabolic equation

$$\mathscr{L}v = x_n a_{ij} v_{ij} + b_i v_i + cv - x_n v_t$$

in $K_{R_0,T}$. Here the coefficients of $\mathscr{L}$ satisfy

$$\lambda|\xi|^2 \le a_{ij}\xi_i\xi_j \le \Lambda|\xi|^2, \qquad 0<\lambda\le\Lambda<\infty, \quad \forall\xi\in\mathbb{R}^n \tag{7}$$

$$|b_i| \le \Lambda (1\le i\le n), \qquad -\Lambda \le c \le 0, \tag{8}$$

$$2a_{nn} - b_n + cx_n \le 0. \tag{9}$$

Denote $\{(x, t): |x_i| < R\ (1 \le i \le n-1),\ 0 < x_n < R,\ -R^2 < t < 0\}$ by K_R, $M(R) = \max_{\overline{K_R}} v$, $m(R) = \min_{\overline{K_R}} v$, and $\omega(R) = M(R) - m(R)$.

LEMMA 15. *Let the coefficients of $\mathscr{L}$ satisfy* (7), (8), *and* (9) *in K_R. Suppose that $v \in C^{2,1}(K_R)\cap C(\overline{K}_R)$ satisfies*

$$-C(1+x_n^2|D_xv|^2) \le \mathscr{L}v \le 0, \qquad 0 \le v \le M_0,$$
$$|\{(x,t)\in \widetilde{K}_R: v(x,t)\ge 1\}| \ge \beta|\widetilde{K}_R|$$

for some $\beta\in(0,1)$, $C\ge 1$, $M_0>0$, and $\widetilde{K}_R = K_R\cap\{t<-R^2/4\}$. There exist constants $R_1>0$ and $\gamma>0$ such that for all $R<R_1$, $\inf_{K_{R/2}} v \ge \gamma$. Here R_1 and γ depend only on n, λ, Λ, C, M_0, and β.

PROOF. Let η be a small number satisfying

$$\eta < \min\{1/6, \beta/3C_1\} \qquad (C_1 \text{ will be determined below}),$$

and

$$\left|\left\{(x,t)\in\widetilde{K}_R\cap\left\{|x_i|<(1-\eta)R(1\le i\le n-1),\ \frac{\eta}{2}R<x_n<\left(1-\frac{\eta}{2}\right)R,\right.\right.\right.$$
$$\left.\left.\left.\left(\frac{\eta}{2}-1\right)R^2<t<-\left(\frac14+\frac{3\beta}{4}\right)R^2\right\}: v\ge 1\right\}\right|$$
$$\ge \frac{\beta}{6}|\widetilde{K}_R|.$$

In the region $|x_i| < (1-\eta/2)R$ $(1\le i\le n-1)$, $\eta R/2 < x_n < (1-\eta/2)R$, and $(\eta/2-1)R^2 < t < -(1/4+3\beta/4)R^2$ there is at least a point (x^*, t^*) with $v(x^*, t^*) \ge 1$. We want to establish a Harnack inequality in the region $|x_i| < (1-\eta/2)R$ $(1\le i\le n-1)$, $\eta R/2 < x_n < (1-\eta/2)R$, $(\eta/2-1)R^2 < t < 0$. In other words, there is $\rho > 0$ such that $v\ge\rho$ on $|x_i| \le (1-\eta)R$ $(1\le i\le n-1)$, $\eta R\le x_n\le(1-\eta)R$, $-(1/4+\beta/2)R^2\le t\le 0$. The proof of the Harnack inequality basically is the same as in Theorem 6 of the previous chapter, except for a few changes stated below.

The term $u(0,0)$ there should be replaced by $v(x^*, t^*)\ge 1$. Hence $v(x^0, t^0) \ge v(x^*, t^*) \ge 1$. When

$$\left|K_{(1-R_0)\eta R/4}(x^0, t^0)\cap\left\{\frac{2^{m-1}v(x^0,t^0)-v}{(2^{m-1}-1/2)v(x^0,t^0)}\ge 1\right\}\right| \ge \mu|K_{(1-R_0)\eta R/4}|,$$

we take

$$w = \frac{1-\exp[-C(2^{m-1}v(x^0,t^0)-v)/\lambda] + (C^2+C\Lambda M_0/\lambda^2)[x_n^0-x_n+(1-R_0)\eta R/4]}{1-\exp[-C(2^{m-1}-1/2)v(x^0,t^0)/\lambda]}.$$

Since $b_n \geq 2a_{nn} - |c|x_n \geq 2\lambda - \Lambda\frac{\lambda}{\Lambda} = \lambda$ for $R \leq \lambda/\Lambda$, we have $w \geq 0$, $\mathscr{L}w \leq 0$ in $K_{(1-R_0)\eta R/4}(x^0, t^0)$. Therefore

$$w|_{B_{(1-R_0)\eta R/4}(x_0)\times\{t=t^0\}} \geq 1 - C_2(1-\mu)^{1/2}.$$

Setting $(x, t) = (x^0, t^0)$ in this inequality yields

$$\frac{\exp[C(2^{m-1}-1/2)v(x^0,t^0)/\lambda]-\exp[C(2^{m-1}-1)v(x^0,t^0)/\lambda] - (C^2+C\Lambda M_0)(1-R_0)\eta R/4\lambda^2}{1-\exp[-C(2^{m-1}-1/2)v(x^0,t^0)/\lambda]}$$
$$\leq C_2(1-\mu)^{1/2}.$$

Since $v(x^0, t^0) \geq 1$, this leads to a contradiction when $R \leq R_1$, R_1 is sufficiently small, and μ is sufficiently close to 1. Now the rest of the proof runs as the proof of Theorem 6 in the previous chapter. The Harnack inequality holds in the present situation.

Consider the function

$$z(x, t) = \rho\left[1 - \beta - 9\sum_{i=1}^{n-1}\frac{(x_i - x_i^0)^2}{R^2} + \frac{C_1x_n}{R} + \frac{4t}{(1+2\beta)R^2} - \frac{R\varepsilon}{x_n}\right],$$

where ε is a small positive number and x_i^0 satisfies $|x_i^0| \leq R/2$ ($i = 1, \ldots, n-1$) in $Q_0 = \{(x, t)\colon \sum_{i=1}^{n-1}(x_i - x_i^0)^2 \leq R^2/9,\ 0 < x_n < \eta R,\ -(1/4+\beta/2)R^2 < t < 0\}$. We have

$$z|_{\partial Q_0\cap\{\sum_{i=1}^{n-1}(x_i-x_i^0)^2=R^2/9\}} \leq 0,$$
$$z|_{\partial Q_0\cap\{t=-(1/4+\beta/2)R^2\}} \leq 0,$$
$$z|_{\partial Q_0\cap\{x_n=\eta R\}} \leq \rho;$$

$$\begin{aligned}\mathscr{L}z = \rho&\left\{-\frac{18}{R^2}\sum_{i=1}^{n-1}[x_na_{ii} + b_i(x_i - x_i^0) + \frac{b_nC_1}{R} - \frac{4x_n}{(1+2\beta)R^2}\right\}\\ &+ c\left(z + \frac{\rho\varepsilon R}{x_n}\right) - \frac{\rho\varepsilon R}{x_n^2}(2a_{nn} - b_n + cx_n)\\ \geq \rho&\left[\frac{C_1}{R}(b_n + cx_n) - \frac{C_2}{R}\right],\end{aligned}$$

where

$$C_2 = \sup_{K_R}\left\{\frac{18}{R}\sum_{i=1}^{n-1}[x_na_{ii} + b_i(x_i - x_i^0) + \frac{4x_n}{(1+2\beta)R} + |c|(1-\beta)\right\}.$$

As

$$b_n + cx_n \geq 2(a_{nn} - |c|x_n) \geq 2(\lambda - \Lambda x_n) \geq 2\lambda(1-\eta) \geq \frac{5}{3}\lambda,$$

we have $\mathscr{L} z > 0$ for $C_1 = C_2/\lambda$. Therefore $z(x, t) - v(x, t)$ cannot attain a positive maximum in Q_0. As a result $z(x, t) - v(x, t) \le 0$ in Q_0 since this is true on $\partial^* Q_0$. Letting $\varepsilon \to 0$ we get

$$v(x, t) \ge \rho \left[1 - \beta - 9 \sum_1^{n-1} \frac{(x_i - x_i^0)^2}{R^2} + \frac{C_1 x_n}{R} + \frac{4t}{(1+2\beta)R^2} \right].$$

So, $v \ge \beta(1 - 8\beta)\rho/4(1 + 2\beta)$ for $-R^2/4 \le t < 0$, $0 \le x_n \le \eta R$, $\sum_1^{n-1}(x_i - x_i^0)^2 \le \beta(1 - 8\beta)/36(1 + 2\beta)$ (we have restricted $\beta < 1/8$). This implies that

$$v \ge \frac{\beta(1 - 8\beta)}{4(1 + 2\beta)} \rho$$

for $|x_i| \le R/2$ $(1 \le i \le n-1)$, $0 \le x_n \le R/2$, and $-R^2/4 \le t \le 0$. □

LEMMA 16. *Suppose for* $l = 1, \ldots, N$, $f_l \in C^1(\mathbb{R})$, $f'_l > 0$ *in* $\mathbb{R}$; $v_l \in C^{2,1}(K_{R_0}) \cap C(\overline{K}_{R_0})$, $|v_l| \le M_0$,

$$-K_1(1 + x_n^2 |Dv_l|^2) \le \mathscr{L} v_l \le K_2.$$

Further there exist $\beta \in (0, 1)$, $\delta_1 > 0$, $K_3 > 0$ *such that for any* $P_1, P_2 \in K_{R_0}$,

$$K_3 \frac{d^\beta(P_1, P_2)}{R_0^\beta} \ge \delta_1 \sum_{l=1}^N (f_l(v_l(P_2)) - f_l(v_l(P_1)))_+ - \sum_{l=1}^N (f_l(v_l(P_2)) - f_l(v_l(P_1)))_-. \qquad (*)$$

Then there are β_1, C *such that*

$$\sum_{l=1}^N |v_l(P_2) - v_l(P_1)| \le C R_0^{-\beta_1} d^{\beta_1}(P_1, P_2)(K_3 + M_0)$$

for all $P_1, P_2 \in K_R \cap \{x_n = 0\}$, $(0 < R < R_0)$. *Here* $0 < \beta_1 < \beta$, *and* β_1, C *depend only on* $n, N, \lambda, \Lambda, \delta_1, K_1, K_2$ *and two-sided bounds of* f'_l *in* $[-M_0, M_0]$. *When* $N = 1$ *the conclusion of the theorem remains valid without* $(*)$.

PROOF. We may take $K_2 = 0$, otherwise replace v_l by $K_2 x_n/\lambda + v_l$. The rest of the proof is contained in Theorem 7 and Theorem 8 of the previous chapter. □

LEMMA 17. *Let* $k_{R_0} = \{x: |x_i| < R_0 \ (1 \le i \le n-1),\ 0 < x_n < R_0\}$ *and* $Q_{R_0, T} = k_{R_0} \times (0, T]$. *Suppose* $\mathscr{L}$ *satisfies* (7), (8), (9) *in* $Q_{R_0, T}$; $v \in C^{2,1}(Q_{R_0, T}) \cap C(\overline{Q}_{R_0, T})$, $|v| \le M_0$, $|\mathscr{L} v| \le K_1(1 + x_n^2 |D_x v|^2)$, *and*

$v(x, 0) = v_0(x) \in C^\alpha$ *for* $x \in \overline{k}_{R_0}$. *Then for* $(x, t) \in Q_{R,T} \cap \{x_n = 0\}$ *and* $R \le \min\{R_0/2, \lambda/4\Lambda\}$,

$$|v(x, t) - v_0(x)| \le MR^{-\alpha}t^{\alpha/2(1+\alpha)}(1 + \|v_0\|_{C^\alpha(\overline{k}_{R_0})})$$

for some M *which depends only on* $n, \lambda, \Lambda, K_1, M_0, R_0$, *and* T.

PROOF. We only consider the special case $x = 0$. The general situation can be reduced to this one by a translation in x. Replacing v by $v - v_0(0)$, we may assume $v_0(0) = 0$. Let $w = e^{K_1 v/\lambda} - 1$. We have

$$\mathscr{L}w \ge e^{K_1 v/\lambda}(\mathscr{L}v - cv + c + K_1|D_x v|^2) \ge -K_2.$$

Consider the auxiliary function

$$W = \varepsilon^\alpha + \frac{1}{\varepsilon^2}\sum_{i=1}^{n}|x_i|^2 + \frac{\widetilde{C}t}{\varepsilon\delta} + \frac{\widetilde{C}}{\varepsilon\lambda}\int_{x_n}^{\varepsilon}\sigma(y)\,dy + \frac{\varepsilon_1}{x_n}$$

in $Q_{\varepsilon,T}$ where $\widetilde{C}$ is positive, $\varepsilon, \delta, \varepsilon_1$ are small, positive $\delta < \varepsilon$, and

$$\sigma(y) = \begin{cases} 1, & y \le \delta, \\ \left(\dfrac{\delta}{y}\right)^{3/2}, & y > \delta. \end{cases}$$

We have

$$\begin{aligned}\mathscr{L}(W) = {} & \frac{2}{\varepsilon^2}\sum_{i=1}^{n}(a_{ii}x_n + b_i x_i) - \frac{\widetilde{C}x_n}{\varepsilon\delta} - \frac{\widetilde{C}}{\varepsilon\lambda}(a_{nn}x_n\sigma'(x_n) + b_n\sigma(x_n)) \\ & + \frac{\varepsilon_1}{x_n^2}(2a_{nn} - b_n + cx_n) + c\left(W - \frac{\varepsilon_1}{x_n}\right).\end{aligned}$$

When $x_n \le \delta$,

$$\mathscr{L}(W) \le \frac{2}{\varepsilon}\sum_{i=1}^{n}(a_{ii} + |b_i|) - \frac{\widetilde{C}}{\varepsilon}\frac{b_n}{\lambda} \le -\frac{\widetilde{C}}{\varepsilon}.$$

This is because $b_n \ge 2a_{nn} - \Lambda\varepsilon \ge \frac{3}{2}a_{nn} \ge \frac{3}{2}\lambda$ when ε is small. So for a larger $\widetilde{C}$ this inequality holds. When $x_n > \delta$,

$$a_{nn}x_n\sigma'(x_n) + b_n\sigma(x_n) = \left(b_n - \frac{3}{2}a_{nn}\right)\left(\frac{\delta}{x_n}\right)^{3/2} \ge 0.$$

Hence

$$\mathscr{L}(W) \le \frac{2}{\varepsilon}\sum_{i=1}^{n}(a_{ii} + |b_i|) - \frac{\widetilde{C}}{\varepsilon} \le -\frac{\widetilde{C}}{2\varepsilon}.$$

Setting $C = \sqrt{n}\|e^{K_1 v_0/\lambda}\|_{C^\alpha} + e^{K_1 M_0/\lambda} + K_2$, we find that $\mathscr{L}(W - w/C) \le -\widetilde{C}/2\varepsilon + 1 < 0$. So $W - w/C$ cannot attain a negative minimum inside $Q_{\varepsilon,T}$.

We claim that it is nonnegative on $\partial^* Q_{\varepsilon,T}$. On $\partial^* Q_{\varepsilon,T} \cap \{x_n = 0\}$ this is true since $W = \infty$. On $t = 0$ using $w(0, 0) = 0$ we get $|w(x, 0)| \le \|w\|_{C^\alpha}|x|^\alpha \le \sqrt{n}\|e^{K_1 v_0/\lambda}\|_{C^\alpha}\varepsilon^\alpha$. Therefore $W - w/c \ge 0$. On the other portion of $\partial^* Q_{\varepsilon,T}$ it is even easier to see that $W - w/c \ge 0$. So $W - w/c \ge 0$ in $\overline{Q}_{\varepsilon,T}$. Letting $\varepsilon_1 \to 0$ and then taking $x = 0$ we get

$$w(0, t) \le CW(0, t) = C\left(\varepsilon^\alpha + \frac{\tilde{C}t}{\varepsilon\delta} + \frac{\tilde{C}}{\varepsilon\lambda}\int_0^\varepsilon \sigma(y)\,dy\right)$$
$$\le C\left(\varepsilon^\alpha + \frac{\tilde{C}t}{\varepsilon\delta} + \frac{3\tilde{C}\delta}{\varepsilon\lambda}\right),$$

which implies

$$w(0, t) \le C_1 t^{\alpha/2(1+\alpha)}, \quad \text{for } t \le R^{2(1+\alpha)},$$

after choosing $\delta = t^{1/2}$ and $\varepsilon = t^{1/2(1+\alpha)}$. In other words,

$$w(0, t) \le C_2 R^{-\alpha} t^{\alpha/2(1+\alpha)} \quad \text{for all } 0 < t \le T,$$

and

$$v(0, t) \le C_3 R^{-\alpha} t^{\alpha/(2(1+\alpha)} \quad \text{for all } 0 < t < T.$$

Similarly using $w_1 = 1 - e^{-K_1 v/\lambda}$ one can show

$$v(0, t) \ge -C_4 R^{-\alpha} t^{\alpha/2(1+\alpha)}. \quad \square$$

THEOREM 18. *Let $\partial\Omega \in C^2$ and let $u \in C^{2,1}(Q) \cap C(\overline{Q})$ be a solution of (1) and (2) with $D_x u \in C(\overline{Q})$, $\varphi \in C^{2,1}(\partial^* Q)$, $|u| \le M_0$, and $|D_x u| \le M_1$ in $\overline{Q}$. Suppose that F satisfies the conditions specified in Theorem 2. Then $\partial u/\partial N \in C^{\alpha_1, \alpha_1/2}(\partial\Omega \times [0, T])$ and α_1 and the Hölder constant of $\partial u/\partial N$ depend only on n, λ, Λ, μ_1, M_0, M_1, α,* diam Q, *$C^{1+\alpha}$-norm of φ on $\partial\Omega \times [0, T]$, and the C^2-modulus of $\partial\Omega$.*

PROOF. By a change of variables locally $\partial\Omega$ is mapped onto $\{x: |x_i| < R_0\ (1 \le i \le n-1),\ x_n = 0\}$ and Ω is contained in $\{x_n > 0\}$. Let $K_{R_0,T} = \{(x, t): |x_i| < R_0\ \ (1 \le i \le n-1),\ 0 < x_n < R_0,\ 0 < t \le T\} \subseteq Q$. Using $u - \varphi$ to replace u, we may assume $u|_{K_{R_0,T}} = 0$. We write (1) as

$$\mathcal{L}u = a_{ij}u_{ij} - u_t = -F(x, t, u, D_x u, 0).$$

Consider the function $v = u/x_n$. Since $u|_{x_n=0} = 0$,

$$v = \frac{\partial u}{\partial x_n}(x_1, \dots, x_{n-1}, \theta x_n, t) \qquad (0 < \theta < 1),$$

$$v|_{x_n=0} = \left.\frac{\partial u}{\partial x_n}\right|_{x_n=0}, \quad \text{and} \quad v|_{t=0} \in C^\alpha$$

in $K_{R_0,T} \cap \{t=0\}$. Furthermore,

$$\widetilde{\mathscr{L}}v = x_n a_{ij} v_{ij} + b_i v_i + cv - x_n v_t = -F(x, t, u, D_x u, 0),$$

where $b_i = -2a_{in}$ $(1 \le i \le n)$ and $c = 0$. We have $|\widetilde{\mathscr{L}}v| \le \mu_1(1+x_n^2|D_x v|^2)$ and $\widetilde{\mathscr{L}}v$ satisfies (7), (8), (9). The function $w = e^{R_0\mu_1 v/\lambda}$ satisfies

$$-\mu_3(1 + x_n^2|D_x w|^2) \le \widetilde{\mathscr{L}}w \le \mu_2.$$

By virtue of Lemma 16 we have

$$|w(P_2) - w(P_1)| \le C d_{P_1P_2}^{-\beta_1} d^{\beta_1}(P_1, P_2),$$
$$\forall P_1, P_2 \in K_{R,T} \cap \{x_n = 0\}, \qquad R < R_0.$$

Consequently $|v(P_2) - v(P_1)| \le C_1 d_{P_1P_2}^{-\beta_1} d^{\beta_1}(P_1, P_2)$, $\forall P_1, P_2 \in K_{R,T} \cap \{x_n = 0\}$, $R < R_0$. When $x^1, x^2 \in K_{R_0/2,T} \cap \{x_n = 0\}$ and $\min\{t^1, t^2\} > d(P_1, P_2)$, $|v(P_2) - v(P_1)| \le C_2 d^{\beta_1/2}(P_1, P_2)$, $\forall P_1, P_2 \in Q_{R_0/2,T} \cap \{x_n = 0\}$. When $x^1, x^2 \in K_{R_0/2,T} \cap \{x_n = 0\}$ and $\min\{t^1, t^2\} = t^1 \le d(P_1, P_2)$,

$$\begin{aligned} t^2 = t^2 - t^1 + t^1 &\le d^2(P_1, P_2) + d(P_1, P_2) \\ &\le 2d(P_1, P_2) \quad \text{if } d(P_1, P_2) \le 1. \end{aligned}$$

By Lemma 17

$$\begin{aligned} |v(x^2, t^2) - v(x^1, t^1)| &\le |v(x^2, t^2) - v(x^2, 0)| + |v(x^1, t^1) - v(x^1, 0)| \\ &\quad + |v(x^2, 0) - v(x^1, 0)| \\ &\le C\frac{(t^1)^{\alpha/2(1+\alpha)} + (t^2)^{\alpha/2(1+\alpha)}}{R_0^\alpha} + C_1|x^2 - x^1|^\alpha \\ &\le C_2 d^{\alpha/2(1+\alpha)}(P_1, P_2). \end{aligned}$$

Combining these two cases we conclude

$$|v(P_2) - v(P_1)| \le C_3 d^\gamma(P_1, P_2) \quad \text{for } P_1, P_2 \in K_{R_0/2,T} \cap \{x_n = 0\},$$

where $\gamma = \min\{\beta_1/2, \alpha/2(1+\alpha)\}$. In view of $v|_{x_n=0} = \partial u/\partial x_n|_{x_n=0}$, the theorem follows by a finite covering argument. □

THEOREM 19. *Under the assumptions of Theorem* 18 $D_x u \in C^{\alpha_1,\alpha_1/2}(\overline{Q})$ *where* α_1 *and the Hölder constant have the same dependence as the* α_1 *in Theorem* 18.

PROOF. Write

$$\mathscr{L} = \frac{\partial F}{\partial r_{ij}} \frac{\partial^2}{\partial x_i \partial x_j} - \frac{\partial}{\partial t}$$

and

$$v^{\pm l} = \pm u_l + \alpha \sum_{m=1}^n u_m^2 \qquad (1 \le l \le n),\ \alpha \in (0, 1) \text{ constant}.$$

We have

$$\begin{aligned}
\mathscr{L}v^{\pm l} &= \pm\mathscr{L}u_l + 2\alpha u_m \mathscr{L}u_m + 2\alpha\frac{\partial F}{\partial r_{ij}}u_{mi}u_{mj} \\
&= -\frac{\partial F}{\partial x_i}v_i^{\pm l} - \frac{\partial F}{\partial u}(\pm u_l + 2\alpha\sum u_m^2) - \frac{\partial F}{\partial x_l} - 2\alpha u_m\frac{\partial F}{\partial u_m} \\
&\quad + 2\alpha\frac{\partial F}{\partial r_{ij}}u_{mi}u_{mj} \\
&\geq -C(1+|r|)(|D_x v^{\pm l}|+1) + 2\alpha\lambda|r|^2 \\
&\geq -\frac{C_1}{\alpha}(1+|D_x v^{\pm l}|^2).
\end{aligned}$$

By Lemmas 11 and 12 we obtain an one-sided Hölder estimate for $v^{\pm l}$ $(1 \leq l \leq n)$ near the boundary. On the other hand, it is easy to see that

$$\begin{aligned}
\sum_{l=1}^n |u_l(P_2) - u_l(P_1)| \leq \sum_{l=1}^n \max(v^l(P_2) - v^l(P_1))_+, -(v^{-l}(P_2) - v^{-l}(P_1))_-\} \\
+ 2\alpha M_1 \sum_{m=1}^n |u_m(P_2) - u_m(P_1)|,
\end{aligned}$$

where $M_1 = \sup_{\overline{Q}}|D_x u|$. So a Hölder estimate for u_l $(1 \leq l \leq n)$ near the boundary follows if we take $\alpha = 1/4M_1$.

Combining with the interior Hölder estimate for $D_x u$ (Theorem 6), we can follow the proof of Theorem 13 to conclude $D_x u \in C^{\alpha_1, \alpha_1/2}(\overline{Q})$. □

When $\varphi \in C^{2,1}(\partial^* Q)$ and $u_t \in C(\overline{Q})$, a global uniform bound for $|u_t|$ can be derived as in the proof of Theorem 6; only now $d_{\overline{P}}$, $d_{\overline{P}}$ are replaced by 1 and the auxiliary function may attain its maximum on $\partial^* Q$. However, these changes pose no serious difficulties.

THEOREM 20. *Let $u \in C^{2,1}(Q) \cap C(\overline{Q})$ be a solution of* (1), (2) *which satisfies $\varphi \in C^{2,1}(\partial^* Q)$, $D_{\tilde{x}}\varphi \in C^{2,1}(\partial^* Q)$ ($\tilde{x}$ is a tangential direction at $\partial\Omega$), $D_x^2 u \in C(\overline{Q})$, $|u| \leq M_0$, and $|D_x u| \leq M_1$ in $\overline{Q}$. Suppose that F satisfies the conditions specified in Theorem* 4. *Then*

$$|D_x^2 u|_{\partial^* Q} \leq C,$$

where the dependence of C is the same as those stated in Theorem 18.

PROOF. Consider the auxiliary function

$$v^{\pm l} = \pm(u_l - u_l|_{x_n=0}) + \alpha\sum_{m=1}^{n-1}(u_m - u_m|_{x_n=0})^2 \qquad (1 \leq l \leq n-1).$$

Since

$$\mathscr{L}v^{\pm l} \geq -C(1+|r|)(|D_x v^{\pm l}|+1) + \alpha\lambda\sum_{(i,j)\neq(n,n)} u_{ij}^2,$$

and from equation (1)

$$u_t = a_{ij}u_{ij} + F(x, t, u, D_x u, 0),$$

using the uniform boundedness of $|u_t|$ we derive

$$|u_{nn}| \le C\left(\sum_{(i,j)\ne(n,n)} |u_{ij}| + 1\right),$$

or

$$u_{nn}^2 \le C_1\left(\sum_{(i,j)\ne(n,n)} |u_{ij}|^2 + 1\right).$$

Consequently

$$\mathscr{L}v^{\pm l} \ge -C(1+|r|)(|D_x v^{\pm l}|+1) - \frac{\alpha\lambda}{2C_1}(|r|^2 - C_2) \ge -\frac{C_3}{\alpha}(1+|D_x v^{\pm l}|^2).$$

As in the proof for Theorem 14 we consider the function

$$W^{\pm l} = K\frac{v}{R} - (v^{\pm l} - v^{\pm l}|_{x_n=0}),$$

where the barrier function v is the same as the one in Theorem 14. By imitating the proof of this theorem a bound for $|\partial v^{\pm l}/\partial x_n|_{x_n=0}$ $(1 \le l \le n-1)$ can be derived. Using

$$\sum_{l=1}^{n-1}\left|\frac{\partial u_l}{\partial x_n}\right|_{x_n=0} \le \sum_{l=1}^{n-1}\max\left\{\left(\frac{\partial v^l}{\partial x_n}\right)_+, -\left(\frac{\partial v^{-l}}{\partial x_n}\right)_-\right\}_{x_n=0} + 2\alpha M_1\sum_{m=1}^{n-1}\left|\frac{\partial u_m}{\partial x_n}\right|_{x_n=0}$$

we obtain $\sum_{l=1}^{n-1}|\partial u_l/\partial x_n|_{x_n=0} \le C$ if we take $\alpha = 1/4M_1$. Finally a bound for $|u_{nn}|_{x_n=0}$ can be obtained from the equation itself. □

§6. Near boundary uniform and Hölder estimates for the second derivatives

Next we consider the global upper bound for $D_x^2 u$. It can be derived as in the proof of Theorem 5, only now $d_{\overline{P}}$, $d_{\overline{P}}$ are replaced by 1 and the auxiliary function under consideration may attain a maximum on $\partial^* Q$. However, these complications are not hard to overcome. Hence one has the estimate $|D_x^2 u| \le C$, $\forall (x, t) \in \overline{Q}$, at hand.

Under the assumption $\varphi \in C^{2+\alpha, 1+\alpha/2}(\partial^* Q)$, the estimate for $u_t \in C^{\alpha_1, \alpha_1/2}(\overline{Q})$ is a consequence of combining the interior estimate in Theorem 8 and near boundary estimate for linear parabolic equations with bounded coefficients. In fact, they are special simple cases of Lemmas 11, 12, and 13.

Concerning the boundary Hölder estimation and global Hölder estimation for $D_x^2 u$ we have the following two theorems.

THEOREM 21. *Let* $\partial\Omega \in C^2$ *and let* $u \in C^{2,1}(Q) \cap C(\overline{Q})$ *be a solution of* (1), (2) *with* $|u| \le M_0$, $|D_x u| \le M_1$, *and* $|D_x^2 u| \le M_2$ *in* $\overline{Q}$. *Suppose that* $\varphi \in C^{2,1}(\partial^* Q)$, $D_{\tilde{x}}\varphi \in C^{2,1}(\partial^* Q)$ ($\tilde{x}$ *is a tangential direction at* $\partial\Omega$), *and* F *satisfies all conditions specified in Theorem* 8. *Then*

$$\frac{\partial^2 u}{\partial N \partial \tilde{x}_i} \quad (1 \le i \le n-1), \qquad \frac{\partial^2 u}{\partial N^2} \in C^{\alpha_1, \alpha_1/2}(\partial^* Q).$$

Moreover, α_1 *and the Hölder constant depend only on* n, λ, Λ, M_0, M_1, M_2, μ_6, $\operatorname{diam} Q$, *and the* C^2*-modulus of* $\partial\Omega$.

PROOF. By a change of variables we assume a portion of $\partial\Omega$ lies on $x_n = 0$ and $\Omega \subseteq \{x_n > 0\}$ locally. The function u_l satisfies

$$\mathscr{L} u_l = \frac{\partial F}{\partial r_{ij}} u_{lij} - u_{lt} = -\frac{\partial F}{\partial p_i} u_{li} - \frac{\partial F}{\partial u} u_l - \frac{\partial F}{\partial x_l} \qquad (1 \le l \le n-1).$$

The function $w^l = (u_l - u_l|_{x_n=0})/x_n$ satisfies

$$\widetilde{\mathscr{L}} w^l = x_n \frac{\partial F}{\partial r_{ij}} w^l_{ij} + 2\frac{\partial F}{\partial r_{in}} w^l_i - x_n \mathscr{L} u_l,$$

$$w^l|_{x_n} = 0, \quad \text{and} \quad |\widetilde{\mathscr{L}} w^l| \le K_3.$$

By applying the argument in Theorem 20 to w^l and $w^{l+n-1} = -w_l$ $(1 \le l \le n-1)$ one concludes a Hölder estimate for $u_{ij}|_{x_n=0}$ $((i, j) \ne (n, n))$ and, as can be solved from the equation itself, a Hölder estimate for $u_{nn}|_{x_n=0}$. □

THEOREM 22. *Let* $\partial\Omega \in C^2$ *and let* $u \in C^{2,1}(\overline{Q})$ *be a solution of* (1), (2) *with* $|u| \le M_0$, $|D_x u| \le M_1$, *and* $|D_x^2 u| \le M_2$ *in* $\overline{Q}$. *Suppose that* $\varphi \in C^{2+\alpha, 1+\alpha/2}(\partial^* Q)$, $D_{\tilde{x}}\varphi \in C^{2,1}(\partial^* Q)$ ($\tilde{x}$ *is a tangential direction at* $\partial\Omega$), *and* F *satisfies all conditions in Theorem* 10. *Then*

$$D_x^2 u \in C^{\alpha_1, \alpha_1/2}(\overline{Q})$$

and both α_1 *and the Hölder constant depend only on* n, λ, M_0, M_1, M_2, μ_7, $M_{t\beta}$, β, α, $C^{2+\alpha, 1+\alpha/2}$*-norm of* φ *on* $\partial^* Q$, $\operatorname{diam} Q$, *and the* C^2*-modulus of* $\partial\Omega$.

PROOF. Consider again the auxiliary function v^k in Theorem 10

$$v^k = u_{\gamma_k \gamma_k} + \varepsilon \sum_{l=1}^{N} (M_2 + u_{\gamma_l \gamma_l})^2,$$

which satisfies

$$\mathscr{L} v^k \ge -C_\varepsilon - \frac{\partial F}{\partial p_i} v^k_i \ge -\widetilde{C}_\varepsilon (1 + |D_x v^k|^2) \qquad (1 \le k \le N),$$

where

$$\mathscr{L} = \frac{\partial F}{\partial r_{ij}} \frac{\partial^2}{\partial x_i \partial x_j} - \frac{\partial}{\partial t}.$$

By Lemma 11 we have

$$(v^k(x, t) - v^k(x, 0))_+ \leq C d_x^{-\beta} t^{\beta_1} \qquad (1 \leq k \leq N). \tag{10}$$

We have learned already from Theorem 10 that

$$\delta \sum_{k=1}^{N} (v^k(P_2) - v^k(P_1))_+ - \sum_{k=1}^{N} (v^k(P_2) - v^k(P_1))_- \leq C d^{\beta_1}(P_1, P_2).$$

Notice that the factor $d_{P_1 P_2}^{-\beta}$ in the right-hand side of the original inequality is out because now we have Hölder estimates for u_i and u_t in the whole domain. Taking $P_1 = (x, 0)$ and $P_2 = (x, t)$ in this inequality it follows from (10) that

$$\begin{aligned} \delta \sum_{k=1}^{N} (v^k(x, 0) - v^k(x, t))_+ &\leq K d^{\beta_1}(P_1, P_2) + \sum_{k=1}^{N} (v^k(x, t) - v^k(x, 0)) \\ &\leq C_1 d_x^{-\beta} t^{\beta_1}. \end{aligned}$$

Putting this into (10) yields

$$|v^k(x, t) - v^k(x, 0)| \leq C_2 d_x^{-\beta} t^{\beta_1}, \qquad (1 \leq k \leq N).$$

Arguing similarly based on Theorem 12 we have

$$|v^k(x, t) - v^k(x, t)|_{x_n=0}| \leq C \left(\frac{x_n}{R}\right)^{\beta_1},$$

for $x_n > 0$, $|x| < R/2$, $t^0 + R^2/2 \leq t \leq t^0 + R^2$, and

$$|v^k(x, t) - v^k(x, t)|_{x_n=t=0}| \leq C \left(\frac{x_n^2 + t}{R^2}\right)^{\beta_1/2}$$

for $x_n > 0$, $|x| < R/2$, $t \leq R^2$. Therefore as in the proof of Theorem 13 we have

$$\begin{aligned} (1 - NC\varepsilon)|u_{\gamma_k \gamma_k}(P_2) - u_{\gamma_k \gamma_k}(P_1)| &\leq |v^k(P_2) - v^k(P_1)| \\ &\leq C d^{\gamma}(P_1, P_2), \qquad (1 \leq k \leq N), \end{aligned}$$

which means

$$|D_x^2 u(P_2) - D_x^2 u(P_1)| \leq C d^{\gamma}(P_1, P_1). \quad \square$$

§7. Uniqueness and existence of a solution for the first boundary value problem under the natural structure conditions

It can be shown that the first boundary value problem (1), (2) possesses a unique solution provided $\partial\Omega$, φ, and the coefficients of the equation are adequately smooth. We formulate this result in the following theorem.

Theorem 23. *Let Ω be a bounded domain in $\mathbb{R}^n$ with $\partial\Omega \in C^{4+\alpha}$ ($\alpha > 0$), $Q = \Omega \times (0, T]$, and $\varphi \in C^{4+\alpha, 2+\alpha/2}(\partial^* Q)$. The function $F(x, t, z, p, r)$ is $C^{2+\alpha}$ in (x, z, p, r) and $C^{1+\alpha}$ in t for $(x, t) \in \overline{Q}$, $|z| \le M_0 + \varepsilon$, $|p| \le M_1 + \varepsilon$, and $|r| \le M_2 + \varepsilon$, where M_0, M_1, and M_2 are determined by the apriori estimates and ε is any positive number. Suppose F satisfies the following conditions*

$$\lambda|\xi|^2 \le F_{r_{ij}}\xi_i\xi_j \le \Lambda|\xi|^2, \quad \forall\xi \in \mathbb{R}^n, \forall(x, t, z, p, r) \in Q \times \mathbb{R}^{n+1} \times \mathcal{S}^n;$$

$$zF(x, t, z, 0, 0) \le \mu_1 z^2 + \mu_2, \quad \forall(x, t, z) \in Q \times \mathbb{R}^{n+1};$$

$$|F(x, t, z, p, 0)| \le \mu_3(1 + |p|^2), \quad \forall(x, t, z, p) \in Q \times \{|z| \le M_0 + \varepsilon\} \times \mathbb{R}^n;$$

$$|F| + |F_z| + |F_p|(1 + |p|) + |F_x|(1 + |p|)^{-1} \le \mu_4(1 + |p|^2 + |r|),$$
$$\forall(x, t, z, p, r) \in Q \times \{|z| \le M_0 + \varepsilon\} \times \mathbb{R}^n \times \mathcal{S}^n;$$

$$|F_t| \le \mu_5(1 + |r|^2),$$
$$\forall(x, t, z, p, r) \in Q \times \{|z| \le M_0 + \varepsilon\} \times \{|p| \le M_1 + \varepsilon\} \times \mathcal{S}^N;$$

$$\begin{aligned} F_{\xi\xi} \le \mu_6\{&(1 + \textstyle\sum |u_{ij}|^2)^{3/2}(|\tilde{x}|^2 + |\tilde{u}|^2) + (1 + \textstyle\sum |u_{ij}|)\textstyle\sum |\tilde{u}_i|^2 \\ &+ [(1 + \textstyle\sum |u_{ij}|)(|\tilde{x}| + |\tilde{u}| + \textstyle\sum |\tilde{u}_i|] \textstyle\sum |\tilde{u}_{ij}|\}, \end{aligned}$$

$$\forall(x, t, z, p, r) \in Q \times \{|z| \le M_0 + \varepsilon\} \times \{|p| \le M_1 + \varepsilon\} \times \mathcal{S}^n$$

and

$$\forall(\tilde{x}, \tilde{u}, \tilde{u}_i, \tilde{u}_{ij}) \in \mathbb{R}^{2n+1} \times \mathcal{S}^n;$$

and

$$\begin{aligned} &\varphi_t = F(x, t, \varphi, \varphi_i, \varphi_{ij}), \\ &\varphi_{tt} = F_t + F_z F + F_{p_i} F_{x_i} + F_{r_{ij}} F_{x_i x_j} \quad \text{for } x \in \partial\Omega \text{ and } t = 0. \end{aligned}$$

Then there exists a unique solution of (1), (2) *in* $C^{4+\alpha, 2+\alpha/2}(\overline{Q})$.

Remark. It is necessary to assume $\partial\Omega \in C^{4+\alpha}$, for otherwise $\varphi \in C^{4+\alpha, 2+\alpha/2}(\partial^* Q)$ does not make sense. Also we point out that the two conditions for $x \in \partial\Omega$ and $t = 0$ are necessary for u to belong to $C^{4+\alpha, 2+\alpha/z}(\overline{Q})$. This can be seen from applying the compatability condition $L\psi = f$ on $x \in \partial\Omega$, $t = 0$ for any $C^{2+\alpha, 1+\alpha/2}(\overline{Q})$-solution for the linear parabolic equation

$$\begin{cases} Lu = a_{ij}u_{ij} + b_i u_i + cu - u_t = f, \\ u|_{\partial^* Q} = \psi \end{cases}$$

to u and u_t.

Proof. Replacing u by $u - \varphi$ we can assume $u|_{\partial^* Q} = 0$ and the compatability conditions become $F(x, 0, 0, 0, 0) = 0$, $F_t(x, 0, 0, 0, 0) = 0$ for $x \in \partial\Omega$.

We shall use the method of continuity to establish the existence of a solution. Consider, for $\kappa \in [0, 1]$,

$$\begin{cases} u_t = (1-\kappa)\Delta u + \kappa F(x, t, u, u_i, u_{ij}), & (x, t) \in Q, \quad (11) \\ u|_{\partial^* Q} = 0. & (12) \end{cases}$$

Let I be the subset of κ in $[0, 1]$ for which (11), (12) has a solution u^κ in $C^{4+\alpha}$ (for simplicity we write $C^{\alpha, \alpha/2}(\overline{Q})$ as C^α). Clearly $0 \in I$. We want to show that I is open and closed in $[0, 1]$.

Based on the a priori estimates we have derived in the last section, we are able to show that I is closed. For, applying them to (11), (12) we know that when $u^\kappa \in C^{4+\alpha}$, $u^\kappa \in C^{2+\alpha_1}$ and both the exponent α_1 and the $C^{2+\alpha_1}$-norm are uniform in κ. Consequently, the $C^{\alpha\alpha_1}$-norms of F_{uu}, F_{up_i}, $F_{p_ip_j}$, $F_{ux_i}, \dots, F_{x_ix_j}$, F_t are independent of κ. Applying the Schauder's estimate to the equation obtained by differentiating (1) we see that the $C^{4+\alpha\alpha_1}$-norm of u^κ is independent of κ. Therefore, F_{uu}, $F_{up_i}, \dots, F_{x_ix_j}$, F_t have their C^α-norms bounded independent of κ. Finally by Schauder's estimate again, the $C^{4+\alpha}$-norm of u^κ is independent of κ.

As a result, the limit function u^κ of a sequence of solutions $\{u^{\kappa_i}\}$ of (11), (12) as $\kappa_i \to \kappa$ $(i \to \infty)$ again belongs to $C^{4+\alpha}$ and solves (11), (12). So I is closed.

Next we prove the openness of I. Let $\kappa_0 \in I$ and $u^0 = u^{\kappa_0}$. We define a mapping $\Psi_\kappa w = v$ by solving

$$\begin{cases} v_t = (1-\kappa)\Delta v + \kappa(a_{ij}v_{ij} + b_i v_i + cv) \\ \qquad + \kappa(F(x, t, w, w_i, w_{ij}) - a_{ij}w_{ij} - b_i w_i - cw), \qquad (x, t) \in Q, \\ v|_{\partial^* Q} = 0, \end{cases} \tag{13}$$

where $(a_{ij}, b_i, c) = (F_{r_{ij}}, F_{p_j}, F_z)(x, t, u^0, u_2^0, u_{ij}^0)$. Clearly a fixed point of Ψ_κ solves (11), (12). We are going to prove that Ψ_κ is a contraction on the subset

$$S_\delta = \{w : \|w - u^0\|_{C^{2+\alpha}} \le \delta\}$$

of $C^{2+\alpha}$ for suitably small δ.

First we prove that Ψ_κ is well-defined. For small δ, $|w| \le M_0 + \varepsilon$, $|D_x w| \le M_1 + \varepsilon$ and $|D_x^2 w| \le M_2 + \varepsilon$ in Q. Hence a_{ij}, b_i, c, and the last term in the right-hand side of (13) all belong to C^α. By the basic result for the first boundary value problem for linear parabolic equations we know that $v \in C^{2+\alpha}$. Hence Ψ_κ is well-defined.

Next we claim that for κ sufficiently close to κ_0, Ψ_κ is a contraction under the $C^{2+\alpha}$-norm. Subtracting (13) from $u_t^0 = (1-\kappa_0)\Delta u^0 +$

$\kappa_0 F(x, t, u^0, u_i^0, u_{ij}^0)$ yields

$$(v - u^0)_t = (1-\kappa)\Delta(v - u^0) + \kappa_0[a_{ij}(v-u^0)_{ij} + b_i(v-u^0)_i + c(v-u^0)] + \kappa G + (\kappa_0 - \kappa)H,$$

where

$$H = \Delta u^0 - F(x, t, u^0, u_i^0, u_{ij}^0)$$

(clearly $\|H\|_{C^\alpha} \le M_3$) and

$$G = F(x, t, w, w_i, w_{ij}) - F(x, t, u^0, u_i^0, u_{ij}^0) - a_{ij}(w - u^0)_{ij} - b_i(w-u^0)_i - c(w-u^0).$$

Denote (w, w_i, w_{ij}) and (u^0, u_i^0, u_{ij}^0) by vectors W and W^0 respectively. We have

$$\begin{aligned}|G(x, t, W^2) - G(x, t, W^1)| &= |(W^2 - W^1)\int_0^1 \operatorname{grad} G(x, t, \theta W^2 + (1-\theta)W^1)\, d\theta| \\ &\le |W^2 - W^1| \max\{|W^2 - W^0|, |W^1 - W^0|\}, \qquad (0<\theta<1).\end{aligned}$$

Let $\rho = (|x-\tilde{x}|^2 + |t - \tilde{t}|)^{1/2}$ and $\widetilde{W} = W(\tilde{x}, \tilde{t})$. We have

$$\begin{aligned}&|(G(x, t, W^2) - G(x, t, W^1)) - (G(\tilde{x}, \tilde{t}, \widetilde{W}^2) - G(\tilde{x}, \tilde{t}, \widetilde{W}^1))| \\ &\quad\le |G(x, t, W^2) - G(x, t, W^1)| + |G(\tilde{x}, \tilde{t}, \widetilde{W}^2) - G(\tilde{x}, \tilde{t}, \widetilde{W}^1)| \\ &\quad\le \|W^2 - W^1\|_{C^\alpha} \frac{1}{p^\alpha} \max\{\|W^2 - W^0\|_C, \|W^1 - W^0\|_C\}.\end{aligned}$$

Also

$$\begin{aligned}&|(G(x, t, W^2) - G(x, t, W^1)) - (G(\tilde{x}, \tilde{t}, \widetilde{W}^2) - G(\tilde{x}, \tilde{t}, \widetilde{W}^1))| \\ &\quad\le |G(x, t, W^2) - G(x, t, W^1) - G(x, t, \widetilde{W}^2) + G(x, t, \widetilde{W}^1)| \\ &\qquad + |G(x, t, \widetilde{W}^2) - G(x, t, \widetilde{W}^1) - G(\tilde{x}, \tilde{t}, \widetilde{W}^2) + G(\tilde{x}, \tilde{t}, \widetilde{W}^1)| \\ &\quad\le |(W^2 - \widetilde{W}^2 - W^1 + \widetilde{W}^1)\int_0^1 \operatorname{grad} G(x, t, \theta W^2 + (1-\theta)\widetilde{W}^2)\, d\theta| \\ &\qquad + |(W^1 - \widetilde{W}^1)\int_0^1 \operatorname{grad}(G(x, t, \theta W^2 + (1-\theta)\widetilde{W}^2) \\ &\qquad\qquad - G(x, t, \theta W^1 + (1-\theta)\widetilde{W}^1))\, d\theta| \\ &\qquad + |(\widetilde{W}^2 - \widetilde{W}^1)\int_0^1 \operatorname{grad}(G(x, t, \theta\widetilde{W}^2 + (1-\theta)\widetilde{W}^1) \\ &\qquad\qquad - G(\tilde{x}, \tilde{t}, \theta\widetilde{W}^2 + (1-\theta)\widetilde{W}^1))\, d\theta|.\end{aligned}$$

Writing $W^1 - \widetilde{W}^1$ as $W^1 - W^0 - (\widetilde{W}^1 - \widetilde{W}^0) + W^0 - \widetilde{W}^0$, the right-hand side of this inequality satisfies

$$\begin{aligned}
&\le C[\|W^2 - W^1\|_{C^\alpha}\rho^\alpha\|W^2 - W^0\|_C + (\|W^1 - W^0\|_{C^\alpha}\rho^\alpha + \rho)\|W^2 - W^1\|_C \\
&\qquad + \|W^2 - W^1\|_C\rho^\alpha \max\{\|W^2 - W^0\|_C, \|W^1 - W^0\|_C\}] \\
&\le C\|W^2 - W^1\|_{C^\alpha}\rho^\alpha(\max\{\|W^2 - W^0\|_C, \|W^1 - W^0\|_C\} + \rho^{1-\alpha}),
\end{aligned}$$

from which we get

$$\begin{aligned}
&\frac{\|G(x, t, W^2) - G(x, t, W^1)\|_{C^\alpha}}{\|W^2 - W^1\|_{C^\alpha}} \\
&\quad \le C_2 \min\left\{\frac{1}{\rho^\alpha}\max\{\|W^2 - W^0\|_C, \|W^1 - W^0\|_C\}, \max\{\|W^2 - W^0\|_C,\right. \\
&\qquad\qquad \left.\|W^1 - W^0\|_C + \rho^{1-\alpha}\}\right\} \\
&\quad \le C_3 \max\{\|W^2 - W^0\|_C, \|W^1 - W^0\|_C\}^{1-\alpha}.
\end{aligned}$$

Since $G(x, t, W^0) = 0$, $\|W\|_{C^\alpha} = \|w\|_{C^{2+\alpha}}$. Taking $\|w - w^0\|_{C^{1+\alpha}} \le \delta$ we get

$$\begin{aligned}
\|\kappa G(x, t, W) + (\kappa_0 - \kappa)H\|_{C^\alpha} &\le (C_\varepsilon\delta^{1-\alpha} + |\kappa - \kappa_0|M_3)\|W - W^0\|_{C^\alpha} \\
&\le (C_3 + M_3)\delta^{1-\alpha}\|w - u^0\|_{C^{2+\alpha}},
\end{aligned}$$

when $|\kappa - \kappa_0| \le \delta^{1-\alpha}$. From linear parabolic theory

$$\|v - u^0\|_{C^{2+\alpha}} \le K_1(C_3 + M_3)\delta^{1-\alpha}\|w - u^0\|_{C^{2+\alpha}} \le \frac{1}{2}\|w - u^0\|_{C^{2+\alpha}} \le \frac{\delta}{2},$$

when δ satisfies $K_1(C_3 + M_3)\delta^{1-\alpha} = 1/2$ and $\|w - u^0\|_{C^{2+\alpha}} \le \delta$ holds. On the other hand,

$$\begin{aligned}
\|v^2 - v^1\|_{C^{2+\alpha}} &\le K_1\kappa\|G(x, t, W^2) - G(x, t, W^1)\|_{C^\alpha} \\
&\le K_1C_3\delta^{1-\alpha}\|W^2 - W^1\|_{C^\alpha} \le \frac{1}{2}\|u^2 - u^1\|_{C^{2+\alpha}}
\end{aligned}$$

which implies Ψ_κ is a contraction in $C^{2+\alpha}$ for $|\kappa - \kappa_0| \le \delta^{1-\alpha}$. By the contraction mapping principle I is open. Since I is nonempty, open, and closed, it is equal to $[0, 1]$.

Thus we have established the existence of a solution for (11), (12). Uniqueness can be proved easily by a standard argument. Actually it holds in a larger space. □

To weaken the requirement on the regularity of the boundary of the domain, the initial-boundary value, and F as well as to remove the compatibility condition (the solution is understood in weak sense) require further treatment. However, basically this is not new material and it can be found in [44] and [45]. We shall not discuss it here.

Problems worth further investigation include Neumann boundary value problem, relaxing the convexity condition, and applications of the results of this chapter.

CHAPTER X

Fully Nonlinear Parabolic Equations (Continued)

In this chapter we establish a density theorem for the quasilinear parabolic equations described below. Based on this result we obtain the unique solvability for the first boundary value problem of fully nonlinear parabolic equations satisfying the natural structure condition of the second kind [46], [47].

Let Ω be a bounded domain in $\mathbb{R}^n$, $T > 0$, and $Q = \Omega \times (0, T]$. Consider the equation

$$\mathscr{L}u = A_{ij}(x, t, u, u_x)u_{ij} - u_t = B(x, t, u, u_x),$$

where A_{ij} $(1 \le i, j \le n)$ and B are continuous functions in $\overline{Q}\times[-M, M]\times \mathbb{R}^n$ satisfying

$$\nu(1 + |p|^{\sigma-2})|\xi|^2 \le A_{ij}(x, t, u, p)\xi_i\xi_j \le \mu(1 + |p|^{\sigma-2})|\xi|^2, \quad \forall\xi \in \mathbb{R}^n, \tag{1}$$

$$|B(x, t, u, p)| \le \mu(1 + |p|^{\sigma}), \tag{2}$$

for some positive constants μ, ν, σ. Here $\sigma > 2$. The case $\sigma = 2$ has been covered in the previous chapter.

Let $\partial\Omega \in C^2$ and let u be a solution of $\mathscr{L}u = B$ in $C^{2,1}(Q) \cap C(\overline{Q})$ which satisfies $u|_{\partial^* Q} = u^* \in C^{\beta, \beta/\sigma}(\partial^* Q)$ $(\partial^* Q = \partial\Omega \times [0, T] \cup \Omega \times \{t = 0\}$ is the parabolic boundary of Q). Under the assumption that the equation $\mathscr{L}u = B$ satisfies the *natural structure condition of the second kind*, that is, (1) and (2) hold for $2 < \sigma < 3$, we shall derive a density theorem for u and consequently a Hölder estimate for u in $\overline{Q}$. Notice that the case $\sigma = 2$ has been discussed in the previous chapter.

Different constants will be denoted by $C_1, C_2, \dots$ and $c_1, c_2, \dots$ respectively.

§1. The density theorem for quasilinear parabolic equations with natural structure condition of the second kind

As in the previous chapter, a couple of lemmas are required for the proof of the density theorem. The distinction between $\sigma = 2$ and the present situation is that when $2 < \sigma < 3$, the density theorem as well as the lemmas hold for those u whose "initial values" lie in a certain range.

The following two lemmas hold for any $\sigma > 2$.

LEMMA 1. *Let $X_1 > 0$ and $t_1 < t_2$. Suppose $u \in C^{2,1}$ in the cylinder $\{|x| \le X_1\} \times \{t_1 \le t \le t_2\}$ satisfies $u \ge 0$, $\mathscr{L}u \le 0$, and $\inf_{|x|\le X_1, t=t_1} u \ge u_0 > 0$. There exist constants C_1, C_2, C_3 depending only on n, μ, ν, σ (> 2) and a parameter $\Lambda \ge 1$ such that*

$$\inf_{\substack{|x|\le 2X_1/3 \\ t=t_2}} u \ge C_3 \Lambda^{-1/(\sigma-2)} u_0$$

provided

$$\frac{u_0^{\sigma-2}(t_2-t_1)}{X_1^\sigma} \le C_1\Lambda \quad \textit{and} \quad X_1 \le C_2 \Lambda^{-1/(\sigma-2)} u_0 .$$

PROOF. Let

$$v = v_0 \left(1 - \frac{|x|^2}{X_1^2}\right)^{1+\delta} \frac{2t_2 - t_1 - t}{2(t_2-t_1)},$$

where $0 < v_0 \le u_0$ and $\delta > 0$. Let $w = u - v$. When $|x| = X_1$, $t_1 \le t \le t_2$, $w = u \ge 0$. When $t = t_1$, $|x| \le X_1$,

$$w = u - v_0 \left(1 - \frac{|x|^2}{X_1^2}\right)^{1+\delta} \ge u_0 - v_0 > 0 .$$

If w attains a minimum in the cylinder $\{|x| < X_1\} \times \{t_1 < t \le t_2\}$, then at the minimum point $w_i = 0$ $(1 \le i \le n)$ and $\mathscr{L}(w) \ge 0$. Therefore

$$0 \ge \mathscr{L}(u) \ge \mathscr{L}(v)|_{u_i = v_i} = A_{ij}(x, t, u, v_i) - v_t .$$

The right-hand side of this inequality is the linearized operator of $\mathscr{L}$. Without causing ambiguity we shall still denote it by $\mathscr{L}v$. Using (1) we have

$$\begin{aligned} \mathscr{L}(v) \ge v_0 \Bigg\{ \Bigg[& 4\delta(1+\delta)\nu \left(1 - \frac{|x|^2}{X_1^2}\right)^{\delta-1} \frac{2t_2 - t_1 - t}{2(t_2 - t_1)} \frac{|x|^2}{X_1^4} \\ & - 2n\mu(1+\delta)\left(1 - \frac{|x|^2}{X_1^2}\right)^{\delta} \frac{2t_2 - t_1 - t}{2(t_2-t_1)} \frac{1}{X_1^2} \Bigg] (1 + |v_x|^{\sigma-2}) \\ & + \left(1 - \frac{|x|^2}{X_1^2}\right)^{1+\delta} \frac{1}{2(t_2 - t_1)} \Bigg\} \\ = v_0 \Bigg\{ & 2(1+\delta) \frac{1 + |v_x|^{\sigma-2}}{X_1^2} \frac{2t_2 - t_1 - t}{2(t_2-t_1)} \left(1 - \frac{|x|^2}{X_1^2}\right)^{\delta-1} \\ & \cdot \left[(2\delta\nu + n\mu)\cdot\frac{|x|^2}{X_1^2} - n\mu\right] + \left(1 - \frac{|x|^2}{X_1^2}\right)^{1+\delta} \frac{1}{2(t_2-t_1)} \Bigg\} . \end{aligned} \tag{3}$$

In case $|x|^2/X_1^2 \ge n\mu/(n\mu + 2\delta\nu)$, $\mathscr{L}v > 0$ which contradicts $\mathscr{L}v \le 0$.

On the other hand, when $|x|^2/X_1^2 < n\mu/(n\mu + 2\delta\nu)$, $1 - |x|^2/X_1^2 \geq 2\delta\nu/(n\mu + 2\delta\nu)$. Taking $\delta = 1/(\sigma - 2)$ we have

$$\begin{aligned}
\mathscr{L}v &\geq v_0 \left\{ \left(1 - \frac{|x|^2}{X_1^2}\right)^{1+\delta} \frac{1}{2(t_2 - t_1)} - 2n\mu(1+\delta)\left(1 - \frac{|x|^2}{X_1^2}\right)^{\delta} \right. \\
&\qquad \left. \cdot \frac{1}{X_1^2}\left\{1 + \left[2(1+\delta)v_0 \cdot \left(1 - \frac{|x|^2}{X_1^2}\right)^{\delta} \frac{|x|}{X_1^2}\right]^{\sigma-2}\right\}\right\} \\
&\geq v_0 \left(1 - \frac{|x|^2}{X_1^2}\right)^{\delta} \left\{ \left(1 - \frac{|x|^2}{X_1^2}\right) \left[\frac{1}{4(t_2 - t_1)} - n\mu(2(1+\delta))^{\sigma-1} v_0^{\sigma-2} \right.\right. \\
&\qquad \left. \cdot \left(\frac{n\mu}{n\mu + 2\delta\nu}\right)^{\sigma/2-1} \frac{1}{X_1^{\sigma}}\right] \\
&\qquad \left. + \frac{2\delta\nu}{n\mu + 2\delta\nu}\frac{1}{4(t_2 - t_1)} - \frac{2n\mu(1+\delta)}{X_1^2}\right\} \\
&> 0 \qquad (4)
\end{aligned}$$

provided the following two conditions hold:

$$t_2 - t_1 \leq \frac{1}{4n\mu(2(1+\delta))^{\sigma-1}} \left(\frac{n\mu + 2\delta\nu}{n\mu}\right)^{\sigma/2-1} v_0^{2-\sigma} X_1^{\sigma} \equiv c_1 v_0^{2-\sigma} X_1^{\sigma}, \qquad (5)$$

and

$$X_1 < \left[\frac{\delta\nu}{4n\mu(1+\delta)(n\mu + 2\delta\nu)c_1}\right]^{1/(\sigma-2)} v_0 \equiv c_2 v_0. \qquad (6)$$

Therefore, under (5) and (6), (4) is in contradiction with $\mathscr{L}v \leq 0$. We conclude that w cannot attain its minimum in $\{|x| < X_1\} \times \{t_1 < t \leq t_2\}$. For $t_1 \leq t \leq t_2$, we have

$$u \geq v_0 \left(1 - \frac{|x|^2}{X_1^2}\right)^{1+\delta} \frac{2t_2 - t_1 - t}{2(t_1 - t_1)} \geq \frac{v_0}{2}\left(1 - \frac{|x|^2}{X_1^2}\right)^{1+\delta}. \qquad (7)$$

So the lemma holds for $\Lambda = (u_0/v_0)^{\sigma-2}$, $C_1 = c_1$, $C_2 = c_2$, and $C_3 = \frac{1}{2}(5/9)^{(\sigma-1)/(\sigma-2)}$. □

LEMMA 2. *Suppose $u \in C^{2,1}$ in the cylinder $\{x : |x| \leq C_4 X_2\} \times \{t_1 \leq t \leq t_2\}$ satisfies $u \geq 0$, $\mathscr{L}u \leq 0$, and $\inf_{|x|\leq X_1, t=t_1} u \geq u_0 > 0$, where $0 < X_1 \leq X_2$, $t_1 < t_2$, and $C_4 = C_4(n, \mu, \nu, \sigma) > 1$. There exist constants C_5, C_6, C_7, C_8 depending only on n, μ, ν, σ, and a parameter $\Lambda \geq 1$ such that*

$$\inf_{\substack{|x|\leq X_2 \\ t=t_2}} u \geq C_8 \Lambda^{-1/(\sigma-2)} \left(\frac{X_1}{X_2}\right)^{C_6} u_0$$

provided

$$C_5 \le \frac{u_0^{\sigma-2}(t_2 - t_1)}{X_2^\sigma \left(\frac{X_2}{X_1}\right)^{C_6(\sigma-2)}} \le C_5\Lambda$$

and

$$X \le C_7\Lambda^{-1/(\sigma-2)} \left(\frac{X_1}{X_2}\right)^{C_6} u_0 .$$

PROOF. Let $t_3 \le t_2$. From (5) and (6) we know that $t_3 - t_1 \le C_1 v_0^{2-\sigma} X_1^\sigma$ when $X_1 < C_2 v_0$. Let

$$v = \tilde{v}_0 \left[1 - F^{1/2}\left(\frac{t_3 - t_1}{t - t_1}\right)^{1/(\sigma-2)}\right],$$

where

$$F = \frac{x_1 - (1-\varepsilon)X_1}{2\varepsilon X_1} + \frac{\beta}{\varepsilon^2 X_1^2} \sum_{i=2}^n x_i^2 ,$$

and $\tilde{v}_0$, ε, β are positive constants, $\varepsilon^2 < \beta < 1$. Let V be the region bounded by $t_1 < t < t_3$, $v = 0$, and $x_1 = (1-\varepsilon)X_1$. The t-slice of V is the intersection of a rotationally symmetric paraboloid and a hyperplane. When $t = t_1$, it degenerates into a single point $((1-\varepsilon)X_1, 0, \dots, 0, t_1) \in \partial V$. When $t = t_3$, the t-slice is maximal and its largest distance to the t-axis, which is attained at the point $((1+\varepsilon)X_1, 0, \dots, 0, t_3) \in \partial V$, is $(1+\varepsilon)X_1$. Thus for $1 + \varepsilon \le C_4$,

$$V \subseteq \{x : |x| < C_4 X_2\} \times \{t_1 < t \le t_2\}.$$

By a suitable choice of $\tilde{v}_0$, β, and ε we claim that $u \ge v > 0$ on ∂V and in V. Let us look at ∂V first. On $\partial V \cap \{v = 0\}$, $w = u - v = u \ge 0$.

On $\partial V \cap \{(x, t) : x_1 = (1-\varepsilon)X_1\} \cap \{(x, t) : v \ge 0\}$,

$$|x|^2 \le [(1-\varepsilon)^2 + \varepsilon^2/\beta^2]X_1^2 .$$

For ε, β satisfying $\varepsilon < 2\beta/(1+\beta)$

$$\begin{aligned} w = u - v &\ge \frac{v_0}{2}\left(1 - \frac{|x|^2}{X_1^2}\right)^{1+\delta} - \tilde{v}_0 \ge \frac{v_0}{2}\left\{1 - \left[(1-\varepsilon)^2 + \frac{\varepsilon^2}{\beta}\right]\right\}^{1+\delta} - \tilde{v}_0 \\ &= 2^\delta \left(1 - \frac{1+\beta}{2\beta}\varepsilon\right)^{1+\delta} \varepsilon^{1+\delta} v_0 - \tilde{v}_0 \end{aligned}$$

on this region in view of (7). We can choose $\tilde{v}_0$ so that the right-hand side of this inequality vanishes, that is,

$$\tilde{v}_0 = 2^\delta \left(1 - \frac{1+\beta}{2\beta}\varepsilon\right)^{1+\delta} \varepsilon^{1+\delta} v_0 .$$

So $w = u - v \geq 0$ on the lateral boundary of ∂V. Next we look at V. Using (1) and the hypotheses of the lemma, $\mathscr{L}u \leq 0$, and $F \leq 1$ we have

$$\mathscr{L}v \geq (1 + |v_x|^{\sigma-2})\left[\frac{\tilde{v}_0\nu}{16\varepsilon^2 X_1^2}F^{-3/2}\left(\frac{t_3 - t_1}{t - t_1}\right)^{2/(\sigma-2)} - \frac{\tilde{v}_0\beta}{\varepsilon^2 X_1^2}n\mu F^{-1/2}\left(\frac{t_3 - t_1}{t - t_1}\right)^{1/(\sigma-2)}\right]$$
$$- \frac{\tilde{v}_0}{\sigma - 2}F^{1/2}\left(\frac{t_3 - t_1}{t - t_1}\right)^{1/(\sigma-2)}\frac{1}{t - t_1}$$
$$\geq (1 + |v_x|^{2-\sigma})\left[\frac{\tilde{v}_0^{\sigma-2}}{(4\varepsilon X_1)^{\sigma-2}}F^{1-\sigma/2}\frac{t_3 - t_1}{t - t_1}\frac{\tilde{v}_0}{(\varepsilon X_1)^2}\cdot\left(\frac{\nu}{16} - \beta n\mu\right)F^{-3/2} - \frac{\tilde{v}_0}{\sigma - 2}F^{1/2}\frac{1}{t - t_1}\right]\left(\frac{t_3 - t_1}{t - t_1}\right)^{1/(\sigma-2)}.$$

Fix $\beta = \nu/32n\mu$ and restrict ε to $\varepsilon \leq \beta/(1+\beta)$. From $F \leq 1$ in V, the choice for $\tilde{v}_0$ and the condition on t_3 we infer

$$\tilde{v}_0^{-1}(t - t_1)\left(\frac{t - t_1}{t_3 - t_1}\right)^{1/(\sigma-2)}F^{(\sigma+1)/2}\mathscr{L}v \geq \frac{\tilde{v}_0^{\sigma-2}(t_3 - t_1)}{(4\varepsilon X_1)^\sigma}\frac{\nu}{2} - \frac{1}{\sigma - 2}F^{\sigma/2+1}$$

$$\geq \frac{\nu}{2}\left(\frac{\varepsilon^{1+\delta}}{2}v_0\right)^{\sigma-2}\frac{t_3 - t_1}{(4\varepsilon X_1)^\sigma} - \frac{1}{\sigma - 2} = \frac{\nu(t_3 - t_1)v_0^{\sigma-2}}{2^{3\sigma-1}\varepsilon X_1^\sigma} - \frac{1}{\sigma - 2}.$$

If $t_3 - t_1 \geq c_3\varepsilon v_0^{2-\sigma}X_1^\sigma$, $c_3 = 2^{3\sigma}/\nu(\sigma - 2)$, then the right-hand side of this inequality becomes positive.

By comparing the two requirements on $t_3 - t_1$ we obtain $c_3\varepsilon \leq C_1$. Hence for $\varepsilon = \min\{\frac{\beta}{1+\beta}, \frac{c_1}{c_3}\}$, w cannot attain its minimum in V when $t_3 - t_1 = c_3\varepsilon v_0^{2-\sigma}X_1^\sigma$. Since $w \geq 0$ on ∂V we conclude $w \geq 0$ in V. In other words,

$$u \geq \frac{1}{2}\varepsilon^{(\sigma-1)/(\sigma-2)}v_0\left[1 - F^{1/2}\left(\frac{t_3 - t_1}{t - t_1}\right)^{1/(\sigma-2)}\right].$$

In particular,

$$u \geq \frac{1}{16}\varepsilon^{(\sigma-1)/(\sigma-2)}v_0 \tag{8}$$

for $(1 - \varepsilon)X_1 \leq x_1 \leq (1 + \frac{\varepsilon}{2})X_1$, $x_i = 0$ $(2 \leq i \leq n)$ and $t = t_3$. After rotating the axes we conclude that (8) holds for $(1 - \varepsilon)X_1 \leq |x| \leq (1 + \frac{\varepsilon}{2})X_1$ and $t = t_3$ when (6) is valid. On the other hand, by (7) we know that

$$u \geq \frac{1}{2}v_0[1 - (1 - \varepsilon)^2]^{(\sigma-1)/(\sigma-2)} \geq \frac{1}{2}\varepsilon^{(\sigma-1)/(\sigma-2)}v_0.$$

Therefore (8) holds for $|x| \leq (1 + \varepsilon/2)X_1$, $t = t_4 = t_1 + c_3\varepsilon v_0^{2-\sigma}X_1^\sigma$.

Replacing X_1 and v_0 by $(1+\frac{\varepsilon}{2})^{k-1}X_1$ and $(\frac{1}{16}\varepsilon^{(\sigma-1)/(\sigma-2)})^{k-1}v_0$ ($k = 1, 2, \ldots$) respectively yields that for $(1+\frac{\varepsilon}{2})^{k-1}(1+\varepsilon)X_1 < C_4X_2$, $(1+\frac{\varepsilon}{2})^{k-1}X_1 < c_2(\frac{1}{16}\varepsilon^{(\sigma-1)/(\sigma-2)})^{k-1}v_0$ and

$$t_2 = t_1 + c_3\varepsilon\frac{\eta^k - 1}{\eta - 1}v_0^{2-\sigma}X_1^\sigma, \tag{9}$$

where $\eta = 16^{\sigma-2}\varepsilon^{1-\sigma}(1+\frac{\varepsilon}{2})^\sigma$, $u(x, t_2) \geq (\frac{1}{16}\varepsilon^{(\sigma-1)/(\sigma-2)})^k v_0$ for all x, $|x| \leq (1+\frac{\varepsilon}{2})^k X_1$. In order to have k satisfying $(1+\frac{\varepsilon}{2})^{k-1}X_1 < X_2$, from $(1+\frac{\varepsilon}{2})^{k-1}(1+\varepsilon)X_1 < C_4X_2$ we see that it suffices to choose $C_4 = 1+\varepsilon$. Now we can fix k such that

$$\left(1+\frac{\varepsilon}{2}\right)^{k-1}X_1 < X_2 \leq \left(1+\frac{\varepsilon}{2}\right)^k X_1.$$

Let

$$C_5 = \frac{c_3\varepsilon\eta}{\eta - 1}, \qquad C_6 = \frac{\log(16\varepsilon^{(\sigma-1)/(\sigma-2)})}{\log(1+\varepsilon/2)}, \quad \text{and} \quad c_4 = \left(\frac{1}{\eta} - \frac{1}{\eta^2}\right)^{1/(\sigma-2)}.$$

From (9)

$$\begin{aligned} v_0^{\sigma-2} &= \frac{X_1^\sigma}{t_2 - t_1}\frac{c_3\varepsilon}{\eta - 1}(\eta^k - 1) \leq \frac{X_1^\sigma}{t_2 - t_1}C_5\eta^{k-1} \\ &< \frac{X_1^\sigma}{t_2 - t_1}C_5\exp\left(\frac{\log(X_2/X_1)\log\eta}{\log(1+\varepsilon/2)}\right) \\ &= \frac{X_2^\sigma C_5}{t_2 - t_1}\left(\frac{X_2}{X_1}\right)^{\log\eta/\log(1+\varepsilon/2)-\sigma} = \frac{X_2^\sigma C_5}{t_2 - t_1}\left(\frac{X_2}{X_1}\right)^{C_6(\sigma-2)}, \end{aligned} \tag{10}$$

and

$$\begin{aligned} v_0^{\sigma-2} &\geq \frac{X_1^\sigma}{t_2 - t_1}c_3\varepsilon\eta^{k-1} \geq \frac{X_2^\sigma}{t_2 - t_1}C_5\left(\frac{\eta - 1}{\eta^2}\right)\left(\frac{X_2}{X_1}\right)^{C_6(\sigma-2)} \\ &= \frac{X_2^\sigma}{t_2 - t_1}C_5c_4^{\sigma-2}\left(\frac{X_2}{X_1}\right)^{C_6(\sigma-2)}. \end{aligned} \tag{11}$$

Therefore $c_4^{\sigma-2}\Lambda^{-1}u_0^{\sigma-2} \leq v_0^{\sigma-2} \leq u_0^{\sigma-2}$ when

$$C_5 \leq \frac{(t_2 - t_1)u_0^{\sigma-2}}{X_2^\sigma(X_2/X_1)^{C_6(\sigma-2)}} \leq C_5\Lambda \qquad (\Lambda \geq 1).$$

We can find v_0 in $[c_4\Lambda^{-1/(\sigma-2)}u_0, u_0]$ so that (10) and (11) hold. For this v_0 we have

$$\left(1+\frac{\varepsilon}{2}\right)^{k-1}X_1 < c_2\left(\frac{1}{16}\varepsilon^{(\sigma-1)/(\sigma-2)}\right)^{k-1}v_0,$$

and

$$X_2 < c_2c_4\Lambda^{-1/(\sigma-2)}u_0\left(\frac{X_1}{X_2}\right)^{C_6}.$$

Taking $C_7 = c_2 c_4$,

$$u(x, t_2) \geq \left(1 + \frac{\varepsilon}{2}\right)^{-C_6} \left(\frac{X_1}{X_2}\right)^{C_6} c_4 \Lambda^{-1/(\sigma-2)} u_0 .$$

So the lemma is valid for $C_8 = (1 + \frac{\varepsilon}{2})^{-C_6}$ and $c_4 = \frac{1}{16}\varepsilon^{(\sigma-1)/(\sigma-2)} c_4$. □

Denote the ball $|x - x^0| < R$ by $B_R(x^0)$, the square $|x_i - x_i^0| < R$ $(1 \leq i \leq n)$ by $k_R(x^0)$, $k_R(x^0) \times \{t^0 < t < t^0 + \eta R^\sigma\}$ by $K_{R,\eta}(x^0, t^0)$ and write $B_R(0)$, $k_R(0)$, $K_{R,\eta}(0, 0)$ simply as B_R, k_R, $K_{R,\eta}$. Also, the measure of a measurable set A is written as $|A|$.

LEMMA 3. *Given constants $0 < R < 1$, $\tilde{\eta} > 0$, $u_0 > 0$, and $\mu_1 > 0$. Suppose that $u \in C^{2,1}$ satisfies $u \geq 0$, $\mathscr{L}u \leq -\mu_1 u_0 |u_x|^\sigma$ in $K_{R,\eta}$ ($\eta = \tilde{\eta} u_0^{2-\sigma}$). There exists a constant τ $(0 < \tau < 1)$ which depends only on $n, \mu, \nu, \sigma, \mu_1, \tilde{\eta}$ such that for $\forall R, u_0$,*

$$u_{B_{3R/4} \times \{(\eta/2)R^\sigma \leq t \leq \eta R^\sigma\}} \geq c u_0$$

if $2 < \sigma < 3$ and $|K_{R,\eta} \cap \{u \geq u_0\}| \geq \tau |K_{R,\eta}|$. Here c is an absolute constant.

PROOF. Let $x = RX$, $t = \delta \tilde{\eta} R^\sigma u_0^{2-\sigma} T$ $(1/2 \leq \delta < 1)$, $u = u_0 U$, and $(A_{ij} u_{ij} - u_t)/(\delta \tilde{\eta} R^\sigma u_0^{1-\sigma}) = \widetilde{A}_{ij} U_{ij} - U_T$, $U = U(X, T)$.

To simplify notation further, let's write $X, T, U, \nu/\tilde{\eta}, 2\mu/\tilde{\eta}, 2\mu_1/\tilde{\eta}$, and $\widetilde{A}_{ij}$ as x, t, u, ν, μ, μ_1 and A_{ij}. We have

$$\nu \left[\left(\frac{R}{u_0}\right)^{\sigma-2} + |p|^{\sigma-2}\right] |\xi|^2 \leq A_{ij}\xi_i\xi_j$$

$$\leq \mu \left[\left(\frac{R}{u_0}\right)^{\sigma-2} + |p|^{\sigma-2}\right] |\xi|^2, \quad \forall \xi \in \mathbb{R}^n,$$

and $\mathscr{L}u \equiv A_{ij}u_{ij} - u_t \leq -\mu_1 |p|^\sigma$ in $Q \equiv \{(x, t) : |x| < 1, 0 < t \leq 1\}$. Also

$$|Q \cap \{u \geq 1\}| \geq \left[1 - 2\frac{|k_1|}{|B_1|}(1 - \tau)\right] |Q| \tag{12}$$

holds.

Choose an approximate function $w(x, t, N)$ $(0 \leq w \leq 1)$ where N is a large parameter and w only depends on $|x|, t, N$. w should also satisfy

$$[w, w_{|x|}, w_t]_{|x|=7t/8} = 0 \qquad (0 \leq t \leq 1).$$

Let

$$v = \begin{cases} u - w & \text{in } Q_1 \equiv \{(x, t) : |x| < 7t/8, \ 0 \leq t \leq 1\} \\ u & \text{in } Q \backslash Q_1 . \end{cases}$$

Then $v \in C^{1,1}(Q) \cap W_{n+1}^{2,1}(Q)$ and v_{ij} $(1 \leq i, j \leq n)$ has finite jumps in $\{(x, t) : |x| = 7t/8, 0 \leq t \leq 1\}$. Let $\tilde{v}(x, t) = \inf_{0 \leq \gamma \leq 1}\{0, v(x, \gamma)\}$ and let $\check{v}(x, t)$ be the lower convex envelope of $\tilde{v}$ in x. Then $\check{v} \le \tilde{v} \le v$. Since $\check{v}$ is Lipschitz continuous, nonincreasing in t, and convex in x (see Lemma 19 of Chapter 7), $\check{v}_t$ and $\check{v}_{ij}$ $(1 \le i, j \le n)$ exist almost everywhere in $\overline{Q}$. By Theorem 23 of Chapter 7 we have

$$[-\inf_{x\in B_1} v(x, 1)]^{n+1} \le \frac{n+1}{|B_1|}\int_{Q_3} -v_t \det(v_{ij})\, dx\, dt, \tag{13}$$

where Q_3 is the negative contact set of v in Q.

In Q_3 one has

$$0 \le -\check{v} = -v = w - u \le w \le 1, \qquad |\check{v}_{|x|}| \le -\check{v}/\left(1 - \frac{7}{8}\right) \le 8, \tag{14}$$

(Note that $Q_3 \subseteq Q \cap \{|x| \le 7/8\}$), $u - w = v = \check{v} \le 0$, and $u \le w \le 1$. By (12)

$$|Q_3| \le |\{(x, t) \in Q : u \le 1\}| \le 2\frac{|k_1|}{|B_1|}(1-\tau)|Q| \equiv c(1-\tau),$$

and $-v_t \ge 0$, $(v_{ij}) \ge 0$. Consequently

$$A_{ij}v_{ij} \le \mathscr{L}v = \mathscr{L}u - \mathscr{L}w \le -\mu_1|p|^\sigma - A_{ij}w_{ij} + w_t$$

in Q_3. Denote the eigenvalues of (v_{ij}) in Q_3 by $(0 \le)$ $\lambda_1 \le \lambda_2 \le \cdots \le \lambda_n$. We have

$$\begin{aligned} 0 \le \Delta v = \lambda_1 + \cdots + \lambda_n &\le \frac{1}{\nu}\frac{A_{ij}v_{ij}}{[(R/u_0)^{\sigma-2} + |p|^{\sigma-2}]} \\ &\le \frac{\mu}{\nu}\sum |w_{ij}| + \frac{1}{2}\frac{(-\mu_1|p|^\sigma + w_t)}{[(R/u_0)^{\sigma-2} + |p|^{\sigma-2}]}. \end{aligned} \tag{15}$$

Now, if we choose w so that

$$|w_t| \le c_2 N, \qquad |w_{ij}| \le c_3 N^2 \qquad (1 \le i, j \le N),$$

it follows from (15) that

$$\begin{aligned} 0 \le \Delta v &\le \frac{\mu}{\nu}n^2 c_3 N^2 + \frac{1}{\nu}\max\left\{N\alpha(N)^{2-\sigma},\ \max_{|p|\ge\alpha(N)}(-\mu_1|p|^2 + c_2 N|p|^{2-\sigma})\right\} \\ &\le c_4 \max\{N^2, N\alpha(N)^{2-\sigma}\}, \\ -v_t &\le \mathscr{L}v \le -\mu_1|p|^\sigma - A_{ij}w_{ij} + w_t \le c_5 N^2 \end{aligned}$$

for $|p| \ge \alpha(N)$ or $R/u_0 \ge \alpha(N)$. As a result,

$$\begin{aligned} &\int_{Q_3\cap\{p\ge\alpha(N) \text{ or } R/u_0\ge\alpha(N)\}} -v_t \det v_{ij}\, dx\, dt \\ &\qquad \le c_6 N^2 \max\{N^2, N\alpha(N)^{2-\sigma}\}^n(1-\tau) \end{aligned} \tag{16}$$

for $c_6 = c_1 c_4 c_5$.

Taking $\alpha(N) = N^{-2/(\sigma-2)}$, then for $|p| \le \alpha(N)$ and $R/u_0 \le \alpha(N)$ one has

$$|A_{ij}w_{ij}| \le \mu n^2[(R/u_0)^{\sigma-2} + |p|^{\sigma-2}]c_3 N^2 \le 2\mu c_3 n^2.$$

Since we can choose w satisfying $w_t \le c_7$ when $|p| \le \alpha(N)$ and $R/u_0 \le \alpha(N)$, $-v_t \le \mathscr{L}v \le -\mu_1|p|^\sigma - A_{ij}w_{ij} + w_t \le c_8$, and

$$\begin{aligned}
&\int_{Q_3\cap\{|p|\le\alpha(N),\, R/u_0\le\alpha(N)\}} -v_t \det v_{ij}\,dx\,dt \\
&\quad\le c_8 \int_{Q\cap\{|x|<7/8\}} \det \check{v}_{ij}\,dx\,dt \\
&\quad\le \frac{64}{15}c_8 \int_Q (1-|x|^2)\det \check{v}_{ij}\,dx\,dt \\
&\quad\le \frac{64}{15}c_8\left(-\int_{B_1}\mu(x,t)\det\check{v}_{ij}\,dx\Big|_{t=0}^{t=1} + \int_Q \mu(x,t)\check{v}_{ijt}\frac{\partial}{\partial \check{v}_{ij}}\det v_{ij}\,dx\,dt\right),
\end{aligned} \tag{17}$$

where $\mu(x,t) = (1-t)(1-|x|^2)$. As $\mu(x,1) = 0$ and $v(x,0) = 0$, the first term in the right-hand side of the above inequality vanishes. Performing integration by parts, by virtue of $\mu|_{\partial B_1} = 0$, $v_t|_{\partial B_1} = 0$, and $(\partial/\partial x_i)(\partial/\partial \check{v}_{ij})\det\check{v}_{ij} = 0$, (see Lemma 2 of Chapter 7) we obtain

$$\begin{aligned}
\int_Q \mu\check{v}_{ijt}\frac{\partial}{\partial\check{v}_{ij}}\det\check{v}_{kl}\,dx\,dt &= \int_Q \check{v}_t\mu_{ij}\frac{\partial}{\partial\check{v}_{ij}}\det\check{v}_{kl}\,dx\,dt \\
&= 2\int_Q (1-t)(-\check{v}_t)\sum_{k=1}^n \frac{\partial}{\partial\check{v}_{kl}}\det\check{v}_{ij}\,dx\,dt \\
&\le 2\int_Q -\check{v}_t\sum_{k=1}^n\frac{\partial}{\partial\check{v}_{kk}}\det\check{v}_{ij}\,dx\,dt.
\end{aligned}$$

Notice that in order to justify the twice integration by parts above, one needs to follow what has been done in Chapter 7. First we extend $\check{v}$ as zero for $t < 0$ and $\check{v}(x,t) = \check{v}(x,1)$ for $t > 1$. Then we smooth out $\check{v}$ by a regularization with the same radius in x and t. Finally we extend the function to be a smooth one, $\check{v}_l$, which is convex in x, in $\mathbb{R}^{n+1}$. By this construction $\check{v}_l$ is independent of t outside $B_1 \times \mathbb{R}$. We perform integration by parts in $\Omega_l \times \{0 \le t \le 1\}$ and notice that $\check{v}_l|_{\partial\Omega_l} = 0$. After the integration by parts, by letting $l \to \infty$ we can follow the proof of Theorem 22 in Chapter 7 to show the above inequality is valid. However, now the terms $-\check{v}_t\,dt$ and $\sum_{k=1}^n \frac{\partial}{\partial\check{v}_{kk}}\det(v_{ij})\,dx$ should be replaced by $\tilde{\omega}(\check{v}, dt)$ and $\sum_{k=1}^n \omega(\check{v}, dx/dx_k)\,dx_k$, where $\omega(\check{v}, dx/dx_k)$ is the normal measure corresponding to $\frac{\partial}{\partial\check{v}_{kk}}\det(\check{v}_{ij})\frac{dx}{dx_k}$, respectively.

In $Q_2 = \{(x,t) : \check{v} = \tilde{v} < v\}\tilde{\omega}(\check{v}, dt) = 0$; in $Q_3 = \{(x,t) : \check{v} = \tilde{v} = v\}\tilde{\omega}(\check{v}, dt) = -\frac{\partial v}{\partial t}dt$, $\omega(\check{v}, dx/dx_k) = \frac{\partial}{\partial v_{kk}}\det v_{ij}\frac{dx}{dx_k}$; and in

$Q_1 = \{(x, t) : \check{v} < \tilde{v}\}\tilde{\omega}(\check{v}, dt) = -\frac{\partial \check{v}}{\partial t} dt$. a.e. We need to show that for a fixed t,

$$\int_{Q_1 \cap \{x_1 = x_1^0\}} -\frac{\partial \check{v}}{\partial t} \omega\left(\check{v}, \frac{dx}{dx_1}\right) = 0 \tag{18}$$

holds for almost every x_1^0. In case this integral is positive at some x_1^0, $\omega(\check{v}, dx/dx_1)|_{x_1=x_1^0} > 0$ and the convex function $\check{v}$ has a $(n-1)$-dimensional corner at $x_1 = x_1^0$. Since $\check{v}$ is the lower convex envelope of $\tilde{v}$, there exist two points $(x_1^0, x_2^i, \dots, x_n^i)$ $(i = 1, 2)$ such that

$$\begin{aligned}\check{v}(x_1^0, x_2, \dots, x_n) = \theta\tilde{v}(x_1^0, x_2^1, \dots, x_n^1) \\ + (1-\theta)\tilde{v}(x_1^0, x_2^2, \dots, x_n^2) \qquad (0 < \theta < 1),\end{aligned}$$

and $\tilde{v}(x_1^0, x_2^i, \dots, x_n^i)$ is an absolute minimum of $\check{v}$ near $(x_1^0, x_2^i, \dots, x_n^i)$ $(i = 1, 2)$ (for otherwise it cannot be a $(n-1)$-dimensional corner). Therefore x_1^0 is isolated and all these x_1^0's form a countable set. (18) holds. Applying the same argument to $x_2, \dots, x_n$ instead of x_1 yields

$$\int_{Q_1} -\frac{\partial \check{v}}{\partial t} \sum_{k=1}^{k} \omega\left(\check{v}, \frac{dx}{dx_k}\right) dx_k\, dt = 0. \tag{19}$$

Combining (16), (17), and (19) yields

$$\int_{Q_3} -v_t \det v_{ij}\, dx\, dt \le c_6 N^{3n+2}(1-\tau) + c_9 \int_{Q_3} -v_t \sum_{k=1}^{n} \frac{\partial}{\partial \check{v}_{kk}} \det v_{ij}\, dx\, dt,$$

$c_9 = \frac{128}{15} c_8$. Analogously we have

$$\begin{aligned}\int_{Q_3} -v_t \sum_{k=1}^{n} \frac{\partial}{\partial v_{kk}} \det v_{ij}\, dx\, dt \\ \le n c_6 N^{3n-1}(1-\tau) + n c_9 \int_{Q_3} -v_t \sum_{k,l=1}^{n} \frac{\partial^2}{\partial v_{kk} \partial v_{ll}} \det v_{ij}\, dx\, dt.\end{aligned}$$

Therefore

$$\begin{aligned}\int_{Q_3} -v_t \det v_{ij}\, dx\, dt &\le c_{10} N^{3n+2}(1-\tau) + c_{11} \int_{Q_3} -v_t \Delta v\, dx\, dt \\ &\le c_{12} N^{3n+2}(1-\tau) + c_{13} \int_{Q \cap \{|p| \le \alpha(N),\, R/u_0 \le \alpha(N)\}} \Delta v\, dx\, dt.\end{aligned} \tag{20}$$

Let us choose w so that $w_{rr} \ge \beta(N) > 0$ $(r = |x|)$ in

$Q\backslash\{(x, t) : |x|/t \le \gamma(N)\}$. Then

$$\begin{aligned}\int_{Q_3\cap\{|x|\le\gamma(N)\}} \Delta v\,dx\,dt &\le \int_{Q_3\cap\{|x|\le 2\gamma(N)\}} \left(2 - \frac{|x|}{\gamma(N)}\right) \Delta v\,dx\,dt \\ &= \int_{Q\cap\{|x|\le 2\gamma(N)\}} \left(2 - \frac{|x|}{\gamma(N)}\right) \Delta \check{v}\,dx\,dt \\ &= \int_{Q\cap\{|x|\le 2\gamma(N)\}} |\check{v}_{|x|}| \frac{dx\,dt}{\gamma(N)} \\ &\le \frac{8}{\gamma(N)} \int_{Q\cap\{|x|\le 2\gamma(N)\}} dx\,dt = 8\cdot 2^n \gamma(N)^{1-n}. \end{aligned} \tag{21}$$

Let $\check{u}_{rr} = \check{v}_{rr} + w_{rr}$ $(r = |x|)$. As $\check{v}_{rr} = v_{ij}x_i x_j/r^2 \ge 0$ in the region $w_{rr} \ge \beta(N)$,

$$\check{u}_{rr} \ge w_{rr} \ge \beta(N) > 0.$$

So $\check{u}_r$ is strictly increasing. This implies that for any $t \in [0, 1]$ and any direction $\theta = x/|x|$, $\check{u}_r$ has at most one zero $r_0(\theta, t)$, i.e., $u_r(r_0, \theta, t) = 0$. Since

$$\check{u}_r \ge \int_{r_0}^{r} \beta(N)\,dr = \beta(N)(r - r_0), \qquad r \ge r_0,$$

and

$$-\check{u}_r \ge \int_{r}^{r_0} \beta(N)\,dr = \beta(N)(r_0 - r), \qquad r \le r_0,$$

$$|\check{u}_r| \ge \beta(N)|r - r_0|.$$

In particular $|p| \ge |u_r| \ge \beta(N)|r - r_0|$ in the lower contact set Q_3. Together with (15) this implies

$$\begin{aligned} 0 \le \Delta v &\le \frac{\mu}{\nu}\sum |w_{ij}| + \frac{1}{\nu}\frac{(-\mu|p|^\sigma + w_t)}{(R/u_0)^{\sigma-2} + |p|^{\sigma-2}} \\ &\le \frac{\mu}{\nu} n^2 c_3 N^2 + \frac{1}{\nu} c_2 N \beta(N)^{2-\sigma}|r - r_0|^{2-\sigma}, \end{aligned}$$

and

$$\begin{aligned} &\int_{Q_3\cap\{|p|\le\alpha(N),\ R/u_0\le\alpha(N),\ w_{rr}\ge\beta(N)\}} \Delta v\,dx\,dt \\ &\quad\le \left[\int_{Q_3\cap\{|p|\le\alpha(N),\ R/u_0\le\alpha(N),\ w_{rr}\ge\beta(N)\}} (\Delta v)^{\frac{(\sigma-1)}{(2\sigma-4)}}\,dx\,dt\right]^{\frac{(2\sigma-4)}{(\sigma-1)}} |Q_3|^{\frac{(3-\sigma)}{(\sigma-1)}} \\ &\quad\le \left[\int_Q (c_{14}N^{\frac{(\sigma-1)}{(\sigma-2)}} + c_{15}N^{\frac{(\sigma-1)}{(2\sigma-4)}}\beta(N)^{\frac{(1-\sigma)}{2}}||x| - r_0|^{\frac{(1-\sigma)}{2}})\,dx\,dt\right]^{\frac{(2\sigma-4)}{(\sigma-1)}} \\ &\qquad\times [c_1(1-\tau)]^{(3-\sigma)/(\sigma-1)} \\ &\quad\le c_{16}(N^2 + N\beta(N)^{2-\sigma})(1-\tau)^{(3-\sigma)/(\sigma-1)}. \end{aligned} \tag{22}$$

Combining (13), (20), (21), and (22) we obtain

$$(-\inf_{x\in B_1} v(x,1))^{n+1} \le c_{17}N^{3n+2}(1-\tau) + c_{18}(N^2 + N\beta(N)^{2-\sigma})(1-\tau)^{\frac{(3-\sigma)}{(\sigma-1)}} + c_{19}\gamma(N)^{1-n}. \tag{23}$$

$w(x,t)$ is chosen as follows. Let $w(x,t) = (1-e^{-Ntf(\theta)})^2$, where $\theta = 1 - \frac{8|x|}{7t}$ and $f \in C^2[0,1]$ is to be specified. We have

$$\begin{aligned}
w_{|x|} &= -\frac{16}{7}Nf'(\theta)e^{-Ntf(\theta)}(1-e^{-Ntf(\theta)}),\\
w_{|x|\,|x|} &= \frac{128}{49}N\left[\frac{f''(\theta)}{t}e^{-Ntf(\theta)}(1-e^{-Ntf(\theta)})\right.\\
&\qquad\left. + f'(\theta)^2N(2e^{-2Ntf(\theta)} - e^{-Ntf(\theta)})\right]\\
&= \frac{128N^2e^{-Ntf}(1-e^{-Ntf})}{49Ntf}\left(ff'' + (f')^2Ntf\frac{2-e^{-Ntf}}{e^{Ntf}-1}\right)\\
&\ge \frac{128N^2e^{-Ntf}(1-e^{-Ntf})}{49Ntf}\left[ff'' + (f')^2\left(1-\frac{3}{2}Ntf\right)\right]\\
&\ge \frac{192N^2}{49}e^{-2/3}(1-e^{-2/3})\left[ff'' + (f')^2\left(1-\frac{3}{2}Ntf\right)\right]
\end{aligned}$$

when $Nf \le \frac{2}{3}$ and $0 \le t \le 1$. We take

$$f(\theta) = \begin{cases} \theta - \dfrac{2\alpha}{1+\alpha}\dfrac{\theta^2}{\theta_0} + \dfrac{\alpha}{2+\alpha}\dfrac{\theta^3}{\theta_0^2}, & 0 \le \theta \le \theta_0,\\[2ex] \dfrac{2}{(1+\alpha)(2+\alpha)}\theta_0^{\alpha}(1-g(\varphi))^{1-\alpha}, & \theta_0 \le \theta \le 1, \end{cases}$$

where $0 < \alpha < 1/2$, $\varphi = 1-\theta = 8|x|/7t$, θ_0 to be determined, and

$$g(\varphi) = \begin{cases} \varphi, & 3\delta \le \varphi \le 1-\theta,\\[1ex] \varphi + \dfrac{(3\delta-\varphi)^3}{6\delta^2}, & 2\delta \le \varphi \le 3\delta,\\[2ex] 2\delta + \dfrac{(\varphi-\delta)^3}{6\delta^2}, & \delta \le \varphi \le 2\delta,\\[1ex] 2\delta, & 0 \le \varphi \le \delta, \end{cases}$$

$\delta = 1/N$. Obviously $f(\theta) \in C^2[0,1]$. Now we fix $\gamma(N) = 21/8N$.

In the interval $\theta_0 \le \theta \le 1 - 3/N$

$$\begin{aligned}
&ff'' + (f')^2\left(1 - \frac{3}{2}Nf\right) \\
&= \frac{4(1-\alpha)}{(1+\alpha)^2(2+\alpha)^2}\left(\frac{\theta_0}{\theta}\right)^{2\alpha}\left[1 - 2\alpha - \frac{3(1-\alpha)N\cdot 2\theta_0^{\alpha}\theta^{1-\alpha}}{2(1+\alpha)(2+\alpha)}\right] \\
&\ge \frac{4(1-\alpha)}{(1+\alpha)^2(2+\alpha)^2}\left(\frac{\theta_0}{\theta}\right)^{2\alpha}\left[1 - 2\alpha - \frac{3(1-\alpha)N\theta_0^{\alpha}}{(1+\alpha)(2+\alpha)}\right] \\
&= \frac{2(1-\alpha)(1-2\alpha)}{(1+\alpha)^2(2+\alpha)^2}\theta_0^{2\alpha} = \frac{(1-2\alpha)^3}{18(1-\alpha)N^2}.
\end{aligned}$$

If we choose

$$\theta_0 = \left[\frac{(1+\alpha)(2+\alpha)(1-2\alpha)}{6(1-\alpha)N}\right]^{1/2}, \quad \text{then} \quad Nf \le \frac{1-2\alpha}{3(1-\alpha)} \le \frac{2}{3},$$

for $0 \le \theta \le 1 - 3/N$.

For $0 \le \theta \le \theta_0$ we have

$$\begin{aligned}
ff'' + (f')^2 = 1 - \frac{12\alpha}{1+\alpha}\frac{\theta}{\theta_0} + \frac{4(5+16\alpha+9\alpha^2)}{3(2+\alpha)(1+\alpha)^2}\alpha^2\left(\frac{\theta}{\theta_0}\right)^2 \\
+ 15\alpha^2\left[-\frac{4}{3(1+\alpha)}\frac{\theta}{\theta_0} + \frac{1}{2+\alpha}\left(\frac{\theta}{\theta_0}\right)^2\right]^2.
\end{aligned}$$

After restricting α to $0 < \alpha \le 1/11$ we have

$$ff'' + (f')^2 \ge c_{20} \qquad (0 \le \theta \le \theta_0),$$

$$\left[ff'' + (f')^2\left(1 - \frac{3}{2}Nf\right)\right]_{0\le\theta\le\theta_0} \ge c_{20} + O(N^{1-\alpha}).$$

Therefore, for $0 \le \theta \le 1 - 3/N$,

$$w_{|x|\,|x|} \ge \frac{192N^2}{49}e^{-2/3}(1 - e^{-2/3})\left[ff'' + (f')^2\left(1 - \frac{3}{2}Nf\right)\right] \ge c_{21}.$$

Now we choose $\beta(N) = c_{21}$. As

$$w_{|x|} = -\frac{16}{7}Nf'(\theta)e^{-Ntf(\theta)}(1 - e^{-Ntf(\theta)}),$$

and

$$f'(\theta) = \begin{cases} 1 - \dfrac{4\alpha}{1+\alpha}\dfrac{\theta}{\theta_0} + \dfrac{3\alpha}{2+\alpha}\dfrac{\theta^2}{\theta_0^2}, & 0 \le \theta \le \theta_0, \\[2ex] \dfrac{2(1-\alpha)}{(1+\alpha)(2+\alpha)}\theta_0^{\alpha}(1 - g(\varphi))^{-\alpha}g'(\varphi), & \theta_0 \le \theta \le 1, \end{cases}$$

$f'(\theta) = O(1)$, $w'_{|x|} = O(N)$, and

$$w_t = -w_{|x|} - 2Nf(\theta)e^{-Ntf(\theta)}(1 - e^{-Ntf(\theta)}) = O(N).$$

In the set $Q_3 \cap \{|p| \le \alpha(N)\}$

$$|w_{|x|}| = |u_{|x|} - v_{|x|}| \le \alpha(N) + 8 \le 9,$$

$$Nf(\theta) \le \begin{cases} c_{22}N\theta \le c_{23}N\theta_0 = O(N^{1-1/\alpha}), & 0 \le \theta \le \theta_0, \\ c_{24}N\theta_0^\alpha \le c_{25}, & \theta_0 \le \theta \le 1. \end{cases}$$

As a result,

$$w_t \le 9 + c_{26}Nf(\theta) \le c_{27},$$

$$w_{ij} = w_{|x|\,|x|}\frac{x_i x_j}{|x|^2} + \frac{w_{|x|}}{|x|}\frac{\delta_{ij}|x|^2 - x_i x_j}{|x|^2}, \qquad 1 \le i, j \le n,$$

where

$$\begin{aligned} w_{|x|\,|x|} &= \frac{128}{49}N^2\left[ff''e^{-Ntf}\frac{1-e^{-Ntf}}{Ntf} + (f')^2(2e^{-2Ntf} - e^{-Ntf})\right] \\ &= O(N^2). \end{aligned}$$

In view of

$$ff'' = \begin{cases} \left(1 - \dfrac{2\alpha}{1+\alpha}\dfrac{\theta}{\theta_0} + \dfrac{\alpha}{2+\alpha}\dfrac{\theta^2}{\theta_0^2}\right)\left(-\dfrac{4\alpha}{1+\alpha}\dfrac{\theta}{\theta_0} + \dfrac{6\alpha}{2+\alpha}\dfrac{\theta^2}{\theta_0^2}\right) = O(1), \\ \hfill 0 \le \theta \le \theta_0, \\ \dfrac{4(1-\alpha)}{(1+\alpha)^2(2+\alpha)^2}\theta_0^{2\alpha}[(1-g(\varphi))^{1-2\alpha}g''(\varphi) - \alpha(1-g(\varphi))^{-2\alpha}g'(\varphi)^2] \\ \hfill = O(1), \quad \theta_0 \le \theta \le 1. \end{cases}$$

as well as

$$\begin{aligned} \frac{w_{|x|}}{|x|} &= -2N^2\frac{ff'}{\varphi}e^{-Ntf}\frac{1-e^{-Ntf}}{Ntf} \\ &= \begin{cases} O(N^2\theta) = O(N^{2-1/\alpha}), & 0 \le \theta \le \theta_0, \\ O\left[N^2\theta_0^{2\alpha}(1-g(\varphi))^{1-2\alpha}g'(\varphi)\dfrac{1}{\varphi}\right] = O(N), & \dfrac{1}{N} \le \varphi \le 1 - \theta_0, \\ 0, & 0 \le \varphi \le \dfrac{1}{N}, \end{cases} \end{aligned}$$

we conclude

$$w_{ij} = O(N^2), \qquad 1 \le i, j \le n.$$

It follows from (23) that

$$\begin{aligned} -\inf_{B_1} v(x, 1) &\le c_{28}[N^{3n+2}(1-\tau) + N^2(1-\tau)^{(3-\sigma)/(\sigma-1)} + N^{1-n}]^{1/(n+1)} \\ &\le c_{29}[N^{3n+2}(1-\tau)^{(3-\sigma)/(\sigma-1)} + N^{1-n}]^{1/(n+1)} \\ &= c_{30}(1-\tau)^{(n-1)(3-\sigma)/(4n+1)(n+1)(\sigma-1)} \end{aligned}$$

provided we choose $N = (1-\tau)^{(3-\sigma)/(4n+1)(\sigma-1)}$.

If we fix $\alpha = \frac{1}{11}$, we obtain

$$\begin{aligned}\inf_{|x|\le 3/4} u(x, 1) &\ge \inf_{|x|\le 3/4} w(x, 1) + \inf_{|x|\le 3/4} v(x, 1)\\ &\ge (1 - e^{-(3/10)7^{-10/11}})^2 - c_{30}(1-\tau)^{(n-1)(3-\sigma)/(4n+1)(n+1)(\sigma-1)}\\ &\ge 2c - c = c.\end{aligned}$$

Here we have set $c = \frac{1}{2}(1 - e^{-(3/10)7^{-10/11}})^2$ and have chosen $\tau \ge (c/c_0)^{(4n+1)(n+1)(\sigma-1)/(n-1)(3-\sigma)}$. Now the lemma follows after returning to the initial notation. □

LEMMA 4. *Given $0 < R < 1$ and $\mu_1 > 0$. Let $\mu_2 = 2^{[\sigma]+2}$ where $[\sigma]$ is the integral part of σ and $2 < \sigma < 3$. Suppose that $u \in C^{2,1}$ satisfies $u \ge 0$, and $\mathscr{L}u \le -\mu_1 u_0 |p|^\sigma$ $(p = u_x)$ in the cylindrical domain $B_{4n^{1/2}R} \times \{0 \le t \le \mu_2 \eta R^\sigma\}$, then there exist constants τ $(0 < \tau < 1)$, C_9, C_{10}, and C_{11} such that*

$$u|_{k_{3R}\times[\eta R^\sigma, \mu_2\eta R^\sigma]} \ge C_{11} u_0$$

if $C_6 \le \eta u_0^{\sigma-2} \le 2C_9$, $R \le C_{10}u_0$, and $|K_{R,\eta} \cap \{u \ge u_0\}| \ge \tau |K_{R,\eta}|$.

REMARK. In case the length of the cylinder in t-direction is less than $\mu_2 \eta R^\sigma$, the conclusion holds on this shorter cylinder.

PROOF. We choose τ according to Lemma 3. Then

$$u|_{B_{3R/4}\times\{t=\eta R^\sigma/2\}} \ge c u_0 .$$

Letting $t_1 = \eta R^\sigma/2$, $X_1 = 3R/4$, $\eta R^\sigma \le t_2 \le \mu_2 \eta R^\sigma$, $X_2 = 3\eta^{1/2}R$, $\Lambda = 2\mu_2 - 1$, and replacing u_0 by cu_0 in Lemma 2, Lemma 4 holds for $C_9 = 2c^{2-\sigma}(3n^{1/2})^\sigma (4n^{1/2})^{C_6(\sigma-2)} C_5$, $C_{10} = (2\mu_2 - 1)^{-1/(\sigma-2)}(4n^{1/2})^{-C_6} C_7$, and $C_{11} = (2\mu_2 - 1)^{-1/(\sigma-2)}(4n^{1/2})^{-C_6} c C_8$. □

LEMMA 5. *Let $\Gamma \subseteq K_{R_0,\eta^0}$ be a measurable set with positive measure and $\eta^1 \le \eta^0$. Suppose that the measurable set $\tilde{\Gamma}$ satisfies*

$$\tilde{\Gamma} \supseteq \Gamma \bigcup_{\substack{K_{R,\eta}(x^0,t^0)\subseteq K_{R_0,\eta^0}\\ \eta^1\le\eta\le 2\eta^1}} \Big\{k_{3R}(x^0) \times [t^0 + \eta R^\sigma, t^0 + 2^{[\sigma]+2}\eta R^\sigma] \cap K_{R_0,\eta^0}$$
$$: |\Gamma \cap K_{R,\eta}(x^0, t^0)| \ge \tau |K_{R,\eta}(x^0, t^0)|\Big\} .$$

Then there exist C_{12}, C_{13} depending only on n, η^0, η^1, and σ so that at least one in the following three cases is valid:

(i) $|\tilde{\Gamma}| \ge [1 + C_{12}(1-\tau)]|\Gamma|$ *or* $|\tilde{\Gamma}| = |K_{R_0,\eta^0}|$;

(ii) *there is* $K_{R,\eta}(x^0, t^0) \subseteq K_{R_0,\eta^0}$, *whose top lies on* $t = \eta^0 R_0^\sigma$, *satisfying*

$$|\Gamma \cap K_{R,\eta}(x^0, t^0)| \geq \tau |K_{R,\eta}(x^0, t^0)|,$$

$$R \geq C_{13} \left(\frac{|\Gamma|}{|K_{R_0,\eta^0}|} \right)^{1/\sigma} R_0;$$

(iii) *there is* $K_{R,\eta^0}(0, t^0) \subseteq K_{R_0,\eta^0}$ *satisfying*

$$|\Gamma \cap K_{R_0,\eta}(0, t^0)| \geq \tau |K_{R_0,\eta}(0, t^0)|, \qquad \eta^1 \leq \eta \leq \eta^2,$$

where $l^0 = \eta^0/\eta$ *is an integer and* $t^0 = l\eta R^\sigma$, $l \in \{0, 1, \ldots, l^0 - 1\}$.

PROOF. Without loss of generality we take $R_0 = 1$. Let $\log_2(\eta^0/\eta^1) = \lambda$. We divide K_{1,η^0} into $2^{[\lambda]}$ many identical small rectangles $K_{1,\eta_0}(0, t^j)$ ($[\lambda]$ is the integral part of λ) where $\eta_0 = \eta^0/2^{[\lambda]} = \eta^1 2^{\lambda-[\eta]}$, $t^j = j\eta_0$, and $0 \leq j < 2^{[\lambda]}$. For any $K_{1,\eta_0}(0, t^j)$ satisfying $|\Gamma \cap K_{1,\eta_0}(0, t^j)| < \tau |K_{1,\eta_0}(0, t^j)|$, we divide it into $2^{n+[\lambda+\sigma]-[\lambda]}$ many small identical rectangles $K_{1/2,\eta_1}$, $\eta_1 = \eta_0 2^\sigma / 2^{[\lambda+\sigma]} = \eta^1 2^{\lambda+\sigma-[\lambda+\sigma]}$. And for $K_{1/2,\eta_1}$ satisfying $|\Gamma \cap K_{1/2,\eta_1}| < \tau |K_{1/2,\eta_1}|$, again we divide it into $2^{n+[\lambda+2\sigma]-[\lambda+\sigma]}$ many identical rectangles ....

In general, the subrectangles K_{2^{-k},η_k} satisfy

$$\eta_k = \frac{\eta^0}{2^{[\lambda+k\sigma]}} (2^k)^\sigma = \eta^1 2^{\lambda+k\sigma-[\lambda+k\sigma]}.$$

Therefore $\eta^1 \leq \eta_k < 2\eta^1$, $\sigma - 1 < [\lambda + k\sigma] - [\lambda + (k-1)\sigma] < \sigma + 1$, and so

$$[\sigma] \leq [\lambda + k\sigma] - [\lambda + (k-1)\sigma] \leq [\sigma] + 1.$$

In other words, divisions effect in t-direction $2^{[\sigma]+\delta}$ ($\delta = 0$ or 1) many times whenever we divide in the x_i $(1 \leq i \leq n)$-direction once.

First we consider the largest rectangle K_{2^{-k},η_k} $(k = 0, 1, \ldots)$ satisfying

$$|\Gamma \cap K_{2^{-k},\eta_k}| \geq \tau |K_{2^{-k},\eta_k}|.$$

When $k = 0$ Case (iii) holds. Otherwise $k \geq 1$. Suppose that K_{2^{-k},η_k} is divided from $\widetilde{K}^0_{2^{-k+1},\eta_{k-1}}$. Let $\widetilde{K}^l_{2^{-k+1},\eta_{k-1}}$, $l = 0, 1, \ldots, L$, be rectangles in the t-direction with the same property, more precisely,

$$|\Gamma \cap K^l_{2^{-k+1},\eta_{k-1}}| < \tau |\widetilde{K}^l_{2^{-k+1},\eta_{k-1}}|, \qquad 0 \leq l \leq L,$$

and there is at least one $K^{(l)}_{2^{-k},\eta_k} \subseteq \widetilde{K}^l_{2^{-k+1},\eta_{k-1}}$ such that

$$|\Gamma \cap K^{(l)}_{2^{-k},\eta_k}| \geq \tau |K^{(l)}_{2^{-k},\eta_k}|, \qquad 0 \leq l \leq L.$$

Moreover, any $K^j_{2^{-k},\eta_k} \subseteq \widetilde{K}^{-1}_{2^{-k+1},\eta_{k-1}}$ or $K^j_{2^{-k},\eta_k} \subseteq \widetilde{K}^{L+1}_{2^{-k+1},\eta_{k-1}}$, $j=1,2,\dots,$ $2^{n+[\lambda+k\sigma]-[\lambda+(k-1)\sigma]}$ satisfies

$$|\Gamma \cap K^j_{2^{-k},\eta_k}| < \tau |K^j_{2^{-k},\eta_k}|.$$

Denote the small rectangles $K^j_{2^{-k},\eta_k} \subseteq \widetilde{K}^{-1}_{2^{-k+1},\eta_k}$ whose tops touch $\widetilde{K}^0_{2^{-k+1},\eta_{k-1}}$ by $K^{-1,j}_{2^{-k},\eta_k}$ $(1 \le j \le 2^n)$. Similarly $K^{L+1,j}_{2^{-k},\eta_k}$ $(1 \le j \le 2^n)$ are those subrectangles of $\widetilde{K}^{L+1}_{2^{-k+1},\eta_{k-1}}$ whose bottoms touch $\widetilde{K}^L_{2^{-k+1},\eta_{k-1}}$.

Since by definition $|\Gamma \cap K^{(l)}_{2^{-k},\eta_k}| \ge \tau |K^{(l)}_{2^{-k},\eta_k}|$ $(0 \le l \le L)$, $\widetilde{K}^l_{2^{-k+1},\eta_{k+1}} \subseteq \widetilde{\Gamma}$ $(1 \le l \le L)$ and $K^{L+1,j}_{2^{-k},\eta_k} \subseteq \widetilde{\Gamma}$ $(1 \le j \le 2^n)$. Using the following inequalities

$$|\widetilde{\Gamma} \cap \widetilde{K}^0_{2^{-k+1},\eta_{k-1}}| \ge |\Gamma \cap \widetilde{K}^0_{2^{-k+1},\eta_{k-1}}|,$$
$$|\widetilde{\Gamma} \cap \widetilde{K}^l_{2^{-k+1},\eta_{k-1}}| \ge |\Gamma \cap \widetilde{K}^l_{2^{-k+1},\eta_{k-1}}| + (1-\tau)|\widetilde{K}^l_{2^{-k+1},\eta_{k-1}}|, \qquad 1 \le l \le L,$$
$$|\widetilde{\Gamma} \cap K^{l+1,j}_{2^{-k},\eta_k}| \ge |\Gamma \cap K^{l+1,j}_{2^{-k},\eta_k}| + (1-\tau)|K^{l+1,j}_{2^{-k},\eta_k}|, \qquad 1 \le j \le 2^n,$$

and

$$\begin{aligned} &\left|\bigcup_{l=1}^{L} \widetilde{K}^l_{2^{-k+1},\eta_{k-1}} \bigcup_{j=1}^{2^n} K^{L+1,j}_{2^{-k},\eta_k}\right| - \frac{1}{2^{[\sigma]}}\left|\bigcup_{j=1}^{2^n} K^{-1,j}_{2^{-k},\eta_k}\right| \\ &\quad = \left[L + 2^{-[\sigma]-\delta}\left(1 - \frac{1}{2^{[\sigma]}}\right)\right] |\widetilde{K}^0_{2^{-k+1},\eta_{k-1}}| \\ &\quad \ge \frac{L + 2^{-[\sigma]-\delta}(1-2^{-[\sigma]})}{L+1+2^{-[\sigma]-\delta}} \left|\bigcup_{l=0}^{L} \widetilde{K}^l_{2^{-k+1},\eta_{k-1}} \bigcup_{j=1}^{2^n} K^{L+1,j}_{2^{-k},\eta_k}\right| \end{aligned}$$

we obtain

$$\begin{aligned} &\left|\widetilde{\Gamma} \cap \left(\bigcup_{l=0}^{L} \widetilde{K}^l_{2^{-k+1},\eta_{k-1}} \bigcup_{j=1}^{2^n} K^{L+1,j}_{2^{-k},\eta_k}\right)\right| - \frac{1-\tau}{2^{[\sigma]}}\left|\bigcup_{j=1}^{2^n} K^{-1,j}_{2^{-k},\eta_k}\right| \\ &\quad \ge \left|\Gamma \cap \left(\bigcup_{l=0}^{L} \widetilde{K}^l_{2^{-k+1},\eta_{k-1}} \bigcup_{j=1}^{2^n} K^{L+1,j}_{2^{-k},\eta_k}\right)\right| \\ &\qquad + (1-\tau)\left|\bigcup_{l=1}^{L} \widetilde{K}^l_{2^{-k+1},\eta_{k-1}} \bigcup_{j=1}^{2^n} K^{L+1,j}_{2^{-k},\eta_k}\right| - \frac{1-\tau}{2^{[\sigma]}}\left|\bigcup_{j=1}^{2^n} K^{-1,j}_{2^{-k},\eta_k}\right| \\ &\quad \ge \left[1 + \frac{L+2^{-[\sigma]-\delta}(1-2^{-[\sigma]})}{L+1+2^{-[\sigma]-\delta}}(1-\tau)\right] \left|\Gamma \cap \left(\bigcup_{l=0}^{L} \widetilde{K}^l_{2^{-k+1},\eta_{k-1}} \bigcup_{j=1}^{2^n} K^{L+1,j}_{2^{-k},\eta_k}\right)\right|. \end{aligned} \tag{24}$$

Notice that when $\widetilde{K}^0_{2^{-k+1},\eta_{k-1}}$ has its bottom lying on $t=0$, the term involving $K^{-1,j}_{2^{-k},\eta_k}$ $(1 \le j \le 2^n)$ is not needed. Therefore the term $\bigcup_{j=1}^{2^n} K^{-1,j}_{2^{-k},\eta_k}$

should be deleted from (24). On the other hand, when the top of $\widetilde{K}^{L}_{2^{-k+1},\eta_{k-1}}$ belongs to $t=\eta^0$, $K^{L+1,j}_{2^{-k},\eta_k}$ lies outside K_{1,η_0} for $j=1,\dots,2^n$. In this case we replace (24) by

$$\left|\widetilde{\Gamma}\cap\left(\bigcup_{l=0}^{L}\widetilde{K}^{l}_{2^{-k+1},\eta_{k-1}}\right)\right|+(1-\tau)\left|\bigcup_{j=1}^{2^n}K^{L+1,j}_{2^{-k},\eta_k}\right|-\frac{1-\tau}{2^{[\sigma]}}\left|\bigcup_{j=1}^{2^n}K^{-1,j}_{2^{-k},\eta_k}\right|$$
$$\geq\left[1+\frac{L+2^{-[\sigma]-\delta}(1-2^{-[\sigma]})}{L+1}(1-\tau)\right]\left|\Gamma\cap\left(\bigcup_{l=0}^{L}\widetilde{K}^{L}_{2^{-k+1},\eta_{k-1}}\right)\right|. \tag{25}$$

We consider those $K_{2^{-k},\eta_k}\subseteq K_{1,\eta^0}$ satisfying $|\Gamma\cap K_{2^{-k},\eta_k}|\geq\tau|K_{2^{-k},\eta_k}|$ following the increasing order in k. In general we obtain (24). However, whenever $\bigcup_{j=1}^{2^n}K^{-1,j}_{2^{-k},\eta_k}$ has been handled before or the bottom of $\widetilde{K}^{0}_{2^{-k+1},\eta_{k-1}}$ lies on $t=0$, the corresponding term $\bigcup_{j=1}^{2^n}K^{-1,j}_{2^{-k},\eta_k}$ should be deleted from (24). Similarly, whenever $\bigcup_{j=1}^{2^n}K^{L+1,j}_{2^{-k},\eta_k}$ has been handled or the top of $\widetilde{K}^{L}_{2^{-k+1},\eta_{k-1}}$ belongs to $t=\eta^0$ we replace (24) by (25).

Summing up (24) and (25) with respect to all k, by noticing that when the bottoms of $K^{L+1,j}_{2^{-k_1},\eta_{k_1}}$ lie above the tops of $\bigcup_{j=1}^{2^n}K^{-1,j}_{2^{-k},\eta_k}$ $(k_1<k)$,

$$2^{-[\sigma]}\left|\bigcup_{j=1}^{2^n}K^{-1,j}_{2^{-k},\eta_k}\right|\geq\sum_{k_1<k}\left|\bigcup_{j=1}^{2^n}K^{L+1,j}_{2^{-k_1},\eta_{k_1}}\right|$$

holds and we have

$$(1-\tau)|\overline{\Gamma}|+|\widetilde{\Gamma}|\geq\left[1+\frac{1-2^{-[\sigma]}}{2^{[\sigma]+1}+1}(1-\tau)\right]\left|\Gamma\cap\left(\bigcup_{k=1}^{\infty}\widetilde{K}_{2^{-k+1},\eta_{k-1}}\right)\right|.$$

Here $\overline{\Gamma}$ is the union of all rectangles (lying outside K_{1,η^0}) sitting on $t=\eta^0$.

By the measurability of Γ

$$\frac{|\Gamma\cap K_{R,\eta}(x^0,t^0)|}{|K_{R,\eta}(x^0,t^0)|}\to 1\geq\tau,$$

as $R\to 0$ for $(x^0,t^0)\in\Gamma$ a.e.. Consequently

$$\left|\Gamma\setminus\left(\bigcup_{k=1}^{\infty}\widetilde{K}_{2^{-k+1},\eta_{k-1}}\right)\right|=0\qquad\text{and}\qquad\left|\Gamma\cap\left(\bigcup_{k=1}^{\infty}\widetilde{K}_{2^{-k+1},\eta_{k+1}}\right)\right|=|\Gamma|.$$

When the maximal height of $\overline{\Gamma}$ does not exceed $(1-2^{-[\sigma]})|\Gamma|/2^{n+1}(2^{[\sigma]+1}+1)$, Case (i) holds for $C_{12}=(1-2^{-[\sigma]})/2(2^{[\sigma]+1}+1)$. Otherwise, Case (ii) holds for $C_{13}=[(1-2^{-[\sigma]})/2^{n+2}(2^{[\sigma]+1}+1)]^{1/\sigma}$. The proof of the lemma is complete. □

Now we are ready for the proof of the following density theorem.

THEOREM 6. *Let* $u_0 > 0$, $0 < \tau_1 < 1$, $\mu_1 > 0$, *and* $C = C(n, \mu, \nu, \sigma) > 1$ *be given where* $2 < \sigma < 3$. *There exist positive constants* $\tilde{\xi}$, $\tilde{\mu}$, $\tilde{\zeta}$ *which depend only on* n, μ, ν, σ, μ_1, τ_1, C *such that*

$$u|_{B_{R/2}\times\{t=\eta R^\sigma\}} \geq \tilde{\zeta} u_0,$$

whenever $u \in C^{2,1}$ *satisfies* $u \geq 0$, $\mathscr{L}u \leq -\mu_1 u_0 |p|^\sigma$ *in* $B_{4n^{1/2}R} \times [0, \eta R^\sigma]$ *for* $R \leq \tilde{\xi} u_0$, $\tilde{\eta} \leq \eta u_0^{\sigma-2} \leq C\tilde{\eta}$, *and* $|K_{R,\eta} \cap \{u \geq u_0\}| \geq \tau_1 |K_{R,\eta}|$.

PROOF. We have

$$|K_{R,(1-\tau_1/2)\eta} \cap \{u \geq u_0\}| \geq \frac{\tau_1}{2} |K_{R,(1-\tau_1/2)\eta}|.$$

Choose $\Gamma = K_{R,(1-\tau_1/2)\eta} \cap \{u \geq u_0\}$ and $\widetilde{\Gamma} = K_{R,(1-\tau_1/2)\eta} \cap \{u \geq C_{11} u_0\}$. When $(1 - \tau_1/2)\eta \geq C_9 u_0^{2-\sigma}$, by Lemma 4 ($\eta$ and τ therein should be replaced by $\eta^1 = C_9 u_0^{2\sigma}$ and $\tau_1/2$ respectively) the hypotheses of Lemma 5 hold for R_0 and η^0 being R and $(1 - \tau_1/2)\eta$ respectively. It follows from Lemma 5 that at least one case in Cases (i)–(iii) must hold.

When Case (i) holds,

$$|K_{R,(1-\tau_1/2)\eta} \cap \{u \geq C_{11} u_0\}| \geq [1 + C_{12}(1-\tau)] \frac{\tau_1}{2} |K_{R,(1-\tau_1/2)\eta}|.$$

We can repeat this step, that is, taking

$$\begin{aligned}
\Gamma &= K_{R,(1-\tau_1/2)\eta} \cap \{u \geq C_{11}^{k-1} u_0\}, \\
\widetilde{\Gamma} &= K_{R,(1-\tau_1/2)\eta} \cap \{u \geq C_{11}^{k} u_0\}, \\
\mu_2 &= 2^{[\sigma]+2}, \qquad \eta^1 = C_9 (C_{11}^{k-1} u_0)^{2-\sigma}, \\
\tau &= [1 + C_{12}(1-\tau)]^{k-1} \frac{\tau_1}{2}.
\end{aligned}$$

If Case (i) continues to hold for $R \leq C_{10} C_{11}^{k-1} u_0$ and $(1 - \tau_1/2)\eta \geq C_9 (C_{11}^{k-1} u_0)^{2-\sigma}$, we have

$$|K_{R,(1-\tau_1/2)\eta} \cap \{u \geq C_{11}^{k} u_0\}| \geq [1 + C_{12}(1-\tau)]^{k} \frac{\tau_1}{2} |K_{R,(1-\tau_1/2)\eta}|.$$

In case Case (i) holds for $k = 1, 2, \ldots, k_0$ where

$$k_0 = \left[\frac{\log(2/\tau_1)}{\log[1 + C_{12}(1-\tau)]}\right] + 1,$$

we can apply Lemma 3 to get

$$u|_{B_{3R/4}\times\{t=(1-\tau_1/2)\eta R^\sigma\}} \geq c C_{11}^{k_0} u_0$$

for $R \leq C_{10} C_{11}^{k_0-1} u_0$ and $(1 - \tau_1/2)\eta \geq C_9 (C_{11}^{k_0-1} u_0)^{2-\sigma}$. From Lemma 1 we further infer

$$u|_{B_{R/2}\times\{t=\eta R^\sigma\}} \geq c C_3 \Lambda^{-1/(\sigma-2)} C_{11}^{k_0} u_0$$

for

$$\frac{3}{4}R \le C_2\Lambda^{-1/(\sigma-2)} cC_{11}^{k_0}u_0 \quad \text{and} \quad \eta \le \frac{2}{\tau_1}C_1\left(\frac{3}{4}\right)^\sigma (cC_{11}^{k_0}u_0)^{2-\sigma}\Lambda.$$

If Case (iii) in Lemma 5 holds once, i.e.,

$$R \le C_{10}C_{11}^{k_1-1}u_0 \quad \text{and} \quad \left(1-\frac{\tau_1}{2}\right)\eta \ge C_9(C_{11}^{k_1-1}u_0)^{2-\sigma} \equiv \eta_1$$

for some k, $1 \le k_1 \le k_0$, we have

$$|K_{R,\eta_2}(0, t_0) \cap \{u \ge C_{11}^{k_1}u_0\}| \ge \tau|K_{R,\eta_2}(0, t_0)|$$

for $\eta_1 \le \eta_2 < 2\eta_1$ and $l = t_0/\eta_2R^\sigma$ an integer satisfying $0 \le l \le (1-\tau_1/2)\eta/\eta_2 - 1$. By Lemma 3

$$u|_{B_{3R/4}\times\{t=t_0+\eta_2R^\sigma\}} \ge cC_{11}^{k_1}u_0.$$

Using Lemma 1 again, we conclude

$$u|_{B_{R/2}\times\{t=\eta R^\sigma\}} \ge C_3\Lambda^{-1/(\sigma-2)}C_{11}^{k_1}u_0$$

for

$$R \le C_2\Lambda^{-1/(\sigma-2)}cC_{11}^{k_1}u_0 \quad \text{and} \quad \eta \le \frac{2}{\tau_1}C_1\left(\frac{3}{4}\right)^\sigma (cC_{11}^{k_1}u_0)^{2-\sigma}\Lambda.$$

If Case (ii) in Lemma 5 holds for some k_2, $1 \le k_2 < k_0$, i.e.,

$$|K_{R_1,\eta_2}(x^0, t^0) \cap \{u \ge C_{11}^{k_2}u_0\}| \ge \tau|K_{R_1,\eta_2}(x^0, t^0)|,$$

where

$$\begin{gathered} |x^0| \le n^{1/2}(R-R_1), \qquad t^0+\eta_2R_1^\sigma = \left(1-\frac{\tau_1}{2}\right)\eta R^\sigma, \\ \eta_1 \le \eta_2 < 2\eta_1, \qquad C_{13}\left\{[1+C_{12}(1-\tau)]^{k_2-1}\frac{\tau_1}{2}\right\}^{1/\sigma}R \le R_1 \le R, \end{gathered} \tag{26}$$

for

$$R \le C_{10}C_{11}^{k_2-1}u_0, \qquad \left(1-\frac{\tau_1}{2}\right)\eta \ge C_9(C_{11}^{k_2-1}u_0)^{2-\sigma} \equiv \eta_1,$$

by Lemma 3

$$u|_{B_{3R_1/4}(x^0)\times\{t=(1-\tau_1/2)\eta R^\sigma\}} \ge cC_{11}^{k_2}u_0.$$

Letting

$$X_1 = \frac{3}{4}R_1, \qquad t_1 = \left(1-\frac{\tau_1}{2}\right)\eta R^\sigma, \qquad X_2 = \frac{3}{2}\eta^{1/2}R, \qquad t_2 = \eta R^\sigma,$$

we can use Lemma 2 to conclude that when

$$\begin{aligned} \frac{2}{\tau_1}C_5\left(\frac{3}{2}n^{1/2}\right)^\sigma\left(2n^{1/2}\frac{R}{R_1}\right)^{C_6(\sigma-2)}(cC_{11}^{k_2}u_0)^{2-\sigma} &\le \eta \\ \le \frac{2}{\tau_1}C_5\left(\frac{3}{2}n^{1/2}\right)^\sigma\left(2n^{1/2}\frac{R}{R_1}\right)^{C_6(\sigma-2)}&(cC_{11}^{k_2}u_0)^{2-\sigma}\Lambda, \end{aligned} \tag{27}$$

and

$$R \le \frac{2C_7}{3n^{1/2}} \left(\frac{R_1}{2n^{1/2}R}\right)^{C_6} \Lambda^{-1/(\sigma-2)} c C_{11}^{k_2} u_0,$$

then

$$\inf_{B_{R/2}\times\{t=\eta R^\sigma\}} \ge C_8 \Lambda^{-1/(\sigma-2)} \left(\frac{R_1}{2n^{1/2}R}\right)^{C_6} c C_{11}^{k_2} u_0$$

holds. Notice that we can substitute (26) into R/R_1 and R_1/R in (27).

All these inequalities for η are valid when $\tilde\eta \le \eta u_0^{\sigma-2} \le C\tilde\eta$ provided we choose

$$\tilde\eta = \max\left\{\frac{\tau}{2-\tau_1} C_9, \frac{2}{\tau_1} c^{2-\sigma} C_5 \left(\frac{3}{2} n^{1/2}\right)^\sigma \left(\frac{2^{1+1/\sigma} n^{1/2}}{C_{13}\tau_1^{1/\sigma}}\right)^{C_6(\sigma-2)}\right\} C_{11}^{k_0(2-\sigma)},$$

$$C\tilde\eta = \min\left\{\frac{2}{\tau_1} C_1 \left(\frac{3}{4}\right)^\sigma c^{2-\sigma}, \frac{2}{\tau_1} c^{2-\sigma} C_5 \left(\frac{3}{2} n^{1/2}\right)^\sigma (2n^{1/2})^{C_6(\sigma-2)}\right\} \Lambda.$$

However, it is always possible to find a parameter Λ (≥ 1) such that this can be done as long as $C \ge 1$. Hence the theorem holds. □

§2. A Hölder estimate for the solution and unique solvability for the first boundary value problem

Since there is no Harnack inequality for the solutions of quasilinear parabolic equations under the second natural structure conditions (1) and (2), we shall derive a Hölder estimate for these solutions based directly on the density theorem.

Denote $d(P_1, P_2) = |x^2 - x^1| + |t^2 - t^1|^{1/\sigma}$ $(P_1 = (x^1, t^1), P_2 = (x^2, t^1))$, $\overline{d}(P_1) = \inf_{P_2\in\partial^* Q} d(P_1, P_2)$, and $\overline{d}_{P_1P_2} = \min\{\overline{d}(P_1), \overline{d}(P_2)\}$.

THEOREM 7. *Let $u \in C^{2,1}(Q)$ satisfy $\max_{\overline{Q}}|u| \le M$ and $\mathscr{L}u = B(x, t, u, u_x)$ in Q. Suppose that* (1) *and* (2) *holds for some σ, $2 < \sigma < 3$. There exist γ, C_{14}, C_{15} which depend only on n, μ, ν, σ, M and* diam Ω *such that*

$$|u(P_2) - u(P_1)| \le C_{15}\left(\frac{d(P_1, P_2)}{\overline{d}_{P_1P_2}^{1+\gamma}}\right)^\gamma,$$

when $d(P_1, P_2) \le C_{14}\overline{d}_{P_1P_2}$.

PROOF. After a translation of axes we may momentarily take $P_2 = (0, \dots, 0)$. Consider the standard rectangles

$$K_j = \{(x, t) : |x_i| < R_j, \ (1 \le i \le n), \ -S_j < t < 0\}, \qquad j = 1, 2, \dots,$$

where $R_j = \xi^{-j}$, $\xi = 8n^{3/2}$, $S_j = \tilde\eta(\frac{\beta}{2}\eta^j)^{2-\sigma}(2nR_j)^\sigma$, and $\beta = M\eta^{-j_0}$. The constants $\tilde\eta$, η $(\frac{3}{4} < \eta < 1)$ and j_0 will be determined as we proceed.

We would like to use induction to establish

$$\operatorname*{osc}_{K_j} u \le \beta\eta^j, \qquad (j \ge j_0). \tag{28}$$

Suppose on the contrary that

$$\operatorname*{osc}_{K_{j-1}} u \le \beta\eta^{j-1}, \qquad \operatorname*{osc}_{K_j} u > \beta\eta^j. \tag{29}$$

We are going to show that this leads to a contradiction for appropriate choices of $\tilde{\eta}$, η, and j_0. Let

$$\max_{\overline{K}_j} u = u(x^1, t^1), \qquad \min_{\overline{K}_j} u = u(x^2, t^2),$$

where $(x^1, t^1), (x^2, t^2) \in \overline{K}_j$ and $u_0 = \frac{1}{2}(\max_{\overline{K}_j} u + \min_{\overline{K}_j} u)$. Consider

$$K_j^1 = \{(x, t) : |x_i| < 2nR_j,\ 1 \le i \le n,\ -S_j < t < 0\},$$

which satisfies

$$|K_j^1 \cap \{u - u_0 \le 0\}| \ge \frac{1}{2}|K_j^1|.$$

Notice we can assume this, otherwise we replace $u - u_0$ by $u_0 - u$ and write the equation $\mathscr{L}u = B$ as $\mathscr{L}(-u) = -B$. Let

$$K_j^2 = \{(x, t) : |x_i| < 2nR_j,\ 1 \le i \le n,\ -S_j < t < t^1\}.$$

Then

$$\begin{aligned} |K_j^2 \cap \{u - u_0 \le 0\}| &\ge |K_j^1 \cap \{u - u_0 \le 0\}| - \frac{S_j}{S_{j-1}}|K_j^1| \\ &\ge \frac{1}{2}|K_j^1| - \frac{\eta^2}{(8n^{3/2}\eta)^\sigma}|K_j^1| \ge \frac{1}{4}|K_j^1|. \end{aligned} \tag{30}$$

The function

$$U = \frac{1}{h}[e^{h(u-u_0)} - 1], \qquad h = \frac{2\mu}{\nu},$$

satisfies

$$\begin{aligned} \widetilde{\mathscr{L}}U &\equiv A_{ij}(x, t, u, e^{-h(u-u_0)}P)U_{ij} - U_t \\ &= e^{h(u-u_0)}B(x, t, u, e^{-h(u-u_0)}P) + he^{-h(u-u_0)}A_{ij}(x, t, u, e^{-h(u-u_0)}P)U_iU_j \\ &\ge e^{-h(u-u_0)}\mu[1 + e^{-h\sigma(u-u_0)}|P|^\sigma] + he^{-h(u-u_0)}\nu e^{-h(\sigma-2)(u-u_0)}|P|^\sigma \\ &= \mu e^{h(\sigma-1)(u-u_0)}|P|^\sigma - \mu e^{h(u-u_0)} \\ &\ge \mu e^{-2h(\sigma-1)M}|P|^\sigma - \mu e^{2hM}, \qquad (P = U_x) \end{aligned}$$

as well as

$$\begin{aligned} \nu e^{-2h(\sigma-2)M}(1 + |P|^{\sigma-2})|\xi|^2 &\le A_{ij}(x, t, u, e^{-h(u-u_0)}P)\xi_i\xi_j \\ &\le \mu e^{2h(\sigma-2)M}(1 + |P|^{\sigma-2})|\xi|^2, \qquad \xi \in \mathbb{R}^n. \end{aligned}$$

Letting $\widetilde{U} = U - \mu e^{2hM}(S_{j-1} + t)$, we have

$$\widetilde{\mathscr{L}}\widetilde{U} \geq \mu e^{-2h(\sigma-1)M}|\widetilde{P}|^{\sigma}, \qquad (\widetilde{P} = \widetilde{U}_x).$$

On the other hand, it follows from (30) that $|K_j^2 \cap \{\widetilde{U} \leq 0\}| \geq \frac{1}{4}|K_j^2|$.

Now, we are going to apply the density theorem (Theorem 6). We shall take the parameters in $\tilde{\xi}$, $\tilde{\eta}$, $\tilde{\zeta}$—n, μ, ν, σ, μ_1, τ_1 and C—to be n, $\mu e^{2h(\sigma-2)M}$, $\nu e^{-(\sigma-2)M}$, σ, $\mu h e^{-2h(\sigma-1)M}/2(e^{hM} - 1)$, $\frac{1}{4}$, and $(\frac{8}{3}e^{Mh})^{\sigma-2}$. We claim

$$\sup_{K_j^3} U \geq \frac{1+\tilde{\zeta}}{h}\left[\exp\left(\frac{h}{2}\operatorname{osc}_{K_j} u\right) - 1\right] \equiv \widetilde{U}_0, \tag{31}$$

where

$$K_j^3 = \{(x, t) : |x_i| < R_{j-1}, \ 1 \leq i \leq n, \ -S_{j-1} < t < t^1\}.$$

In case (31) is not true, we verify that the density theorem is applicable to the function $w = \widetilde{U}_0 - \widetilde{U}$ in K_j^2 where u_0, η, and R now are taken to be $\widetilde{U}_0$, $\tilde{\eta}(\frac{\beta}{2}\eta^j)^{2-\sigma}$, and $2nR_j$ respectively. As (31) does not hold, $w \geq 0$ and

$$\widetilde{\mathscr{L}}w \leq -\mu e^{-2h(\sigma-1)M}|w_x|^{\sigma} \leq -\frac{\mu}{2}\frac{he^{-2h(\sigma-1)M}}{(e^{hM}-1)}\widetilde{U}_0|w_x|^{\sigma}$$

in K_j^3. By virtue of

$$K_j^2 = K_{2nR_j,\tilde{\eta}}\left(\frac{\beta}{2}\eta^j\right)^{2-\sigma} \times (0, -S_{j-1}),$$

$$\tilde{\eta}\left(\frac{\beta}{2}\eta^j\right)^{2-\sigma}\widetilde{U}_0^{\sigma-2} = \tilde{\eta}\left\{\frac{2}{\beta\eta^j}\frac{1+\tilde{\zeta}}{h}\left[\exp\left(\frac{h}{2}\operatorname{osc}_{K_j} u - 1\right)\right]\right\}^{\sigma-2}$$

$$= \tilde{\eta}\left[\frac{1+\tilde{\zeta}}{\beta\eta^j}\operatorname{osc}_{K_j} u \exp\left(\frac{\theta h}{2}\operatorname{osc}_{K_j} u\right)\right]^{\sigma-2}, \qquad 0 < \theta < 1,$$

and (29) we have

$$\tilde{\eta} \leq \tilde{\eta}\left(\frac{\beta}{2}\eta^j\right)^{2-\sigma}\widetilde{U}_0^{\sigma-2} \leq \left(\frac{8}{3}e^{Mh}\right)^{\sigma-2}\tilde{\eta},$$

and

$$\frac{2nR_j}{\widetilde{U}_0} \leq \frac{4nR_j}{\operatorname{osc}_{K_j} u} \leq \frac{4n(8n^{3/2})^{-j}}{\beta\eta^j} \leq \left(\frac{4n}{\beta}\right)(8n^{3/2}\eta)^{-j_0} = \frac{2n}{M}(8n^{3/2})^{-j_0} \leq \tilde{\xi}$$

for all $j \geq j_0$ where j_0 is sufficiently large. Now all conditions in Theorem 6 are fulfilled, and we infer from it

$$w(x^1, t^1) \geq \frac{\tilde{\zeta}(1+\tilde{\zeta})}{h}\left[\exp\left(\frac{h}{2}\operatorname{osc}_{K_j} u\right) - 1\right],$$

or

$$\widetilde{U}(x^1, t^1) \le \frac{1-\tilde{\zeta}^2}{h}\left[\exp\left(\frac{h}{2}\operatorname*{osc}_{K_j} u\right) - 1\right]. \tag{32}$$

Since $u(x^1, t^1) - u_0 = \frac{1}{2}\operatorname{osc}_{K_j} u$,

$$\widetilde{U}(x^1, t^1) = \frac{1}{h}\left[\exp\left(\frac{h}{2}\operatorname*{osc}_{K_j} u\right) - 1\right] - \mu e^{2hM}(S_{j-1} + t^1).$$

Putting this into (32) we obtain

$$\frac{\tilde{\zeta}^2}{h}\left[\exp\left(\frac{h}{2}\operatorname*{osc}_{K_j} u\right) - 1\right] \le \mu e^{2hM} S_{j-1},$$

or

$$\begin{aligned}\frac{\tilde{\zeta}^2}{2\mu e^{2hM}\tilde{\eta}} &\le \frac{S_{j-1}}{\tilde{\eta}\operatorname{osc}_{K_j} u} \le \frac{1}{\beta\eta^j}\frac{(2nR_{j-1})^\sigma}{(\beta\eta^{j-1}/2)^{\sigma-2}} \\ &\le \frac{[16n^{5/2}(8n^{3/2})^{-j}]^\sigma}{2(\beta\eta^j/2)^{\sigma-1}} \le \frac{[16n^{3/2}(8n^{3/2})^{-j_0}]^\sigma}{2M^{\sigma-1}}.\end{aligned}$$

If we choose j_0 so large that

$$(8n^{3/2})^{-j_0} \le \min\left\{\frac{M\tilde{\xi}}{2n}, \frac{1}{16n^{5/2}}\left[\frac{\tilde{\zeta}^2 M^{\sigma-1}}{\mu e^{2hM}\tilde{\eta}}\right]^{1/\sigma}\right\}, \tag{33}$$

then (32) cannot hold. Consequently (31) must be valid. In other words,

$$\sup_{K_j^2} e^{h(u-u_0)} \ge e^{(h\operatorname{osc}_{K_j} u)/2}(1+\tilde{\zeta}) - \tilde{\zeta} \ge \exp\left[\frac{h}{2}(1+e^{-2Mh}\tilde{\zeta})\operatorname*{osc}_{K_j} u\right].$$

The second inequality in this estimate is derived in the following way. Let $h/2\operatorname{osc}_{K_j} u = \lambda \in [0, Mh]$ and

$$f(\lambda) = 1 + \tilde{\zeta} - \tilde{\zeta}e^{-\lambda} - e^{e^{-2Mh\tilde{\zeta}\lambda}}.$$

Then $f(0) = 0$ and

$$f'(\lambda) = \tilde{\zeta}e^{-\lambda} - e^{-2Mh}\tilde{\zeta}e^{e^{-2Mh\tilde{\zeta}\lambda}} \ge \tilde{\zeta}e^{-\lambda}[1 - e^{-Mh(1-\tilde{\zeta}e^{-2Mh})}] > 0.$$

Thus $f(\lambda) > 0$ in $[0, Mh]$. It follows that

$$\sup_{K_j^2} u \ge u_0 + \frac{1}{2}(1+e^{-2Mh}\tilde{\zeta})\operatorname*{osc}_{K_j} u,$$

$$\operatorname*{osc}_{K_{j-1}} u \ge \sup_{K_j^2} u - \inf_{K_j} u \ge \left(1 + \frac{1}{2}e^{-2Mh}\tilde{\zeta}\right)\operatorname*{osc}_{K_j} u,$$

or

$$\operatorname*{osc}_{K_j} u \le \left(1 + \frac{1}{2}e^{-2Mh}\tilde{\zeta}\right)^{-1}\operatorname*{osc}_{K_{j-1}} u.$$

Taking $\eta = (1+\frac{1}{2}e^{-2Mh\tilde{\zeta}})^{-1}$, we conclude $\operatorname{osc}_{K_j} u \le \eta \operatorname{osc}_{K_{j-1}} u \le \beta\eta^j$, which contradicts (29). So (28) holds for $j = j_0$. By induction (28) is valid for all $j \ge j_0$.

When $j = j_0$ we have

$$R_{j_0} = \xi^{-j_0} = (8n^{3/2})^{-j_0}, \qquad S_{j_0} = \tilde{\eta}M^{2-\sigma}[2n(8n^{3/2})^{-j_0}]^{\sigma}.$$

Now we do not assume that $P_2 = (0, \dots, 0)$. We need to choose j_0 so that $\max_{P_1\in P_2+\overline{K}_{j_0}} d(P_1, P_2) \le \frac{1}{2}\overline{d}(P_2)$ is also satisfied. (Incidentally $P_2 + K_{j_0} \subseteq Q$). Therefore, in addition to (33) j_0 must also satisfy

$$(8n^{3/2})^{-j_0} \le \min\left\{\frac{\overline{d}(P_2)}{2n}, \frac{1}{2n}\left(\frac{\tilde{d}(P_2)M^{\sigma-2}}{2\tilde{\eta}}\right)^{1/\sigma}\right\}.$$

Under this requirement $\operatorname{osc}_{P_2+K_{j_0}} u \le 2M = \beta\eta^{j_0}$, that is, (28) holds for $j = j_0$. It also holds for all $j \ge j_0$.

For $P_1(x^1, t^1), P_2(x^2, t^2) \in Q$, $t^1 \le t^2$, satisfying $d(P_1, P_2) \le \frac{1}{2}\overline{d}(P_2) \le C_{14}\overline{d}_{P_1P_2}$ (we may take $C_{14} = 2$) we can use (28) and the definition of K_j to get

$$\begin{aligned}
|u(x^2, t^2) - u(x^1, t^1)| &\le \frac{\beta}{\eta}\max\left\{\max_{1\le i\le n}|x_i^2 - x_i^1|^{\log(1/\eta)/\log\xi},\right.\\
&\qquad\left.\left[\frac{(\beta/2)^{\sigma-2}}{\tilde{\eta}(2n)^{\sigma}}|t^2 - t^1|^{\log(1/\eta)/\log(\xi^{\sigma}\eta^{\sigma-2})}\right]\right\}\\
&\le \frac{\beta}{\eta}\max\left\{1, \left[\frac{(\beta/2)^{\sigma-2}}{\tilde{\eta}(2n)^{\sigma}}\right]^{\log(1/\eta)/\log(\xi^{\sigma}\eta^{\sigma-2})}\right\}\\
&\qquad\cdot\max\left\{\max_{1\le i\le n}|x_i^2 - x_i^1|, |t^2 - t^1|^{1/\sigma}\right\}^{\log(1/\eta)/\log\xi}\\
&\le c_2\frac{(|x^2 - x^1| + |t^2 - t^1|^{1/\sigma})^{\log(1/\eta)/\log\xi}}{(8n^{3/2})^{j_0}[\log(1/\eta)/\log\xi + (\sigma-2)(\log(1/\eta)/\log\xi)^2/\sigma]}\\
&\le C_{15}\left(\frac{d(P_1, P_2)}{\overline{d}_{P_1P_2}^{1+\gamma}}\right)^{\gamma}, \qquad \gamma = \frac{\log(1/\eta)}{\log\xi}.
\end{aligned}$$

The proof of the theorem is complete. □

In order to obtain a Hölder estimate in $\overline{Q}$ it remains to derive a Hölder estimate near the boundary. This can be achieved by the method in Chapter 9. Let $u|_{\partial^* Q} = u^*$ satisfy a Hölder condition with exponents β and β/σ in x and t respectively. The Hölder constant of u^* is given by

$$c_\beta = c_{\beta,\beta/\sigma} = \sup_{(x^1,t^1),(x^2,t^2)\in\partial^* Q}\frac{|u^*(x^2, t^2) - u^*(x^1, t^1)|}{(|x^2 - x^1| + |t^2 - t^1|^{1/\sigma})^{\beta}}, \qquad 0 < \beta \le 1.$$

LEMMA 8. *There exist constants β_1, C_{16} depending only on n, μ, ν, σ, M, β and $\operatorname{diam}\Omega$ with $\sigma\beta_1 < \beta$ such that whenever $u \in C^{2,1}(Q)\cap C(\overline{Q})$, $u^* \in C^{\beta,\beta/\sigma}(\partial^* Q)$ $(0 < \beta \le 1)$, u satisfies the equation $\mathscr{L}u = B(x, t, u, u_x)$ and the boundary condition $u|_{\partial^* Q} = u^*$, $|u| \le M$ in $\overline{Q}$ and* (1), (2) *hold, one has*

$$|u(x, t) - u(x, 0)| \le C_{16}(1 + c_\beta)\frac{t^{\beta_1}}{d^\beta(x)}, \qquad d(x) = d(x, \partial\Omega).$$

PROOF. Let

$$U = \frac{1}{h}\{\exp[h(u(x, t) - u^*(x, 0))] - 1\}, \qquad h = \frac{2\mu}{\nu}, \qquad x^0 \in \Omega.$$

Proceeding in the proof of Theorem 7 we get

$$\widetilde{\mathscr{L}}U \ge \mu_1|P|^\sigma - \mu_2, \qquad \widetilde{\mathscr{L}} = \widetilde{A}_{ij}\frac{\partial^2}{\partial x_i \partial x_j} - \frac{\partial}{\partial t}, \qquad P = U_x,$$

$$\nu_0(1 + |P|^{\sigma-2})|\xi|^2 \le \widetilde{A}_{ij}\xi_i\xi_j \le \mu_0(1 + |P|^{\sigma-2})|\xi|^2, \qquad \xi \in \mathbb{R}^n$$

for $\nu_0 = \nu e^{-2h(\sigma-2)M}$, $\mu_0 = \mu e^{2h(\sigma-2)M}$, $\mu_1 = \mu e^{-2h(\sigma-1)M}$, and $\mu_2 = \mu e^{2hM}$. Let

$$\eta(x) = \left(1 - \frac{|x - x^0|^2}{R^2}\right)^m, \qquad v = \eta U,$$

where $R < d(x^0)$ and $m > 1$. The constant m will be determined below. We have

$$|\eta_i| \le \frac{2m}{R}\eta^{1-1/m}, \quad \text{and} \quad \widetilde{\mathscr{L}}\eta \ge -4m\mu_0(1 + |P|^{\sigma-2})\frac{\eta^{1-1/m}}{R^2}.$$

In view of the inequality

$$\min(\alpha|p|^a - \beta|p|^b) = -(a-b)\left(\frac{\beta}{a}\right)^{\frac{a}{(a-b)}}\left(\frac{b}{\alpha}\right)^{\frac{b}{(a-b)}} \qquad (a > b > 0, \alpha, \beta > 0),$$

and the fact $R \le \operatorname{diam}\Omega$ we have

$$\begin{aligned}\widetilde{\mathscr{L}}v &\ge \eta\widetilde{\mathscr{L}}U - 2\mu_0(1 + |P|^{\sigma-2})|\eta_x||P| + U\widetilde{\mathscr{L}}\eta \\ &\ge \left[\mu_1\eta|P|^{2+(\sigma-2)/(n+1)} - \mu_2\eta - 2\mu_0(1 + |P|^{(\sigma-2)/(n+1)})c_1\frac{\eta^{1-1/m}}{R}|P| \right. \\ &\qquad \left. - c_2(1 + |P|^{(\sigma-2)/(n+1)})\frac{\eta^{1-1/m}}{R^2}\right] D^{1/(n+1)}\end{aligned}$$

$$\geq \left\{ -\mu_2 - 2\mu_0 \left[1 + \left(\frac{c_3 \eta^{-1/m}}{R} \right)^{(\sigma-2)/(n+1)} \right] c_1 \frac{\eta^{-1/m}}{R} \left(\frac{c_3 \eta^{-1/m}}{R} \right) \right.$$

$$\left. - c_2 \left[1 + \left(\frac{c_3 \eta^{-1/m}}{R} \right)^{(\sigma-2)/(n+1)} \frac{\eta^{-1/m}}{R^2} \right] \right\} \eta D^{1/(n+1)}$$

$$\geq -c_4 R^{-2-(\sigma-2)/(n+1)} D^{1/(n+1)}, \qquad D = \det(\widetilde{A}_{ij}),$$

provided $m \geq 2 + (\sigma - 2)/(n+1)$.

An application of the Aleksandrov maximum principle yields

$$\sup_{\substack{|x-x^0|\leq R \\ 0\leq t\leq \tau}} v(x, t) \leq \sup_{\substack{|x-x^0|\leq R \\ t=0}} U + c_5 \frac{R^{n/(n+1)}}{R^{2+(\sigma-2)/(n+1)}} \left(\int_{\substack{|x-x^0|\leq R \\ 0\leq t\leq \tau}} dx\, dt \right)^{1/(n+1)}$$

$$\leq c_6 c_\beta R^\beta + c_5 \left(\frac{\tau}{R^\sigma} \right)^{1/(n+1)}.$$

Therefore,

$$u(x^0, \tau) - u^*(x^0, 0) \leq c_7(1 + c_\beta) \left[c_6 R^\beta + c_5 \left(\frac{\tau}{R^\sigma} \right)^{1/(n+1)} \right].$$

An analogous estimate for $u^*(x^0, 0) - u(x^0, \tau)$ can be derived replacing U by $-U$. Taking $\tau = R^{\sigma+(n+1)\beta}$ we get

$$|u(x^0, \tau) - u^*(x^0, 0)| \leq c_8(1 + c_\beta) \tau^{\beta/(\sigma+(n+1)\beta)}$$

$$\leq c_9(1 + c_\beta) d^{-\beta}(x_0) \tau^{\beta/(\sigma+(n+1)\beta)}$$

if $\tau \leq d^{\sigma+(n+1)\beta}(x^0)$. However, it is obviously valid when $\tau > d^{\sigma+(n+1)\beta}(x^0)$. Hence, this lemma is true for $C_{16} = \max\{2, c_9\}$ and $\beta_1 = \beta/(\sigma + (n+1)\beta)$. □

When $\partial\Omega \in C^2$, by a suitable change of variables we can assume a portion of $\partial\Omega$ lies on $x_n = 0$. By shifting the origin, we can also assume $\{x : |x| < R,\ x_n = 0\}$ belongs to this portion and $\{x : |x| < R,\ x_n > 0\} \subseteq \Omega$.

LEMMA 9. *Let*

$$u \in C^{2,1}\{x : |x| < R, x_n > 0, 0 < t \leq T\} \cap C\{x : |x| \leq R, x_n \geq 0, 0 \leq t \leq T\}$$

and

$$u^* \in C^{\beta, \beta/\sigma}\{x : |x| \leq R, x_n = 0, 0 \leq t \leq T\} \quad (0 < \beta \leq 1).$$

Suppose that u *satisfies the equation* $\mathscr{L}u = B$ *and* $u|_{\partial^* Q} = u^*$, $|u| \leq M$ *in* $\overline{Q}$ *and* (1), (2) *hold. Then there exist constants* δ, δ_1, C_{17}, C_{18} *depending solely on* n, μ, ν, σ, M, β, *and* $\operatorname{diam}\Omega$ *with* $\delta < \delta_1$ *such that*

$$|u(x, t) - u^*(x_1, \dots, x_{n-1}, 0, t)| \leq C_{17}(1 + c_\beta) \left(\frac{x_n}{R} \right)^\delta \tag{34}$$

for $x_n > 0$, $|x| \le R/2$, *and* $t^0 + R^\sigma/2 \le t \le t^0 + R^\sigma$ $(t^0 \ge 0)$. *Also*

$$|u(x, t) - u^*(x_1, \dots, x_{n-1}, 0, 0)| \le C_{18}(1 + c_\beta)\left(\frac{x_n + t^{1/\sigma}}{R}\right)^\delta \tag{35}$$

for $x_n + t^{1/\sigma} > 0$, $|x| \le R/2$, *and* $t \in R^\sigma$.

PROOF. Let

$$\eta(x, t) = \left(1 - \frac{|x|^2}{R^2}\right)^m \left(\frac{t - t_0}{R^\sigma}\right)^m, \qquad m > 1.$$

For $t^0 + R^\sigma/2 \le t^1 \le t^0 + R^\sigma$ we write $u^* = u^*(0, \dots, 0, t^1)$,

$$s = \frac{x_n}{R} + \varepsilon\left[\frac{(t - t^1)^2}{R^{\sigma+2}} + \frac{\sum_{i=1}^{n-1} x_i^2}{R^2}\right] + \varepsilon_1,$$

and $v(x, t) = \eta(x, t)s^{-\alpha}(u - u^*)^m$, where ε is a small positive constant, ε_1 a small positive number, $0 < \alpha < \min\{1/2(n + 1), m\beta/2\sigma\}$ and m (> 2) an even number.

For $|x| \le R$, $t^0 \le t \le t^0 + R^0$,

$$\begin{aligned} v|_{x_n=0} &\le c_1\left(\frac{R^2}{\varepsilon}\right)^\alpha \left[\frac{u^*(x_1, \dots, x_{n-1}, 0, t^1) - u^*}{(\sum_{i=1}^{n-1} x_i^2 + |t - t^1|^{2/\sigma})^{\sigma\alpha/m}}\right]^m \\ &\le c_2 c_\beta^m R^{\beta m - 2(\sigma-1)\alpha}, \end{aligned}$$

$$\begin{aligned} \mathscr{L}v &= s^{-\alpha}(u - u^*)^m \mathscr{L}\eta + \eta s^{-\alpha}\mathscr{L}(u - u^*)^m + \eta(u - u^*)^m \mathscr{L}s^{-\alpha} \\ &\quad + 2A_{ij}\left[s^{-2}\frac{\partial(u - u^*)^m}{\partial x_i}\eta_j + \eta\frac{\partial s^{-\alpha}}{\partial x_i}\frac{\partial(u - u^*)^m}{\partial x_j} + (u - u^*)^m\frac{\partial s^{-\alpha}}{\partial x_i}\eta_j\right]. \end{aligned}$$

We have the following estimates:

$$\mathscr{L}\eta \ge -c_3\eta^{1-1/m}\left[\frac{1}{R^2}(1 + |p|^{\sigma-2}) + \frac{1}{R^\sigma}\right],$$

$$\begin{aligned} \mathscr{L}(u - u^*)^m &= m(m-1)(u - u^*)^{m-2}A_{ij}u_iu_j + m(u - u^*)^{m-1}\mathscr{L}u \\ &\ge m(u - u^*)^{m-2}[(m-1)\nu(1 + |p|^{\sigma-2})|p|^2 - 2M\mu(1 + |p|^\sigma)] \\ &\ge m(u - u^*)^{m-2}\left[\frac{(m-1)\nu}{2}|p|^\sigma - 2M\mu\right] \\ &\ge c_3(n - u^*)^{m-2}|p|^\sigma + c_4(u - u^*)^m|p|^\sigma - c_5. \end{aligned}$$

Notice that the last inequality holds for $m \geq 1+2M\mu/\nu$ and $\nu m(m-1)/4 \geq c_3$. We can choose a large m so that this requirement is satisfied.

$$\begin{aligned}\mathscr{L}s^{-\alpha} &\geq \alpha(\alpha+1)s^{-\alpha-2}\nu(1+|p|^{\sigma-2})\frac{1+4\varepsilon^2\sum_{i=1}^{n-1}x_i^2}{R^2} \\ &\quad - \alpha s^{-\alpha-1}\mu(1+|p|^{\sigma-2})\frac{2n\varepsilon}{R^2} \\ &\geq \alpha\frac{1+|p|^{\sigma-2}}{R^2}s^{-\alpha-2}(\nu-2n\mu\varepsilon c_6) \geq \frac{\alpha\nu}{2}\frac{1+|p|^{\sigma-2}}{R^2}s^{-\alpha-2}.\end{aligned}$$

This inequality holds for $\varepsilon \leq \nu/4n\mu c_6$;

$$\begin{aligned}\left|2A_{ij}\frac{\partial(u-u^*)^m}{\partial x_i}\eta_j\right| &\leq 2\mu mn|u-u^*|^{m-1}(1+|p|^{\sigma-2})|p|\frac{2m\eta^{1-1/m}}{R} \\ &\leq \frac{c_3}{2}(u-u^*)^{m-2}\eta(1+|p|^{\sigma})+c_7(1+|p|^{\sigma-2})\frac{\eta^{1-2/m}}{R^2},\end{aligned}$$

$$\begin{aligned}\left|2A_{ij}\frac{\partial s^{-\alpha}}{\partial x_i}\frac{\partial(u-u^*)^m}{\partial x_j}\right| &\leq c_8\frac{\alpha s^{-\alpha-1}}{R}(1+|p|^{\sigma-2})|p|\,|u-u^*|^{m-1} \\ &\leq \frac{c_3}{2}(u-u^*)^{m-2}s^{-\alpha}(1+|p|^{\sigma}) \\ &\quad + \frac{c_9\alpha^2}{R^2}s^{-\alpha-2}(1+|p|^{\sigma-2}),\end{aligned}$$

and

$$\begin{aligned}\left|2A_{ij}\frac{\partial s^{-\alpha}}{\partial x_i}\eta_j\right| &\leq \frac{c_{10}\alpha}{R^2}s^{-\alpha-1}(1+|p|^{\sigma-2})\eta^{1-1/m} \\ &\leq \frac{1+|p|^{\sigma-2}}{R^2}\left(\frac{\alpha\nu}{4}\eta s^{-\alpha-2}+c_{11}\alpha\eta^{1-2/m}\right).\end{aligned}$$

Now let us fix m and ε as above and take $\alpha = \nu/4c_9$. Substituting all these inequalities into the inequality for $\mathscr{L}v$ yields, for $m \geq 2+(\sigma-2)/(n+1)$,

$$\begin{aligned}\mathscr{L}v &\geq s^{-\alpha}(u-u^*)^m\left(c_4\eta|p|^{\sigma}-c_{12}\eta^{1-2/m}\frac{|p|^{\sigma-2}}{R^2}-\frac{c_{13}}{R^{\sigma}}\right) \\ &\geq s^{-\alpha}(u-u^*)^m c_{14}\Bigg(c_4\eta|p|^{2+(\sigma-2)/(n+1)} \\ &\qquad\qquad - c_{12}\frac{\eta^{1-2/m}}{R^2}|p|^{(\sigma-2)/(n+1)}-\frac{c_{13}}{R^{\sigma}}\Bigg)D^{1/(n+1)} \\ &\geq -\frac{c_{15}s^{-\alpha}}{R^{\sigma}}D^{1/(n+1)} \\ &\geq -\frac{c_{15}s^{-\alpha}}{\varepsilon^{\alpha}R^{\sigma}}\left(\frac{R}{r}\right)^{2\alpha}D^{1/(n+1)},\end{aligned}$$

where $r = (\sum_{i=1}^n x_i^2)^{1/2}$ and $D = \det(A_{ij})$. By the Aleksandrov maximum principle we have

$$\begin{aligned}\sup_{\substack{|x|\le R,\, x_n\ge 0\\ t^0\le t\le t^0+R^\sigma}} v(x, t) &\le c_2 c_\beta^m R^{\beta m-2(\sigma-1)\alpha}\\ &\quad + c_{16} R^{n/(n+1)+2\alpha-\sigma}\left(\int_{\substack{|x|\le R,\, x_n>0\\ t^0\le t\le t^0+R^\sigma}} r^{-2(n+1)\alpha}\,dx\,dt\right)^{1/(n+1)}\\ &\le c_{17}(1+c_\beta^m)R^{n(2-\sigma)/(n+1)}.\end{aligned}$$

Therefore

$$|u(0, \dots, 0, x_n, t^1) - u^*(0, \dots, 0, t^1)| \le \left[c_{18}(1+c_\beta)\frac{(x_n + R\varepsilon_1)}{R^{n(\sigma-2)/(n+1)+\alpha}}\right]^{1/m}.$$

Letting $\varepsilon_1 \to 0$ and writing $\alpha/m = \delta$, $(1/m)[n(\sigma-2)/(n+1)+\alpha] = \delta_1$ we conclude that (34) holds when $x_1 = \cdots = x_{n-1} = 0$.

The proof of (35) for $x_1 = \cdots = x_{n-1} = 0$ is more or less the same, only we need to modify v to

$$\begin{aligned}v(x, t) = \left(1 - \frac{|x|^2}{R^2}\right)^m \left[\frac{x_n}{R} + \varepsilon\left(\frac{t^2}{R^{\sigma+2}} + \sum_{i=1}^{n-1}\frac{x_i^2}{R^2}\right) + \varepsilon_1\right]^{-\alpha}&\\ \times (u(x, t) - u^*(0, \dots, 0)).&\end{aligned}$$

Finally we point out that the general case follows by translating the origin. □

Concerning the global Hölder estimation of u the following theorem holds.

THEOREM 10. *Let $u \in C^{2,1}(Q) \cap C(\overline{Q})$ satisfy $\mathscr{L}u = B$, $u|_{\partial^* Q} = u^*$ and let* (1), (2) *hold for $2 < \sigma < 3$. There exist constants α $(0 < \alpha < 1)$ and C_{19} which only depend on $n, \mu, \nu, \sigma, M, \beta, \partial\Omega$, $\operatorname{diam}\Omega$ and the Hölder constant of u^* in $C^{\beta,\beta/\sigma}(\partial\Omega)$ such that*

$$|u(x^2, t^2) - u(x^1, t^1)| \le C_{19}(1 + c_\beta)(|x^2 - x^1| + |t^2 - t^1|^{1/\sigma})^\alpha$$

for all $P_1(x^1, t^1), P_2(x^2, t^2) \in \overline{Q}$.

PROOF. Let $d(x) = d(x, \partial\Omega)$ and $\rho = |x^2 - x^1| + |t^2 - t^1|^{1/\sigma}$. Without loss of generality we may take $\rho \le 1$.

CASE I. $\overline{d}_{P_1P_2} < \rho^{1/\sigma}/C_{14}$. By Theorem 7

$$|u(x^2, t^2) - u(x^1, t^1)| \le C_{20}\rho^{(1-(1+\gamma)/\sigma)\gamma}.$$

CASE II. $\overline{d}_{P_1P_2} \le \rho^{1/\sigma}/C_{14}$, $\overline{d}_{P_1P_2} = \min\{t^1, t^2\}^{1/\sigma}$. Suppose that $\min\{t^1, t^2\} = t^1$. We have

$$t^2 = t^1 + (t^2 - t^1) \le \frac{1}{C_{14}^\sigma}\rho + \rho^\sigma \le 2\rho.$$

When $\min\{d(x^1), d(x^2)\} > \rho^{\beta_1/2\beta}$, it follows from Lemma 8 that

$$\begin{aligned}|u(x^2, t^2) - u(x^1, t^1)| &\le |u(x^2, t^2) - u(x^2, 0)| + |u(x^1, t^1) - u(x^1, 0)| \\ &\quad + |u(x^2, 0) - u(x^1, 0)| \\ &\le C_{16}(1 + c_\beta)\frac{(t^1)^{\beta_1} + (t^2)^{\beta_1}}{\min\{d(x^1), d(x^2)\}^\beta} + c_\beta \rho^\beta \\ &\le c_1(1 + c_\beta)\rho^{\beta_1/2}.\end{aligned}$$

When $\min\{d(x^1), d(x^2)\} \le \rho^{\beta_1/2\beta}$, letting $\min\{d(x^1), d(x^2)\} = d(x^1)$, we have

$$d(x^2) \le d(x^1) + |x^2 - x^1| \le \rho^{\beta_1/2\beta} + \rho \le 2\rho^{\beta_1/2\beta}.$$

Therefore

$$\max\{d(x^1), d(x^2), (t^1)^{1/\sigma}, (t^2)^{1/\sigma}\} \le c_2\rho^{\beta_1/2\beta}. \tag{36}$$

(36) will be handled in Case IV.

CASE III. $\bar{d}_{P_1P_2} \le \rho^{1/\sigma}/C_{14}$, $\bar{d}_{P_1P_2} = \min\{d(x^1), d(x^2)\}$. Let $\min\{d(x^1), d(x^2)\} = d(x^1) = |x^1 - y^1|$ and $d(x^2) = |x^2 - y^2|$, $y^1, y^2 \in \partial\Omega$. We have

$$\begin{aligned}d(x^2) = d(x^2, \partial\Omega) &\le |x^2 - y^1| \le |x^2 - x^1| + |x^1 - y^1| \\ &\le \rho + \frac{1}{C_{14}}\rho^{1/\sigma} \le \frac{2}{C_{14}}\rho^{1/\sigma},\end{aligned}$$

and

$$|y^2 - y^1| \le |y^2 - x^2| + |y^1 - x^1| + |x^2 - x^1| \le \frac{4}{C_{14}}\rho^{1/\sigma}.$$

When $\min\{t^1, t^2\} > \rho^{\delta/2\delta_1}$, letting $R = \rho^{\delta/2\sigma\delta_1}$ we have

$$\begin{aligned}|u(x^2, t^2) - u(x^1, t^1)| &\le |u(x^2, t^2) - u(y^2, t^2)| + |u(x^1, t^1) - u(y^1, t^1)| \\ &\quad + |u(y^2, t^2) - u(y^1, t^1)| \\ &\le C_{17}(1 + c_\beta)\frac{d(x^2)^\delta + d(x^1)^\delta}{R^{\delta_1}} + c_\beta(|y^2 - y^1| + |t^2 - t^1|^{1/\sigma})^\beta \\ &\le c_3(1 + c_\beta)\rho^{\min\{\delta/2, \beta\}/\sigma}.\end{aligned}$$

When $\min\{t^1, t^2\} \le \rho^{\delta/2\delta_1}$, if $\min\{t^1, t^2\} = t^1$, $t^2 = t^1 + (t^2 - t^1) \le \rho^{\delta/2\delta_1} + \rho^\sigma \le 2\rho^{\delta/2\delta_1}$. Therefore we get

$$\max\{d(x^1), d(x^2), (t^1)^{1/\sigma}, (t^2)^{1/\sigma}\} \le C_4\rho^{\delta/2\sigma\delta_1}, \tag{37}$$

which will be treated below.

CASE IV. (36) and (37) are contained in the following inequality.

$$\max\{d(x^1), d(x^2), (t^1)^{1/\sigma}(t^2)^{1/\sigma}\} \le c_5\rho^{\min\{\beta_1/\beta, \delta/\sigma\delta_1\}/2}.$$

Taking R to be a constant, by virtue of (35) in Lemma 9 we get

$$|u(x^2,t^2)-u(x^1,t^1)| \le C_{18}(1+c_\beta)\left(\frac{\max\{d(x^1),\,d(x^2)\}+\max\{t^1,\,t^2\}^{1/\sigma}}{R}\right)^\delta$$
$$\le c_6(1+c_\beta)\rho^{\delta\min\{\beta_1/\beta,\,\delta/\sigma\delta_1\}/2}.$$

So the theorem holds in all cases. □

From the Hölder estimation one deduces the unique solvability for the first boundary value problem for quasilinear parabolic equations satisfying (1), (2). The reader is referred to [4] for details.

§3. Certain apriori estimates for the solutions of a fully nonlinear parabolic equation with natural structure condition of the second kind and the unique solvability of the first boundary value problem

Now let us turn to fully nonlinear parabolic equations. Given μ, μ_1, μ_2, ν, and σ, $\mu>\nu>0$, $2<\sigma<3$, μ_1, $\mu_2>0$. Let Ω be a bounded domain in $\mathbb{R}^n$, $T>0$, $Q=\Omega\times(0,T]$, and let $F(x,t,u,u_i,u_{ij})$ $(1\le i,j\le n)$ be a real function defined in $Q\times\mathbb{R}^{1+n}\times\mathscr{S}^n$. Suppose that F is C^2 in x, u, u_i, u_{ij}, C^1 in t and the following structure conditions hold for $(x,t)\in Q$, $u,\tilde u\in\mathbb{R}$, $\xi,\tilde x,u_i,\tilde u_i\in\mathbb{R}^n$, and $u_{ij},\tilde u_{ij}\in\mathscr{S}^n$

$$uF(x,t,u,0,0)\le\mu_1u^2+\mu_2, \tag{38}$$

$$\nu\left(1+\sum_i|u_i|^{\sigma-2}\right)|\xi|^2\le\sum F_{ij}\xi_i\xi_j\le\mu\left(1+\sum_i|u_i|^{\sigma-2}\right)|\xi|^2, \tag{39}$$

$$|F(x,t,u,u_i,0)|\le M(u)\left(1+\sum_i|u_i|^\sigma\right), \tag{40}$$

$$|F|+|F_u|+\sum_j|F_{u_j}|\left(1+\sum_i|u_i|\right)+\sum_j|F_{x_j}|\left(1+\sum_i|u_i|\right)^{-1}$$
$$\le M_1(u)\left[1+\sum_i|u_i|^\sigma+\sum_{i,j}|u_{ij}|\left(1+\sum_i|u_i|^{\sigma-2}\right)\right], \tag{41}$$

$$F_{(\eta)(\eta)} \le M_2(u, u_k)\left\{\left(1+\sum_{i,j}|u_{ij}|^3\right)(|\tilde{x}|^2+|\tilde{u}|^2) + \left(1+\sum_{i,j}|u_{ij}|\right)\sum_i|\tilde{u}_i|^2 + \left[\left(1+\sum_{i,j}|u_{ij}|\right)(|\tilde{x}|+|\tilde{u}|)+\sum_i|\tilde{u}_i|\right]\sum_{i,j}|\tilde{u}_{ij}|\right\}, \tag{42}$$

$$|F_t| \le M_2(u, u_k)\left(1+\sum_{i,j}|u_{ij}|^2\right), \tag{43}$$

where $M(u)$, $M_1(u)$, $M_2(u, u_k) = M_2(u, u_1, \dots, u_n)$ are bounded, $\eta = (\tilde{x}, \tilde{u}, \tilde{u}_i, \tilde{u}_{ij})$ and $F_{\eta\eta}$ is the second directional derivative of F along η.

Conditions (38)–(43) constitute the *natural structure condition of the second kind* for the second order fully nonlinear parabolic equation

$$u_t = F(x_1, t, u, u_i, u_{ij}). \tag{44}$$

When $\sigma = 2$, they become the natural structure condition encountered in Chapter 9. (In contrast we may call then natural conditions of the first kind.)

Let $\partial^*Q = \partial\Omega \times \{0 \le t \le T\} \cup \Omega \times \{t = 0\}$ be the parabolic boundary of Q and let

$$u|_{\partial^*Q} = u^* \tag{45}$$

be a given function on ∂^*Q.

When $|u|_{C(\partial^*Q)} < \infty$ and $u \in C^{2,1}(Q) \cap C(\overline{Q})$ satisfies (38), (39), (44), and (45), the proof of Theorem 1 in Chapter 9 shows that u is bounded in $\overline{Q}$. Thus we can assume $|u| \le |u|_{C(\overline{Q})} \le M_0$ where M_0 depends only on $\mu, \nu, \mu_1, \mu_2, \sigma, |u^*|_{C(\partial^*Q)}$ in the following discussion. Moreover, (40) is simplified to

$$|F(x, t, u, u_i, 0)| \le \mu\left(1+\sum_i|u_i|^\sigma\right),$$

(take μ to be $\max\{\mu, \max_{|u|\le M_0} M(u)\}$ if necessary).

For $P = (x, t)$ and $\widetilde{P} = (\tilde{x}, \tilde{t})$, write $d(P, \widetilde{P}) = (|x-\tilde{x}|^2+|t-\tilde{t}|)^{1/2}$. We also write $\|u\|_{C^{\alpha,\alpha/2}(\overline{Q})} = \sup_{P,\widetilde{P}\in\overline{Q}} |u(P)-u(\widetilde{P})|/d^\alpha(P, \overline{P})$ so that the Hölder constant is simply $\|u\|_{C^{\alpha,\alpha/2}(\overline{Q})}$. Similarly we do the same for $\|u\|_{C^{\alpha,\alpha/2}(\partial^*Q)}$.

THEOREM 11. *Let $u \in C^{2,1}(Q)\cap C(\overline{Q})$ be a solution of* (44), (45) *in which F satisfies* (39), (40) *and*

$$\|u\|_{C(\overline{Q})} \le M_0, \qquad \|u\|_{C^{\beta,\beta/2}(\partial^*Q)} \le M_\beta, \qquad (0 < \beta < 1).$$

There exist α *and* C, $0 < \alpha < \beta$, $c > 0$, *such that*

$$\|u\|_{C^{\alpha,\alpha/2}(\overline{Q})} \leq C(1 + M_\beta).$$

Here α *and* C *depend only on* n, μ, ν, σ, M_0, T, $\partial\Omega$, *and* $\operatorname{diam}\Omega$.

PROOF. For $(x^1, t^1), (x^2, t^2) \in \partial^* Q$,

$$\frac{|u(x^2, t^2) - u(x^1, t^1)|}{(|x^2 - x^1|^2 + |t^2 - t^1|^{2/\sigma})^\beta} \leq \max\{1, T^{\beta(1-2/\sigma)}\} \frac{|u(x^2, t^2) - u(x^1, t^1)|}{(|x^2 - x^1| + |t^2 - t^1|)^\beta}$$
$$\leq \max\{1, T^{\beta(1-2/\sigma)}\} M_\beta.$$

Expressing (44) as

$$\mathscr{L}u = A_{ij}u_{ij} - u_t = -F(x, t, u, u_i, 0),$$
$$A_{ij} = \int_0^1 \frac{\partial F}{\partial u_{ij}}(x, t, u, u_i, \theta u_{ij})\, d\theta,$$

and applying Lemma 10 we obtain

$$\sup_{(x^1, t^1), (x^2, t^2) \in \overline{Q}} \frac{|u(x^2, t^2) - u(x^1, t^1)|}{(|x^2 - x^1|^2 + (t^2 - t^1)^{2/\sigma})^{\alpha_1}} \leq C_1(1 + M_\beta).$$

So, from

$$\frac{|u(x^2, t^2) - u(x^1, t^1)|}{(|x^2 - x^1|^2 + |t^2 - t^1|)^{2\alpha_1/\sigma}} \leq 2^{(1-2/\sigma)\alpha_1} \max\{1, (\operatorname{diam}\Omega)^{2\alpha_1(1-2/\sigma)}\}$$
$$\cdot \frac{|u(x^2, t^2) - u(x^1, t^1)|}{(|x^2 - x^1|^2 + |t^2 - t^1|^{2/\sigma})^{\alpha_1}}$$

we obtain the theorem by taking $\alpha = 2\alpha_1/\sigma$. □

THEOREM 12. *Let* $u^* \in C^{2,1}(\partial\Omega \times [0, T]) \cup C^1(\overline{\Omega} \times \{t = 0\})$ *and let* $u \in C^{2,1}(Q) \cap C(\overline{Q})$ *be a solution of* (44), (45) *with* $\|u\|_{C(\overline{Q})} \leq M_0$. *Suppose that* F *satisfies* (39) *and* (41). *Then*

$$\left\| \frac{\partial u}{\partial N} \right\|_{C(\partial\Omega \times [0, T])} \leq C(1 + \|u^*\|_{C^{2,1}(\partial\Omega \times [0, T])} + \|u^*\|_{C^1(\Omega \times \{t=0\})}).$$

Here N *is the unit inner normal of* $\partial\Omega$ *and* C *depends only on* n, μ, ν, M_0, σ, T, *and* $\partial\Omega$.

PROOF. Straighten a portion of ∂Q to

$$\Sigma_{R,T} = \{(x, t) : |x_i| < R, \ 1 \leq i \leq n-1, \ x_n = 0\} \times \{0 \leq t \leq T\}$$

by a change of variables. Assume that

$$Q_{R,T} = \{(x, t) : |x_i| < R, \ 1 \leq i \leq n-1, \ 0 < x_n < R\} \times \{0 < t < T\}$$

belongs to the new Q. As in Theorem 11 we write the equation as $\mathscr{L}u = A_{ij}u_{ij} - u_t = -F(x, t, u, u_i, 0)$. Denote

$$U = \frac{\nu}{2\mu}\left\{\exp\left[\frac{2\mu}{\nu}(u(x_1, \dots, x_n, t) - u^*(x_1, \dots, x_{n-1}, 0, t))\right] - 1\right\}$$

and $P = U_x$. We have

$$\begin{aligned}
\widetilde{\mathscr{L}}U &= \widetilde{A}_{ij}(x, t, U, P, D^2U)U_{ij} - U_t \\
&\equiv A_{ij}(x, t, u, u_x^* + e^{-2\mu(u-u^*)/\nu}P, \dots)U_{ij} - U_t \\
&= \frac{2\mu}{\nu}e^{2\mu(u-u^*)/\nu}A_{ij}(u_i - u_i^*)(u_j - u_j^*) + e^{-2\mu(u-u^*)/\nu}\mathscr{L}u \\
&\quad - e^{2\mu(u-u^*)/\nu}(\widetilde{A}_{ij}u_{ij}^* - u_t^*) \\
&\geq \mu e^{-2\mu(\sigma-1)(u-u^*)/\nu}[2|P + u_x^* e^{2\mu(u-u^*)/\nu}||P|^2 - |P + u_x^* e^{2\mu(u-u^*)/\nu}|^\sigma \\
&\qquad - c_1|P|^{\sigma-2} - c_2] \\
&\geq c_3|P|^\sigma - c_4|P|^{\sigma-2} - c_5 \geq -c_6.
\end{aligned}$$

Let

$$v = \log\left(1 + \frac{x_n}{R} + \varepsilon\sum_{i=1}^{n-1}\frac{x_i^2}{R^2}\right) \quad \text{and} \quad w = K\frac{v}{R} - U,$$

where ε is small and K is a positive constant. Taking $K = K_\varepsilon$ so large that $w \geq 0$ in $\partial Q_{R,T} \cap \{0 \leq t \leq T\}$. On $\partial Q_{R,T} \cap \{t = 0\}$

$$\begin{aligned}
\frac{\partial w}{\partial x_n} &\geq \frac{K}{1 + x_n/R + \varepsilon\sum_{i=1}^{n-1}x_i^2/R_2} \\
&\quad - \exp\left[\frac{2\mu}{\nu}(u^*(x_1, \dots, x_n, 0) - u^*(x_1, \dots, x_{n-1}, 0, 0))\right] \\
&\quad \times \left[\frac{\partial u^*}{\partial x_n}(x_1, \dots, x_n, 0) - \frac{\partial u^*}{\partial x_n}(x_1, \dots, x_{n-1}, 0, 0)\right] \\
&\geq \frac{K}{2 + (n-1)\varepsilon} - c_7\|u^*\|_{C^1(\Omega\times\{t=0\})} \geq 0.
\end{aligned}$$

As a result $w|_{\partial Q_{R,T}\cap\{t=0\}} \geq 0$.

If w attains a minimum inside $Q_{R,T}$, then at the minimum $w_x = 0$ and $\widetilde{\mathscr{L}}w \geq 0$. But, on the other hand,

$$\begin{aligned}
\widetilde{\mathscr{L}}w|_{w_x=0} &\leq \left(1 + \left|\frac{K}{R}v_x\right|^{\sigma-2}\frac{K}{R^3}\right)\left[-\nu\frac{1 + 4\varepsilon^2\sum_{i=1}^{n-1}\frac{x_i^2}{R^2}}{(1 + x_n/R + \varepsilon\sum_{i=1}^{n-1}\frac{x_i^2}{R^2})^2}\right. \\
&\qquad \left. + \frac{\mu n\varepsilon}{1 + \frac{x_n}{R} + \varepsilon\sum_{i=1}^{n-1}\frac{x_i^2}{R^2}}\right] + c_7 \\
&\leq -\frac{K}{R^3}\frac{\nu - \mu n(n+1)\varepsilon}{(1 + x_n/R + \varepsilon\sum_{i=1}^{n-1}x_i^2/R^2)^2} + c_7 < 0,
\end{aligned}$$

when $\varepsilon < \nu/2n(n+1)\mu$ and $K = K_\varepsilon$ is sufficiently large. Hence for such ε and K, w cannot attain its minimum in $Q_{R,T}$. Therefore $w \geq 0$ in $\overline{Q}_{R,T}$. We conclude

$$0 \leq \left.\frac{\partial w}{\partial x_n}\right|_{x=0} = \left.\frac{K}{R^2} - u_x\right|_{\substack{x=0\\ 0\leq t\leq T}} .$$

Replacing u by $-u$ the same reasoning leads to $0 < K/R^2 + u_x|_{x=0,0\leq t\leq T}$. Hence $|u_x|_{x=0,0\leq t\leq T} \leq K/R^2$. At points on $x_n = 0$ with $|x_i| \leq R/2$ $(1 \leq i \leq n-1)$ the estimation of $|u_x|$ is similar.

By a finite covering argument we conclude the estimation on $\partial\Omega \times [0, T]$. □

THEOREM 13. *Suppose that F satisfies the structure inequalities* (39), (40), *and* (41). *Let $u \in C^{2,1}(Q) \cap C(Q)$ be a solution of* (44), (45), *and $u_x \in C^{2,1}(\overline{Q})$. If $\|u\|_{C(\overline{Q})} \leq M_0$, $\|u_x\|_{C(\partial^* Q)} \leq M_1$, and $\|u\|_{C^{\beta,\beta/2}(\overline{Q})} \leq M_\beta$, then there exists a constant C depending solely on n, μ, ν, σ, M, M_1, M_β,* diam Ω, *and T such that*

$$\|u_x\|_{C(\overline{Q})} \leq C.$$

PROOF. Let $\max_{x\in\overline{Q}}|u_x| = M_2$. Clearly we may assume $M_2 \geq \sqrt{2}$. Let $M_2 = u_x(P_0)$, $P_0 = (x^0, t^0)$. Denote the cylinder

$$\left\{(x, t) : |x - x^0| < \frac{1}{M_2},\ t^0 - \frac{1}{M_2^2} < t < t^0\right\}$$

by Q_0. We consider the function

$$v(x, t) = \eta^2(x, t)|u_x|^2 + \alpha M_2^2(u(x, t) - u(x^0, t^0))^2,$$

in $\overline{Q_0 \cap Q}$ where α is a constant to be specified below and η is a cut-off function satisfying $0 \leq \eta(x, t) \leq 1$, $\eta(x^0, t^0) = 1$, $\eta|_{\partial^*\Omega_0} = 0$, $|\eta_i| \leq c_1 M_2$, $|\eta_{ij}| \leq c_2 M_2^2$, $|\eta_t| \leq c_3 M_2^2$.

Let $P_1(x^1, t^1)$ be a maximum of v. We have

$$\begin{aligned} M_2^2 = v(P_0) \leq v(P_1) &= \eta^2(P_1)|u_x(P_1)|^2 + \alpha M_2^2(u(P_1) - u(P_0))^2 \\ &\leq \eta^2(P_1)|u_x(P_1)|^2 + \alpha M_2^{2-2\beta} M_\beta^2 . \end{aligned} \tag{46}$$

CASE I. $\alpha M_\beta^2 M_2^{-2\beta} \geq \frac{1}{2}$. We have $M_2 \leq (2\alpha M_\beta^2)^{1/2\beta}$.

CASE II. $\alpha M_\beta^2 M_2^{-2\beta} < \frac{1}{2}$, $P_1 \in \partial^*(Q_0 \cap Q)$. From (46), $M_2 \leq (2M_1^2)^{1/2}$.

CASE III. $\alpha M_\beta^2 M_2^{-2\beta} < \frac{1}{2}$, $P_1 \in Q_0 \cap Q$. From (46)

$$\eta(P_1) \geq \frac{1}{\sqrt{2}}, \qquad \frac{M_2}{\sqrt{2}} \leq |u_x(P_1)| \leq M_2 . \tag{47}$$

Let $\mathscr{L}v = F_{u_{ij}}v_{ij} + F_{u_i}v_i - v_t$. Since v takes a maximal value at P_1, we have

$$0 \le \mathscr{L}v|_{P_1} = |u_x^2|\mathscr{L}(\eta^2) + \eta^2\mathscr{L}(|u_x|^2) + 2F_{u_{ij}}(\eta^2)_i \cdot (|u_x|^2)_j + 2\alpha M_2^2[u(P_1) - u(P_0)]\mathscr{L}u + 2\alpha M_2^2 F_{u_{ij}}u_iu_j,$$

$$|\mathscr{L}\eta^2| = \left|F_{u_{ij}}(\eta^2)_{ij} + F_{u_i}(\eta^2)_i - (\eta^2)_t\right| \le (1+|p|^{\sigma-2})c_4M_2^2 + \frac{1+|p|^\sigma + |r|(1+|p|^{\sigma-2})}{1+|p|}\cdot c_5M_2 + c_6M_2^4,$$

$$(|p| = |u_x|,\ |r| = |u_{xx}|),$$

$$\mathscr{L}(|u_x|^2) = 2u_k\mathscr{L}u_k + 2F_{u_{ij}}u_{ki}u_{kj},$$

$$|\mathscr{L}u_k| = \left|F_{u_{ij}}(u_{ij})_k + F_{u_i}(u_i)_k - \frac{dF}{dx_k}\right| = |F_uu_k + F_k| \le c_7(1+|p|)[1+|p|^\sigma + |r|(1+|p|^{\sigma-2})],$$

$$|\mathscr{L}u| = \left|F_{u_{ij}}u_{ij} + F_{u_i}u_i - F\right| \le c_8[1+|p|^\sigma + |r|(1+|p|^{\sigma-2})].$$

Combining these four inequalities and (47) yields

$$\begin{aligned} &2\alpha\nu(1+|p|^{\sigma-2})|p|^4 + \nu(1+|p|^{\sigma-2})|r|^2 \\ &\le 2\alpha\nu M_2^2(1+|p|^{\sigma-2})|p|^2 + 2\nu\eta^2(1+|p|^{\sigma-2})|r|^2 \\ &\le 2\alpha M_2^2F_{u_{ij}}u_iu_j + 2\eta^2F_{u_{ij}}u_{ki}u_{kj} \\ &\le |u_x|^2|\mathscr{L}\eta^2| + 2\eta^2\,|u_k\mathscr{L}u_k| + 2\left|\Sigma F_{u_{ij}}(\eta^2)_i \cdot (|u_x|^2)_j\right| + 2\alpha M_2^2|u(P_1) - u(P_0)|\,|\mathscr{L}u| \\ &\le c_9(|p|^{\sigma+2} + |r|\,|p|^\sigma) + c_{10}\alpha(|p|^{\sigma+2-\beta} + |r|\,|p|^{\sigma-\beta}). \end{aligned}$$

Therefore

$$\begin{aligned} \alpha|p|^{\sigma+2} + |r|^2|p|^{\sigma-2} &\le |r|^2|p|^{\sigma-2} + c_{11}|p|^{\sigma+2} + c_{12}\alpha^2|p|^{\sigma+2-2\beta} + c_{10}\alpha|p|^{\sigma+2-\beta} \\ &\le |r|^2|p|^{\sigma-2} + c_{11}|p|^{\sigma+2} + c_{13}(\alpha^2+1)|p|^{\sigma+2-\beta}. \end{aligned}$$

Taking $\alpha = 2c_{11}$ we obtain

$$M_2 \le \sqrt{2}|p| \le \sqrt{2}\left[\frac{c_{13}(1+4c_{11}^2)}{c_{11}}\right]^{1/\beta}.$$

In view of Cases I, II, III we conclude

$$M_2 \le \max\{(4c_{11}M_\beta^2)^{1/\beta},\ (2M_1^2)^{1/2},\ \sqrt{2}[c_{13}(1+4c_{11}^2)/c_{11}]^{1/\beta}\}. \quad \square$$

After deriving a uniform estimate $\|u_x\|_{C(\overline{Q})}$, the structure conditions of the second kind and the first kind become the same. Consequently, estimates for $\|u_x\|_{C^{\alpha,\alpha/2}(\overline{Q})}$, $\|u\|_{C^{2,1}(\overline{Q})}$, and $\|u\|_{C^{2+\alpha,1+\alpha/2}(\overline{Q})}$ can be deduced from the results of Chapter 9.

The proof for unique solvability for the first boundary value problem essentially coincides with §7 of Chapter 9. Minor differences include: (11) in §7 of Chapter 9 should be replaced by

$$u_t = (1-\kappa)(1+|u_x|^2)^{\sigma/2-1}\Delta u + \kappa F(x, t, u, u_i, u_{ij}), \qquad (x, t) \in Q;$$

(13) replaced by

$$\begin{aligned}(v-u^0)_t = (1-\kappa)[(1+|u^0_\tau|^2)^{\sigma/2-1}\Delta(v-u^0) \\ + (\sigma-2)(1+|u^0_x|^2)^{\sigma/2-2}u^0_i(v_i-u^0_i)] \\ + \kappa[a_{ij}(v-u^0)_{ij} + b_i(v-u^0)_i + c(v-u^0)] + (1-\kappa)E + \kappa H \\ + (\kappa_0-\kappa)G,\end{aligned}$$

where $(a_{ij}, b_j, c, f) = (F_{r_{ij}}, F_{p_i}, F_z, F)(x, t, u^0, u^0_i, u^0_{ij})$,

$$\begin{aligned}E = (1+|w_x|^2)^{\sigma/2-1}\Delta w - (1+|u^0_x|^2)^{\sigma/2-1}\Delta u^0 - (1+|u^0_x|^2)^{\sigma/2-1}\Delta(w-u^0) \\ - (\sigma-2)(1+|u^0_x|^2)^{\sigma/2-2}u^0_i(w-u^0)_i,\end{aligned}$$

$$G = F(x, t, w, w_i, w_{ij}) - f - a_{ij}(w-u^0)_{ij} - b_i(w-u^0)_i - c(w-u^0),$$

and

$$H = (1+|u^0_x|^2)^{\sigma/2-1}\Delta u^0 - f.$$

G and H can be handled as before. As E is the function G corresponding to $F(x, t, w, w_i, w_{ij}) = (1+|w_x|^2)^{\sigma/2-1}\Delta w$, its estimation is as that of G. Therefore we can derive an estimate for $\|v-u^0\|_{C^{2+\alpha}}$ as before. Finally, as

$$\begin{aligned}\|v^2-v^1\|_{C^{2+\alpha}} \le K_1\|(1-\kappa)(E(W^2)-E(W^1)) \\ + \kappa(G(x, t, W^2) - G(x, t, W^1))\|_{C^\alpha},\end{aligned}$$

it can also be estimated as before. Thus we obtain the unique solvability for the first boundary value problem. □

Symbols

Ω	usually a bounded (connected) open set
$\overline{\Omega}$	the closure of Ω
$\partial\Omega$	the boundary of Ω
$\subset\subset$	$A \subset\subset B$ means $\overline{A} \subset B$
Ω'	$\Omega' \subset\subset \Omega$
$d(x, A)$	the distance between x and the set A, i.e., $d(x, A) = \inf_{y\in A} \overline{xy}$
Ω_ρ	$\Omega_\rho \subseteq \Omega$ and $d(\Omega_\rho, \partial\Omega) = \rho$
d_x	$d_x = d(x, \partial\Omega)$
d_{xy}	$d_{xy} = \min\{d_x, d_y\}$
Q	usually the cylinder $\Omega \times (0, T]$
$\partial^* Q$	the parabolic boundary of Q, i.e., $\partial^* Q = \Omega \times \{t = 0\} \cup \partial\Omega \times [0, T]$
Q'	$Q' \subseteq Q$ and $d(Q', \partial^* Q) > 0$
Q_ρ	$Q_\rho \subseteq Q$ and $d(Q_\rho, \partial^* Q) = \rho$
d_P	the distance between the point $P = (x, t)$ and $\partial^* Q$
$d_{P_1 P_2}$	$d_{P_1 P_2} = \min\{d_{P_1}, d_{P_2}\}$
$K(\rho)$, B_ρ	a ball with radius ρ
$B_\rho(x^0)$	a ball with radius ρ, centered at x^0, i.e., $\{x: \lvert x - x^0\rvert < \rho\}$
$k_R(x^0)$	a square with length $2R$, centered at $x^0 = (x_1^0, \ldots, x_n^0)$, i.e., $$\{x: \lvert x_i - x_i^0\rvert < R,\ 1 \le i \le n\}$$
$K_R(x^0, t^0)$	$k_R(0) \times (0, T)$
$\lvert A\rvert$	the measure of the set A, or $\lvert A\rvert = \operatorname{meas} A$
a.e.	almost everywhere
$\varnothing$	empty set
$\partial u/\partial N$	N is the normal at the boundary, $\partial u/\partial N$ in general is the normal derivative of u along the boundary; usually we take N to be the inner one.
ω_n	the area of the unit sphere in $\mathbb{R}^n$
κ_n	the volume of the unit ball in $\mathbb{R}^n$

$\|u\|_p$	$\|u\|_p = (\int_\Omega \|u\|^p\, dx)^{1/p}$, Ω given
$C^{k+\alpha}$	$u = u(x_1, \dots, x_n)$ is continuously differentiable up to kth order and its kth derivatives satisfy a Hölder condition with exponent α $(0 < \alpha \le 1)$
$C^{2,1}$	$u = u(x_1, \dots, x_n, t)$ twice continuously differentiable in $x_1, \dots, x_n$ and continuously differentiable in t
$C^{2+\lambda, 1+\lambda/2}$	u_{ij} $(1 \le i, j \le n)$ and u_t are Hölder continuous in x with exponent λ and Hölder continuous in t with exponent $\lambda/2$
$\|u\|_{\alpha,\Omega}$	$u \in C^\alpha(\Omega)$ $(0 < \alpha \le 1)$, $\|u\|_{\alpha,\Omega}$ is the Hölder constant
$\|u\|_{\alpha,Q}$	$u \in C^{\alpha,\alpha/2}(Q)$, $Q = \Omega \times (0, T]$ and $$\|u\|_{\alpha,Q} = \sup \frac{\|u(x,t) - u(x',t')\|}{(\|x - x'\|^2 + \|t - t'\|)^{\alpha/2}}$$
$W_p^k(\Omega)$	The function space consisting of all functions in Ω whose weak derivatives up to kth order belong to $L^p(\Omega)$
$\mathring{W}_p^k(\Omega)$	The closure (under the $W_p^k(\Omega)$-norm) of functions in $W_p^k(\Omega)$ which vanish near $\partial\Omega$. In other words, it is the closure of $C_0^\infty(\Omega)$ under $W_p^k(\Omega)$-norm
$W_p^{k,k/2}(Q)$	$Q = \Omega \times (0, T]$, the function space consisting of all functions in Q whose weak derivatives in x up to kth order and weak derivatives in t up to $k/2$ order (k is even) belong to $L^p(Q)$
a^+	$a^+ = \max\{a, 0\}$
a^-	$a^- = -\min\{a, 0\}$

References

1. K. C. Chang [Gong Qing Zhang], *Critical points theory and its applications*, Shanghai Science and Technology Press, Shanghai, 1986. (Chinese)

2. Avner Friedman, *Partial differential equations*, Holt, Rinehart and Winston, Inc., New York-Montreal-London, 1969.

3. A. Pazy, *Semigroups of linear operators and applications to partial differential equations*, Applied Mathematical Sciences, 44. Springer-Verlag, New York, 1983.

4. O. A. Ladyzhenskaya, V. A. Solonnikov and N. N. Ural′tseva, *Linear and quasilinear equations of parabolic type*, "Nauka", Moscow, 1967; English transl., Transl. Math. Monographs, vol. 23, Amer. Math. Soc., Providence, RI, 1968.

5. O. A. Ladyzhenskaya and N. N. Ural′tseva, *Linear and quasilinear equations of elliptic type*, "Nauka", Moscow, 1964; English transl., Academic Press, New York, 1968.

6. D. Gilbarg and N. S. Trudinger, *Elliptic partial differential equations of second order*, 2nd ed., Grundlehren der mathematischen Wissenschaften, 224. Springer-Verlag, Berlin and New York, 1983.

7. G. C. Dong and S. Li, *A boundary value problem for nonlinear telegraph equations*, Nonlinear Anal. **5** (1981), no. 7, 705–711.

8. H. Brezis and L. Nirenberg, *Characterizations of the ranges of some nonlinear operators and applications to boundary value problems*, Ann. Scuola Norm. Sup. Pisa, Cl. Sci. (4) **5** (1978), no. 2, 225–326.

9. E. M. Landesman and A. C. Lazer, *Nonlinear perturbations of linear elliptic boundary value problems at resonance*, J. Math. Mech. **19** (1969/70), 609–623.

10. Ding Weiyue, *Periodic solutions of dissipative wave equations*. Preprint.

11. G. C. Dong and S. Li, *On the initial value problem for a nonlinear Schrödinger equation*, J. Differential Equations **42** (1981), no. 3, 353–365.

12. W. A. Strauss, *Dispersion of low-energy waves for two conservative equations*, Arch. Rational Mech. Anal. **55** (1974), 86–92.

13. S. Klainerman, *Loss of decay for a nonlinear Schrödinger equation*. Preprint.

14. ——, *Long-time behavior of solutions to nonlinear evolution equations*, Arch. Rational Mech. Anal. **78** (1982), no. 1, 73–98.

15. Y. Tsutsumi, *Global solutions of the nonlinear Schrödinger equation in exterior domains*, Comm. Partial Differential Equations **8** (1983), no. 12, 1337–1374.

16. Y. M. Chen, *Global solvability for nonlinear evolution equations*, Ph. D. dissertation. (Chinese)

17. L. Bers, *Existence and uniqueness of subsonic flows past a given profile*, Comm. Pure Appl. Math. **7** (1954), 441–504.

18. R. Finn and D. Gilbarg, *Three-dimensional subsonic flows, and asymptotic estimates for elliptic partial differential equations*, Acta Math. **98** (1957), 265–296.

19. G. C. Dong, *Subsonic flows in space and its higher dimensional generalization*, Zhejiang Univ. J. **1** (1979), 33–63. (Chinese)

20. B. Ou, *On "Subsonic flows in space and its higher dimensional generalization."* Preprint. (Chinese)

21. ——, *Existence, uniqueness and continuous dependence for subsonic flows with "source" in space.* Preprint. (Chinese)

22. Carlo Miranda, *Equazioni alle derivate parziali di tipo ellittico*, Ergebnisse der Mathematik und ihrer Grenzgebiete, Heft 2, Springer-Verlag, Berlin, 1955; English transl., Springer-Verlag, Berlin, 1970.

23. J. Bear, *Dynamics of fluids in porous media*, American Elsevier, New York, 1972.

24. B. H. Gilding, *A nonlinear degenerate parabolic equation*, Ann. Scuola Norm. Sup. Pisa, Cl. Sci. (4) **4** (1977), no. 3, 393–432.

25. Y. Z. Chen, *Uniqueness of the generalized solution of quasilinear degenerate parabolic equations*, Proceedings of the 1982 Changchun Symposium on differential geometry and differential equations, Science Press, 317–332.

26. G. C. Dong, *A higher dimensional nonlinear degenerate parabolic equation*, Proceedings of the 1982 Changchun Symposium on differential geometry and differential equations, Science Press, 373–382.

27. Y. Z. Chen, *Hölder estimates for solutions of uniformly degenerate quasilinear parabolic equations*, Chinese Ann. Math. Ser. B, **5** (1984), no. 4, 661–678.

28. G. C. Dong, *The first boundary value problems for solutions of degenerate quasilinear parabolic equations.* Preprint.

29. Y. Z. Chen, *On finite diffusing speed for uniformly degenerate quasilinear parabolic equations.* Preprint.

30. B. J. Bian, *On the finite propagation for solutions of second order degenerate parabolic equations.* Preprint. (Chinese)

31. A. D. Aleksandrov, *Studies on the maximum principle*, Izv. Vyssh. Uchebn. Zaved. Matematika 1961, no. 1, 3–20. (Russian)

32. C. Pucci, *Limitazioni per soluzioni di equazioni ellittiche*, Ann. Mat. Pura Appl. (4) **74** (1966), 15–30.

33. N. V. Krylov, *Sequences of convex functions, and estimates of the maximum of the solution of a parabolic equation*, Sibirsk. Mat. Zh. **17** (1976), no. 2, 290–303; English transl. in Siberian Math. J. **17** (1976), no. 2.

34. Y. Z. Chen, *Aleksandrov and Bony maximum principles for parabolic equations.* Preprint. (Chinese)

35. A. D. Aleksandrov, *Intrinsic geometry of convex surfaces*, OGIZ, Moscow, 1948; German transl., Akademie-Verlag, Berlin, 1955.

36. S. Y. Cheng and S. T. Yau, *On the regularity of the Monge-Ampère equation* $\det(\partial^2 u/\partial x_i\}\partial x_j) = F(x, u)$, Comm. Pure Appl. Math. **30** (1977), no. 1, 41–68.

37. J. M. Bony, *Principe du maximum dans les espaces de Sobolev*, C. R. Acad. Sci. Paris, Sér. A-B 265 (1967), A333–A336.

38. P.-L. Lions, *A remark on Bony maximum principle*, Proc. Amer. Math. Soc. **88** (1983), no. 3, 503–508.

39. N. V. Krylov and M. V. Safonov, *A certain property of solutions of parabolic equations with measurable coefficients*, Izv. Akad. Nauk SSSR Ser. Mat. **44** (1980), no. 1, 161–175; English transl. in Math. USSR-Izv. **16** (1981), no. 1, 151–164.

40. N. S. Trudinger, *Local estimates for subsolutions and supersolutions of general second order elliptic quasilinear equations*, Invent. Math. **61** (1980), no. 1, 67–79.

41. G. C. Dong, *Local estimates for subsolutions and supersolutions of general second order parabolic quasilinear equations.* Preprint.

42. N. V. Krylov, *On estimates for the derivatives of solutions of nonlinear parabolic equations*, Dokl. Akad. Nauk SSSR **274** (1984), no. 1, 23–26; English transl. in Soviet Math. Dokl. **29** (1984), no. 1, 14–17.

43. Y. Z. Chen, *On Krylov's a priori estimates for fully nonlinear equations*, Adv. Math. **15** (1986), 63–101. (Chinese)

44. N. V. Krylov, *Boundedly nonhomogeneous elliptic and parabolic equations*, Izv. Akad. Nauk SSSR Ser. Mat. **46** (1982), no. 3, 487–523; English transl., Math. USSR Izv. **20** (1983), no. 3, 459–492.

45. ——, *Boundedly nonhomogeneous elliptic and parabolic equations in a domain*, Izv. Akad. Nauk SSSR Ser. Mat. **47** (1983), no. 1, 75–108; English transl. in Math. USSR-Izv. **22** (1984), no. 1, 67–97.

46. G. C. Dong, *A priori Hölder estimates for solutions of quasilinear parabolic equations with natural structure conditions.* Preprint.

47. ——, *On the estimation of derivatives for solutions of fully nonlinear parabolic equations.* Preprint.

48. K. Tso, *On an Aleksandrov-Bakel'man type maximum principle for second-order parabolic equations*, Comm. Partial Differential Equations **10** (1985), no. 5, 543–553.

49. ——, *Linear second order partial differential equations*, Zhejiang Univ. Press, 1987. (Chinese)

50. G. C. Dong, *Initial and nonlinear oblique boundary value problems for fully nonlinear parabolic equations*, J. Partial Differential Equations Ser. A **1** (1988), no. 2, 12–42.

51. ——, *Hölder estimate of quasilinear parabolic equations with nonlinear oblique derivative boundary condition*, J. Partial Differential Equations **3** (1990), no. 3, 49–53.

52. ——, *Viscosity solutions of the first boundary value problem for the second order fully nonlinear parabolic equation under natural structure conditions.* Preprint.

53. G. C. Dong and B. J. Bian, *Regularity properties of viscosity solutions of fully nonlinear equations.* Preprint.

54. ——, *Viscosity solutions of fully nonlinear second-order parabolic partial differential equations.* Preprint.

Epilogue

Our discussion on quasilinear and fully nonlinear equations is far from complete. Many important aspects, such as the oblique derivative problem for fully nonlinear equations, as well as the applications of fully nonlinear equations, are left out. Hopefully they will be covered in a future publication.